INTRODUCTION TO LINEAR ALGEBRA

JONES AND BARTLETT BOOKS IN MATHEMATICS AND COMPUTER SCIENCE

Advanced Calculus, Revised Edition
Lynn H. Loomis and Shlomo Sternberg

Algorithms and Data Structures in Computer Engineering
Christopher H. Nevison, et al., Editors

Calculus with Analytic Geometry, Fourth Edition
Murray H. Protter and Philip E. Protter

Classical Complex Analysis
Liang-shin Hahn and Bernard Epstein

College Geometry
Howard Eves

Complex Analysis for Mathematics and Engineering, Third Edition
John H. Mathews and Russell W. Howell

Differential Equations: Theory and Applications
Ray Redheffer

Discrete Mathematics
James L. Hein

Discrete Structures, Logic, and Computability
James L. Hein

Fundamentals of Modern Elementary Geometry
Howard Eves

An Introduction to Computational Science and Mathematics
Charles F. Van Loan

Introduction to Differential Equations
Ray Redheffer

Introduction to Fractals and Chaos
Richard M. Crownover

Introduction to Linear Algebra
Géza Schay

Introduction to Numerical Analysis
John Gregory and Don Redmond

Lebesgue Integration on Euclidean Space
Frank Jones

The Limits of Computing
Henry M. Walker

Logic, Sets, and Recursion
Robert L. Causey

Mathematical Models in the Social and Biological Sciences
Edward Beltrami

The Poincaré Half-Plane: A Gateway to Modern Geometry
Saul Stahl

Selected Works in Applied Mathematics and Mechanics
Eric Reissner

The Theory of Numbers: A Text and Source Book of Problems
Andrew Adler and John E. Coury

Theory of Computation
James L. Hein

Wavelets and Their Applications
Mary Beth Ruskai, et al.

The Way of Analysis
Robert S. Strichartz

INTRODUCTION TO LINEAR ALGEBRA

Géza Schay

Department of Mathematics and Computer Science
University of Massachusetts—Boston

Jones and Bartlett Publishers
Sudbury, Massachusetts
Boston London Singapore

Editorial, Sales, and Customer Service Offices
Jones and Bartlett Publishers
40 Tall Pine Drive
Sudbury, MA 01776
1-800-832-0034
508-443-5000
info@jbpub.com
http://www.jbpub.com

Jones and Bartlett Publishers International
Barb House, Barb Mews
London W6 7PA
UK

Library of Congress Cataloging-in-Publication Data
Schay, Géza.
Introduction to Linear Algebra / Géza Schay.
p. cm.
Includes index.
ISBN 0-86720-498-2
1. Algebra, Linear. I. Title
QA184.S33 1996
512'.5—dc20 96-13520
CIP

Acquisitions Editor: David Geggis
Senior Production Administrator: Mary Sanger
Manufacturing Manager: Dana L. Cerrito
Design: Mary Gordon
Editorial Production Service: Superscript Editorial Production Services
Typesetting and Illustrations: LM Graphics
Cover Design: Jeannet Leendertse
Prepress: Pure Imaging
Printing and Binding: Courier Companies Inc.
Cover Printing: Henry N. Sawyer Company, Inc.

Printed in the United States of America
00 99 98 97 96 10 9 8 7 6 5 4 3 2 1

CONTENTS

PREFACE

This is a book for a one-semester post-calculus linear algebra course. The selection of topics conforms to a large extent to the recommendations of the Linear Algebra Curriculum Study Group. The main differences are that the book begins with a chapter on Euclidean vector geometry, mostly in three dimensions; determinants are treated more fully and are placed just before eigenvalues, which is where they are needed; the LU factorization is relegated to Chapter 8 on numerical methods; the facts about linear transformations are collected in one chapter and not scattered; and the QR factorization is not discussed.

The present book is considerably shorter than the 400 to 800 pages of most introductory linear algebra books, which are more suitable for two- or three-semester courses. We have avoided unnecessary verbosity and omitted some topics, but have not shortchanged the topics we have included.

Although many applications are presented, their number is kept low by providing only one of each kind. These applications are intended to give only a glimpse of how the subject is used in other fields, and further details are left to texts in those fields. Hopefully, the student's interest will be aroused not only by the possible applications but also by the geometrical background and the beautiful structure of linear algebra.

The more difficult exercises are marked by an asterisk, and exercises with answers in the back of this book are indicated by a bullet. Some exercises are used to develop new topics, whose inclusion in the main text would have disrupted the flow of ideas.

Main Features

- The brevity mentioned above makes the book easier to use. Important points are not drowned in a sea of detail, and instructors and students do not have to search for what to keep and what to omit. In a minimal course the following sections may be omitted entirely: Section 4.3 on computer graphics, Section 5.1 on orthogonal projections and least-squares, Section 6.3 on the cross product, Sections 7.3 and 7.4 on principal axes and complex matrices, and Chapter 8 on numerical methods.
- The geometric content is heavily emphasized.
- The letter symbols are selected so as to reflect the connections between related quantities. This makes, for instance, the notoriously messy topic of change of basis much simpler. Vectors and their components, matrices and their column and row vectors and entries are denoted by the

same letters such as $\boldsymbol{v}$, $\boldsymbol{v}_i$ and A, $\boldsymbol{a}_i$, $\boldsymbol{a}^j$, $\boldsymbol{a}_{ij}$; the main exception being the unit matrix, which, bowing to tradition, is denoted by I, its columns by $\boldsymbol{e}_i$ and its entries by δ_{ij}.

- Important concepts are presented as definitions and theorems. Students are advised to memorize these. It is not enough just to understand the material; the main concepts must be remembered well to be able to build on them.
- Except for the Spectral Theorem in the complex case, the Fundamental Theorem of Algebra, and some elementary facts, all theorems are proved. It is thus left to the instructor to adjust the level of the course from the computational to the fairly theoretical by omitting as many or as few proofs as desired.
- Great care has been taken to motivate every new concept, even those that many books do not, such as dot product, matrix operations, linear independence (not just in two or three dimensions), determinants, eigenvalues, and eigenvectors.
- Only standard notation is used, so that students who go on will have no difficulty in reading applied or more advanced texts. Nonstandard notation, such as the use of a row in parentheses for column vectors and brackets for row vectors, found in some other introductory linear algebra books, is avoided.
- The connections of the theory are emphasized. For example, in Section 1.3 the equations of lines and planes are given in both parametric and nonparametric forms, in Section 2.1 these are related to solutions of linear systems, in Section 3.2 to subspaces being defined as linear combinations of vectors and as solution spaces; and finally, in Section 3.4 to representing subspaces either as column spaces or as nullspaces of matrices.
- MATLAB exercises at the end of most sections reinforce and expand the material.
- The appendix on implication and equivalence introduces the student in an informal way to certain crucial elements of proofs and is highly recommended reading for most.
- The book has two supplements: A *Student Solutions Manual*, coauthored with Dennis Wortman, contains detailed solutions to all the odd-numbered exercises and an *Instructor's Manual* with brief answers to the even-numbered exercises.

Acknowledgments

I would like to express my gratitude to the following colleagues who tested early drafts of this book in their classes and suggested numerous, extremely valuable improvements: Guy T. Hogan, Matthew P. Gaffney, John Lutts, Seymour Kass, and Dennis Wortman. Thanks go also to Robert Seeley for a thorough reading of the first three chapters, which resulted in many helpful corrections as well. Also, I want to thank my son, Peter Schay, for reading the first and last drafts and providing a long list of insightful comments. Finally, I wish to express my appreciation to Mary Sanger of Jones and Bartlett, to Ann H. Knight of Superscript Editorial Production Services, and to their staffs for shepherding the book through production.

INTRODUCTION TO LINEAR ALGEBRA

<<1>> ANALYTIC GEOMETRY OF EUCLIDEAN SPACES

1.1. VECTORS

We begin by describing some geometrical concepts. This approach may seem strange in a book on algebra, but the influence of geometry is fundamental to our subject, for the underlying geometrical ideas provide motivation, examples, and applications for the algebraic constructions. In fact, the adjective "linear" in the title means "pertaining to lines" (which in mathematics always means straight lines), and indicates the geometric origins of linear algebra.

In physics several important notions such as displacement, velocity, and force possess not just a magnitude but a direction as well. These are typical of a large class of quantities called *vectors* that can be depicted by arrows showing the desired directions and representing the vectors' magnitudes by their lengths. In geometry we can use them to locate points and also, as we shall see, to write equations of lines and planes. Let us look at a few such examples before stating formal definitions:

FIGURE 1.1

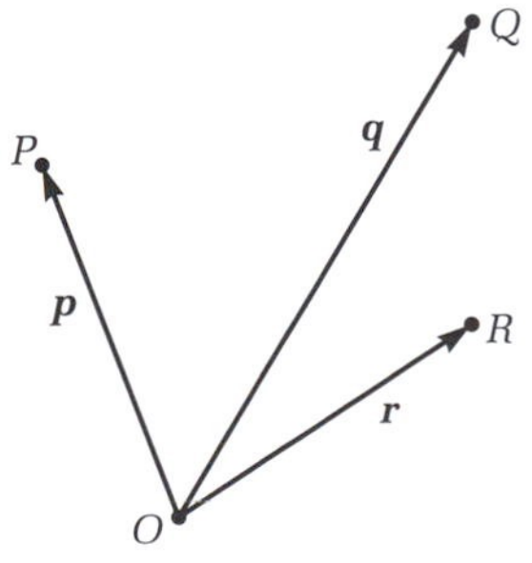

EXAMPLE 1.1.1. Either in the plane or in three-dimensional space, consider a fixed point O and other points P, Q, R, and draw correspondingly labeled arrows $\boldsymbol{p}$, $\boldsymbol{q}$, $\boldsymbol{r}$ from O to the other points (see Figure 1.1). These arrows are called the *position vectors* or *radius vectors* of P, Q, R relative to the point O, which is usually regarded as the origin of a coordinate system. Such vectors are also sometimes called *bound* vectors, for they are bound to the origin, in contrast to free vectors, to be introduced shortly. The position vector of the point O is a special vector $\mathit{0}$, called the zero or null vector, whose length is 0, and whose direction is undefined.

<<>>

Whereas, in print vectors are generally denoted by lowercase boldface letters such as $\boldsymbol{p}$, $\boldsymbol{q}$, $\boldsymbol{r}$ or by symbols like $\overrightarrow{OP}$, $\overrightarrow{OQ}$; in handwriting boldface would be difficult and so $\underline{p}$, $\underline{q}$ or $\vec{p}$, $\vec{q}$, and so on are used instead.

Since position vectors and points are in one-to-one correspondence, you may wonder why we need position vectors at all. The answer is that various arithmetic operations that would make no sense with points can be performed with vectors, and will lend themselves to all kinds of useful constructions. Such operations are also essential for the vectors of physics:

EXAMPLE 1.1.2. If, in Figure 1.2, $\boldsymbol{p}$ and $\boldsymbol{q}$ represent two *forces* acting simultaneously on a point mass at O, then the single force represented by $\boldsymbol{r}$, to be defined as $\boldsymbol{p} + \boldsymbol{q}$, would have the same effect.

<<>>

FIGURE 1.2

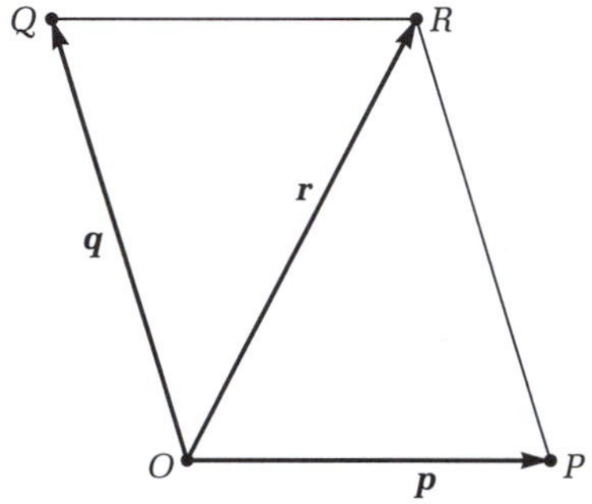

$\boldsymbol{r} = \boldsymbol{p} + \boldsymbol{q}$

EXAMPLE 1.1.3. On the other hand, if $\boldsymbol{p}$ and $\boldsymbol{q}$ represent simultaneous *displacements,* then $\boldsymbol{r}$ represents their combined effect. For example, if a person on a boat at O walks to Q while the point O of the boat moves, together with the boat, to P (and the point Q of the boat to R), then, as seen from shore, the person ends up at R.

<<>>

FIGURE 1.3

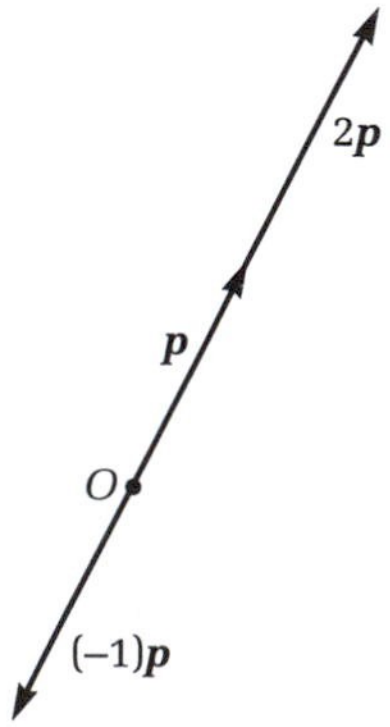

The last two examples illustrate how addition of such vectors is defined: Given any pair $\boldsymbol{p}$ and $\boldsymbol{q}$ as in Figure 1.2, the corresponding points O, P, Q determine a parallelogram $OPRQ$, and the sum $\boldsymbol{p} + \boldsymbol{q}$ is defined as the diagonal vector $\boldsymbol{r} = \overrightarrow{OR}$. This definition is called the *parallelogram law* of vector addition.

A second operation we consider is multiplication of vectors by scalars. (In this context real numbers are usually called scalars, since they can be pictured on a scale, unlike vectors.) Let c be any scalar and $\boldsymbol{p}$ any vector, as in the previous examples. The vector $c\boldsymbol{p}$ is defined as the vector whose length is $|c|$ times the length of $\boldsymbol{p}$ and whose direction is the same as that of $\boldsymbol{p}$ if $c > 0$, and opposite if $c < 0$. If $c = 0$, then $c\boldsymbol{p}$ is the zero vector. Some examples of this type of multiplication are shown in Figure 1.3.

The discussion has been somewhat informal so far because we have not really specified very precisely the sets of vectors under consideration. It is best to remedy this omission by bringing a coordinate system into the picture:

If we consider the position vector $\boldsymbol{p}$ of a point P in a plane (see Figure 1.4) and introduce a coordinate system, then we can represent the vector $\boldsymbol{p}$, as well as the point P, by the ordered pair (p_1, p_2) of coordinates, and write $\boldsymbol{p} = (p_1, p_2)$. For this representation to be of any use, we then have to recast the parallelogram law and the multiplication of vectors by scalars in terms of the coordinates, as follows.

For any two vectors $\boldsymbol{p} = (p_1, p_2)$ and $\boldsymbol{q} = (q_1, q_2)$, a good look at Figure 1.5 shows that if $\boldsymbol{r} = \boldsymbol{p} + \boldsymbol{q}$ is the diagonal of the parallelogram spanned by $\boldsymbol{p}$ and $\boldsymbol{q}$, then $\boldsymbol{p} + \boldsymbol{q} = (p_1 + q_1, p_2 + q_2)$ must hold; that is, we must simply add the corresponding coordinates. Similarly, we must have $c\boldsymbol{p} = (cp_1, cp_2)$ for any scalar c.

FIGURE 1.4

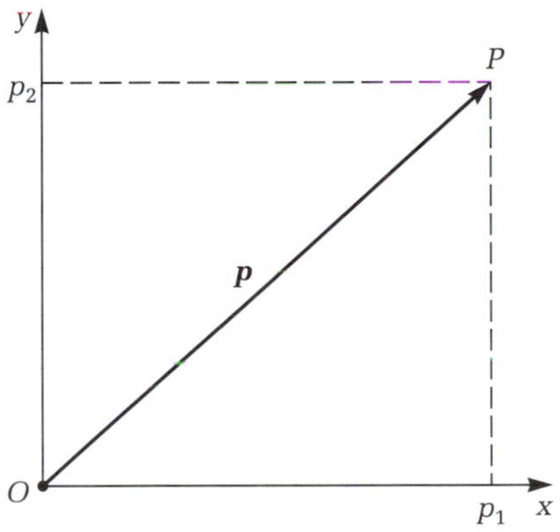

FIGURE 1.5

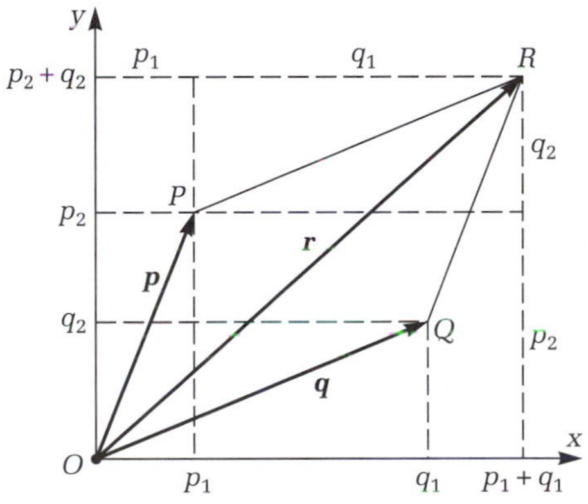

In light of the discussion above we now make this formal definition:

DEFINITION 1.1.1.

The set of all ordered pairs of real numbers is called the *vector space* $\mathbb{R}^2$ and the ordered pairs *vectors* (or *coordinate vectors*) if, for any (p_1, p_2), (q_1, q_2) and any scalar c, we have the operations of vector addition and multiplication of vectors by scalars defined by[1]

$$(p_1, p_2) + (q_1, q_2) = (p_1 + q_1, p_2 + q_2), \tag{1.1.1}$$

and

$$c(p_1, p_2) = (cp_1, cp_2). \tag{1.1.2}$$

The scalars p_1 and p_2 are called the *components* of the vector $\boldsymbol{p} = (p_1, p_2)$.

[1]By definition, equality of two quantities means that they are the same and so two vectors are equal if and only if their corresponding components are equal.

EXAMPLE 1.1.4. In Figure 1.6 let the points $P = (1, 5)$ and $Q = (3, 1)$ be given. Then the corresponding coordinate vectors are $\boldsymbol{p} = (1, 5)$ and $\boldsymbol{q} = (3, 1)$, and the position vector of the point R that makes $OQRP$ into a parallelogram is $\boldsymbol{r} = \boldsymbol{p} + \boldsymbol{q} = (1 + 3, 5 + 1) = (4, 6)$. The midpoint M of the parallelogram has the position vector $\frac{1}{2}\boldsymbol{r} = (2, 3)$.

<<>>

FIGURE 1.6

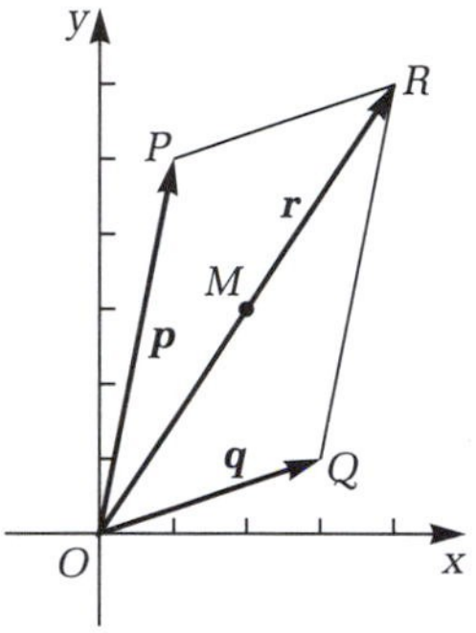

The definition above implies the following simple properties, which will be used in Chapter 3 as the defining rules for general vector spaces:

THEOREM 1.1.1. For all vectors $\boldsymbol{p}, \boldsymbol{q}, \boldsymbol{r}$ in $\mathbb{R}^2$ and all scalars a, b we have:

1. $\boldsymbol{p} + \boldsymbol{q} = \boldsymbol{q} + \boldsymbol{p}$ (commutativity of addition)
2. $(\boldsymbol{p} + \boldsymbol{q}) + \boldsymbol{r} = \boldsymbol{p} + (\boldsymbol{q} + \boldsymbol{r})$ (associativity of addition)
3. There is a vector $\boldsymbol{0}$ such that $\boldsymbol{p} + \boldsymbol{0} = \boldsymbol{p}$ (existence of zero vector)
4. $1\boldsymbol{p} = \boldsymbol{p}$ (rule of multiplication by 1)
5. $a(b\boldsymbol{p}) = (ab)\boldsymbol{p}$ (associativity of multiplication by scalars)[2]
6. $(a + b)\boldsymbol{p} = a\boldsymbol{p} + b\boldsymbol{p}$ (first distributive law)
7. $a(\boldsymbol{p} + \boldsymbol{q}) = a\boldsymbol{p} + a\boldsymbol{q}$ (second distributive law)

Proof

1. $\boldsymbol{p} + \boldsymbol{q} = (p_1 + q_1, p_2 + q_2) = (q_1 + p_1, q_2 + p_2) = \boldsymbol{q} + \boldsymbol{p}$.
2. $(\boldsymbol{p} + \boldsymbol{q}) + \boldsymbol{r} = (p_1 + q_1, p_2 + q_2) + (r_1, r_2) = (p_1 + q_1 + r_1, p_2 + q_2 + r_2) = (p_1, p_2) + (q_1 + r_1, q_2 + r_2) = \boldsymbol{p} + (\boldsymbol{q} + \boldsymbol{r})$.
3. Defining $\boldsymbol{0} = (0, 0)$ we have $\boldsymbol{p} + \boldsymbol{0} = (p_1 + 0, p_2 + 0) = (p_1, p_2) = \boldsymbol{p}$.

We leave the rest to the reader.

<<>>

[2]"Associativity" is nonstandard here; there is no official name for this property.

Subtraction and the negative of a vector can be defined just as for numbers:

DEFINITION 1.1.2.

For any $\boldsymbol{p} = (p_1, p_2)$, $\boldsymbol{q} = (q_1, q_2) \in \mathbb{R}^2$ we define

$$-\boldsymbol{p} = (-1)\boldsymbol{p} \tag{1.1.3}$$

and

$$\boldsymbol{p} - \boldsymbol{q} = \boldsymbol{p} + (-\boldsymbol{q}). \tag{1.1.4}$$

EXAMPLE 1.1.5. Let $\boldsymbol{p} = (1, -3)$ and $\boldsymbol{q} = (-4, 5)$.
Then $-\boldsymbol{p} = (-1)(1, -3) = (-1, 3)$, $-\boldsymbol{q} = (-1)(-4, 5) = (4, -5)$ and $\boldsymbol{p} - \boldsymbol{q} = (1, -3) + (4, -5) = (5, -8)$.

<<>>

As in the example above, the definitions lead at once to the following alternative expressions for the negatives of vectors and for their subtraction in terms of components:

THEOREM 1.1.2.

For any $\boldsymbol{p} = (p_1, p_2)$, $\boldsymbol{q} = (q_1, q_2) \in \mathbb{R}^2$

$$-\boldsymbol{p} = (-p_1, -p_2) \tag{1.1.5}$$

and

$$\boldsymbol{p} - \boldsymbol{q} = (p_1 - q_1, p_2 - q_2) \tag{1.1.6}$$

hold.

We have the following list of further properties of vectors:

THEOREM 1.1.3.

For all vectors $\boldsymbol{p}, \boldsymbol{q}, \boldsymbol{x}$ in $\mathbb{R}^2$ and all scalars c and d we have

1. $0\boldsymbol{p} = \boldsymbol{0}$
2. $c\boldsymbol{0} = \boldsymbol{0}$
3. $\boldsymbol{p} + \boldsymbol{x} = \boldsymbol{q}$ if and only if $\boldsymbol{x} = \boldsymbol{q} - \boldsymbol{p}$
4. If $c\boldsymbol{p} = \boldsymbol{0}$ then either $c = 0$ or $\boldsymbol{p} = \boldsymbol{0}$ or both
5. $(-c)\boldsymbol{p} = c(-\boldsymbol{p}) = -(c\boldsymbol{p})$
6. $c(\boldsymbol{p} - \boldsymbol{q}) = c\boldsymbol{p} - c\boldsymbol{q}$
7. $(c - d)\boldsymbol{p} = c\boldsymbol{p} - d\boldsymbol{p}$

Proof We prove only Part 3: Writing $\boldsymbol{p} = (p_1, p_2)$, $\boldsymbol{q} = (q_1, q_2)$, and $\mathbf{x} = (x_1, x_2)$, if $\mathbf{x} = \boldsymbol{q} - \boldsymbol{p}$, then we have

$$\mathbf{x} = (x_1, x_2) = (q_1 - p_1, q_2 - p_2) \tag{1.1.7}$$

and so

$$\boldsymbol{p} + \mathbf{x} = (p_1 + (q_1 - p_1), p_2 + (q_2 - p_2)) = (q_1, q_2) = \boldsymbol{q} \tag{1.1.8}$$

must hold.

Conversely, if $\boldsymbol{p} + \mathbf{x} = \boldsymbol{q}$, then this equation can be written in components as

$$(p_1 + x_1, p_2 + x_2) = (q_1, q_2) \tag{1.1.9}$$

and, because the equality of two vectors means that the corresponding components must be equal, we must have

$$p_1 + x_1 = q_1 \tag{1.1.10}$$

and

$$p_2 + x_2 = q_2. \tag{1.1.11}$$

Solving these equations for x_1 and x_2 and combining the latter into a vector, we get

$$\mathbf{x} = (x_1, x_2) = (q_1 - p_1, q_2 - p_2) = \boldsymbol{q} - \boldsymbol{p}. \tag{1.1.12}$$

<<>>

There is an additional, important way of associating arrows with ordered pairs of coordinates. If we draw an arrow $\boldsymbol{p}'$ anywhere in a coordinate system (see Figure 1.7), not necessarily at the origin, then we can still project it perpendicularly onto the axes and consider the signed lengths p_1, p_2 of the projections to be the components of a vector in $\mathbb{R}^2$. Of course, any other arrow, such as $\boldsymbol{p}''$, obtained from $\boldsymbol{p}'$ by a parallel shift, will produce the same p_1, p_2 values. Thus, for a given vector $(p_1, p_2) \in \mathbb{R}^2$, there corresponds a class $\boldsymbol{p}$ of infinitely many arrows parallel to each other and equal in length,[3] all having the same signed projections p_1 and p_2. The arrows like $\boldsymbol{p}'$ and $\boldsymbol{p}''$ are equivalent representatives of the class $\boldsymbol{p}$. Such classes of equivalent arrows are called free vectors, since the arrows can be shifted freely. We usually identify the free vector $\boldsymbol{p}$ with the vector $(p_1, p_2) \in \mathbb{R}^2$; that is, we write $\boldsymbol{p} = (p_1, p_2)$. This identification should not

[3] "Equivalence class" is a standard term used for sets whose members constitute all objects equivalent to each other under a type of relation called *equivalence relation*.

lead to confusion, just as referring to a point (x, y) instead of a point P with coordinates (x, y) does not.

FIGURE 1.7

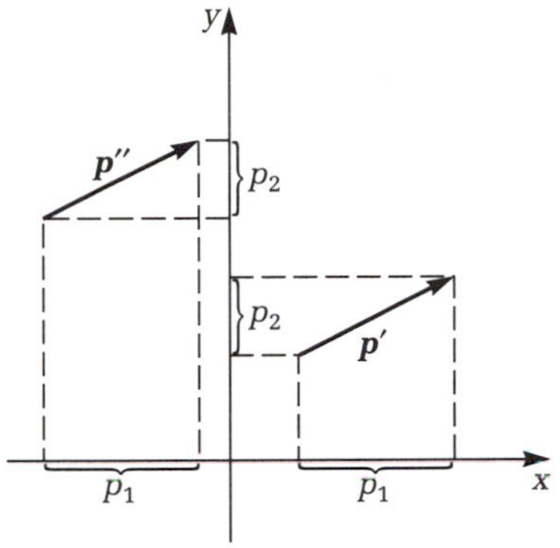

A free vector can be represented by any one of its arrows; that is, the whole class $\boldsymbol{p}$ is known if any member $\boldsymbol{p}'$ is known. Unfortunately, many people confuse the class $\boldsymbol{p}$ with the individual arrows, and call $\boldsymbol{p}'$ and $\boldsymbol{p}''$ equal *vectors*, rather than just equivalent representative *arrows* of the vector $(p_1, p_2) \in \mathbb{R}^2$ or of the free vector $\boldsymbol{p}$.

Why do we use free vectors at all? There are at least three reasons: First, they arise rather naturally as representations of coordinate vectors, as we have just seen. Second, in physical applications some vector quantities are not bound to any fixed point. (For example, the velocity vector of a non-rotating object can reasonably be drawn at any point of the object.) Third, the pictures of many constructions become simpler and less cluttered if we use well-positioned representative arrows, rather than just vectors at O. Many examples of this usage will follow.

FIGURE 1.8

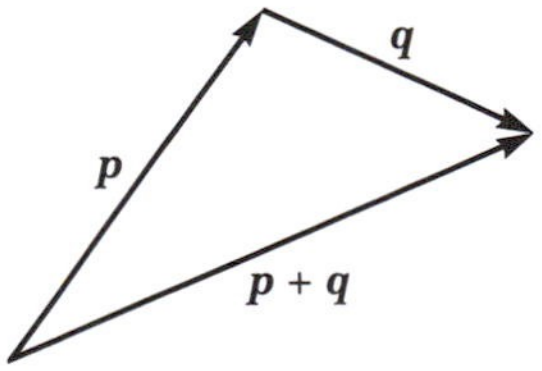

FIGURE 1.9

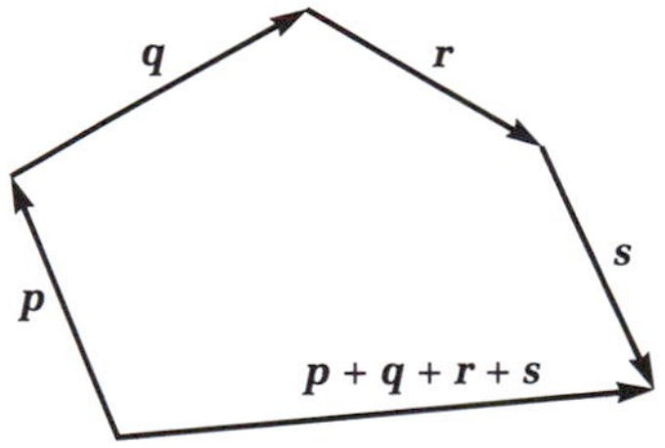

Let us first look at a variation on the addition of two vectors in terms of free vectors. In Figure 1.8, let the arrows marked $\boldsymbol{p}$ and $\boldsymbol{q}$ represent the free vectors $\boldsymbol{p}$ and $\boldsymbol{q}$ (we shall abbreviate this statement from now on as customary to "let $\boldsymbol{p}$ and $\boldsymbol{q}$ be two vectors as shown"). Then, obviously, their sum is represented by the arrow $\boldsymbol{p} + \boldsymbol{q}$. This result is sometimes called the *triangle law* of vector addition. If the arrows represent successive displacements, then it is the natural description of their sum; that is, of one displacement followed by another.

Now let us turn to the addition of several, say four, vectors $\boldsymbol{p}$, $\boldsymbol{q}$, $\boldsymbol{r}$, $\boldsymbol{s}$, as given in Figure 1.9. By repeated application of the triangle law we get the sum as shown. (Because of the associativity of vector addition, just as with numbers, we do not need parentheses in the sum.) Contrast the simplicity of this construction with the mess we would get if all vectors were drawn at O.

Because we have $(\boldsymbol{p} + \boldsymbol{q}) - \boldsymbol{p} = \boldsymbol{q}$, we can relabel Figure 1.8 to illustrate the subtraction of vectors, by writing $\boldsymbol{r}$ for $\boldsymbol{p} + \boldsymbol{q}$ and $\boldsymbol{r} - \boldsymbol{p}$ for $\boldsymbol{q}$ as in Figure 1.10 below. The triangle law applied to Figure 1.10 shows that $\boldsymbol{p} + (\boldsymbol{r} - \boldsymbol{p}) = \boldsymbol{r}$, as it should be, and that $\overrightarrow{PR} = \boldsymbol{r} - \boldsymbol{p}$. This construction is especially useful for obtaining the coordinate vectors of arrows joining given points, as in Example 1.1.6.

FIGURE 1.10

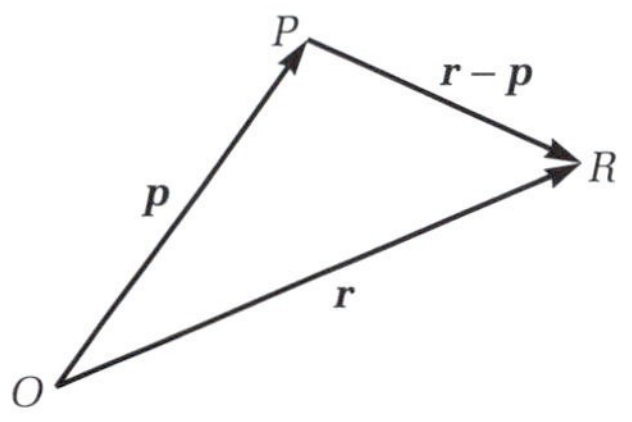

FIGURE 1.11

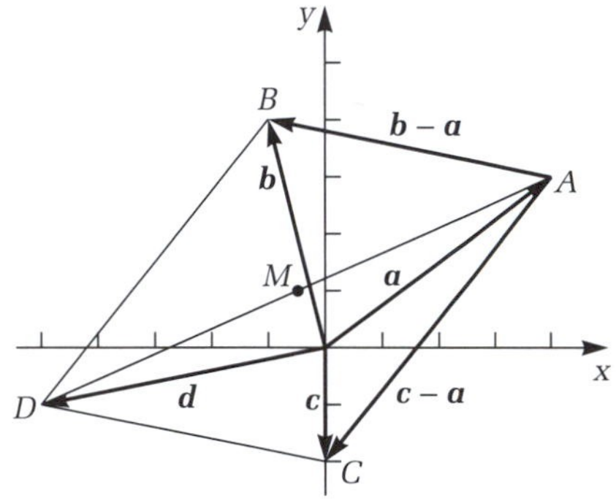

EXAMPLE 1.1.6. Given two points $P = (1, 2)$ and $R = (3, 6)$ in the plane, find the coordinate vector of $\overrightarrow{PR}$.

We can write the position vectors of the given points as $\boldsymbol{p} = (1, 2)$ and $\boldsymbol{r} = (3, 6)$, and so $\overrightarrow{PR} = \boldsymbol{r} - \boldsymbol{p} = (3 - 1, 6 - 2) = (2, 4)$.

<<>>

EXAMPLE 1.1.7. Given three points $A = (4, 3)$, $B = (-1, 4)$ and $C = (0, -2)$ in the plane (see Figure 1.11), find the coordinate vectors of $\overrightarrow{AB}$, $\overrightarrow{AC}$, the coordinates of the point D that makes $ABDC$ a parallelogram, and those of the midpoint M of the parallelogram.

The position vectors of the given points are $\boldsymbol{a} = (4, 3)$, $\boldsymbol{b} = (-1, 4)$, and $\boldsymbol{c} = (0, -2)$. Then $\overrightarrow{AB} = \boldsymbol{b} - \boldsymbol{a} = (-5, 1)$, $\overrightarrow{AC} = \boldsymbol{c} - \boldsymbol{a} = (-4, -5)$ and $\overrightarrow{AD} = (\boldsymbol{b} - \boldsymbol{a}) + (\boldsymbol{c} - \boldsymbol{a}) = (-9, -4)$. Now $\boldsymbol{d} = \boldsymbol{a} + \overrightarrow{AD} = (-5, -1)$ and this ordered pair also gives the coordinates of D. The position vector $\boldsymbol{m}$ of the midpoint M can be obtained as $\boldsymbol{m} = \boldsymbol{a} + \frac{1}{2}\overrightarrow{AD} = (4, 3) + \frac{1}{2}(-9, -4) = \left(-\frac{1}{2}, 1\right)$.

<<>>

We now define the three-dimensional vector space $\mathbb{R}^3$ of coordinate vectors:

DEFINITION 1.1.3.

The vector space $\mathbb{R}^3$ is the set of all ordered triples (p_1, p_2, p_3) of real numbers with the operations defined componentwise, just as before for any (p_1, p_2, p_3), (q_1, q_2, q_3), and any scalar c, by

$$(p_1, p_2, p_3) + (q_1, q_2, q_3) = (p_1 + q_1, p_2 + q_2, p_3 + q_3), \tag{1.1.13}$$

and

$$c(p_1, p_2, p_3) = (cp_1, cp_2, cp_3). \tag{1.1.14}$$

Just as in two dimensions, if we introduce a coordinate system with the origin at O, then every $\boldsymbol{p} \in \mathbb{R}^3$ can be regarded as the position vector of the corresponding point P. (See Figure 1.12.) Thus we identify the arrow $\boldsymbol{p}$ with the coordinate vector (p_1, p_2, p_3).

FIGURE 1.12

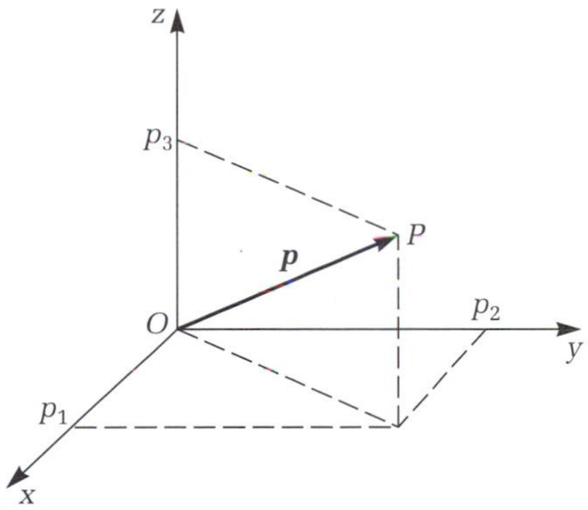

In three-dimensional space we can again represent coordinate vectors also by arrows drawn anywhere, not just at the origin, and we define free vectors much as in the plane.

About Figure 1.12, let us remark that the x-axis is meant to be interpreted as pointing out of the paper toward the reader. (This sense is not obvious: if you stare at the picture hard, you may see it as pointing into the paper.) In three dimensions, two kinds of coordinate systems are possible: the kind pictured here and its mirror image. The one shown is called a *right-handed coordinate system*, for the x, y, z axes point like the thumb, index, and middle finger of the right hand, respectively. (See Figure 1.13.) By convention, left-handed coordinate systems are rarely used.

FIGURE 1.13

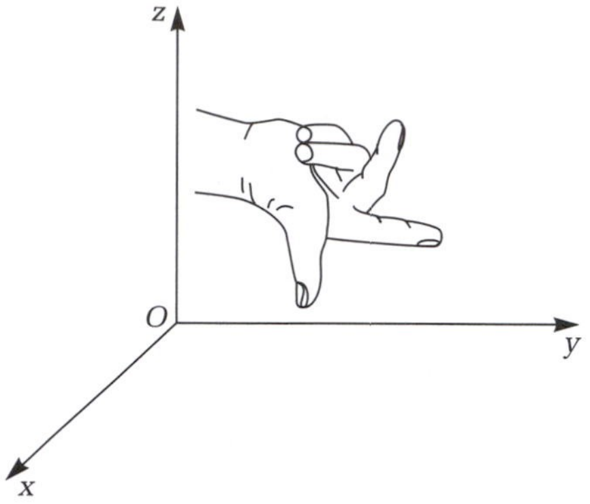

Although there is no way of picturing it when $n > 3$, the n-dimensional vector space $\mathbb{R}^n$ is defined algebraically as follows:

DEFINITION 1.1.4. For any positive integer n, $\mathbb{R}^n$ is the vector space of ordered n-tuples $(x_1, x_2, \ldots, x_n)$ of real numbers, called n-dimensional Euclidean space, with the operations defined by

$$(x_1, x_2, \ldots, x_n) + (y_1, y_2, \ldots, y_n) = (x_1 + y_1, x_2 + y_2, \ldots, x_n + y_n) \tag{1.1.15}$$

and

$$c(x_1, x_2, \ldots, x_n) = (cx_1, cx_2, \ldots, cx_n) \tag{1.1.16}$$

for any vectors $(x_1, x_2, \ldots, x_n)$ and $(y_1, y_2, \ldots, y_n)$ and any scalar c.

Such coordinate vectors with $n > 3$ are used in many applications. For instance, in physics the configuration space of n pointlike particles is defined as the $3n$-dimensional vector space $\mathbb{R}^{3n}$ whose vectors are made up of the particles' coordinates, and the phase space as the $6n$-dimensional vector space $\mathbb{R}^{6n}$ whose vectors' components are the coordinates and the momentum components of the particles. Similarly, in theoretical economics the prices and quantities of n commodities are frequently represented by n-dimensional vectors.

The definition of negatives and of subtraction of vectors can be adopted, for any n, unchanged from the two-dimensional case:

DEFINITION 1.1.5. For any $\boldsymbol{p}, \boldsymbol{q} \in \mathbb{R}^n$ we define

$$-\boldsymbol{p} = (-1)\boldsymbol{p} \tag{1.1.17}$$

and

$$\boldsymbol{p} - \boldsymbol{q} = \boldsymbol{p} + (-\boldsymbol{q}). \tag{1.1.18}$$

Additionally, we adopt the following notational conventions:

$$\boldsymbol{p}c = c\boldsymbol{p} \text{ and } \frac{\boldsymbol{p}}{c} = \frac{1}{c}\boldsymbol{p} \tag{1.1.19}$$

for all vectors and scalars (except for $c = 0$ in the latter, of course).

If $n = 1$, then $\mathbb{R}^1$ denotes the one-dimensional vector space formed by the set $\mathbb{R}$ of real numbers itself, with ordinary addition and multiplication serving as the vector operations. Although $\mathbb{R}$ has more structure than that of a vector space, it is still customary to write $\mathbb{R}$ not just for the field of real numbers but for the vector space $\mathbb{R}^1$ as well.

The theorems we stated for vectors in the plane also remain valid for vectors in $\mathbb{R}^n$, for any n, with the obvious change to n components in Theorem 1.1.2.

Let us remark that all the given operations can be extended to several vectors and scalars much as for numbers, and we shall use such extensions without further ado.

Before closing this section let us consider a three-dimensional example:

EXAMPLE 1.1.8. Given the points $P = (1, 2, 3)$ and $Q = (-1, 6, 5)$, find the midpoint M of the line segment PQ.

Just as in two dimensions, we can write $\overrightarrow{PQ} = \boldsymbol{q} - \boldsymbol{p} = (-2, 4, 2)$ and the position vector $\boldsymbol{m}$ of the point M as

$$\boldsymbol{m} = \boldsymbol{p} + \frac{1}{2}\overrightarrow{PQ} = (1, 2, 3) + \frac{1}{2}(-2, 4, 2) = (0, 4, 4). \tag{1.1.20}$$

<<>>

Notice that in the example above we could also have written

$$\boldsymbol{m} = \boldsymbol{p} + \frac{1}{2}(\boldsymbol{q} - \boldsymbol{p}) = \boldsymbol{p} + \frac{1}{2}\boldsymbol{q} - \frac{1}{2}\boldsymbol{p} = \frac{1}{2}(\boldsymbol{p} + \boldsymbol{q}), \tag{1.1.21}$$

which gives a general formula for the midpoint of a line segment.

In later chapters we discuss many further examples of vector spaces. Most of them do not resemble sets of directed line segments at all, but their vector-space structure allows us to study them together. The set of all polynomials in one variable, the set of all polynomials of degree less than some arbitrary number, the set of all functions continuous on a given interval, the set of all solutions of certain differential equations, and so on, are all vector spaces, just to mention a few.

Exercises 1.1

• **1.1.1.** Referring to Figure 1.2, find expressions in terms of $\boldsymbol{p}$ and $\boldsymbol{q}$ of the free vectors $\overrightarrow{PR}$, $\overrightarrow{PQ}$, $\overrightarrow{QP}$ and of the vectors $\overrightarrow{QC}$, $\overrightarrow{PC}$, and $\overrightarrow{OC}$, where C denotes the center of the parallelogram.

• **1.1.2.** If in Figure 1.5 $\boldsymbol{p} = (1, 3)$ and $\boldsymbol{q} = (4, 2)$, then what are the coordinate vectors of $\boldsymbol{r}$, $\boldsymbol{p} - \boldsymbol{q}$, and $\boldsymbol{q} - \boldsymbol{p}$?

1.1.3. Let $\boldsymbol{p} = (2, 3, -1)$ and $\boldsymbol{q} = (1, 2, 2)$ be two vectors in $\mathbb{R}^3$. Find $\boldsymbol{p} + \boldsymbol{q}$, and draw all three vectors from the origin in the xyz coordinate system to illustrate the parallelogram law in three dimensions.

• **1.1.4.** In $\mathbb{R}^3$ the unit cube is defined as the cube with vertices (0, 0, 0), (0, 0, 1), (0, 1, 0), (1, 0, 0), (0, 1, 1), (1, 0, 1), (1, 1, 0), (1, 1, 1). Find the position vectors (in coordinate form) of the midpoints of the edges, the centers of the faces, and the center of the whole cube.

1.1.5. Draw a diagram to illustrate the second distributive law of vectors (Property 7 of Theorem 1.1.1, page 4) with $a = 2$ and $\boldsymbol{p}$, $\boldsymbol{q}$ any vectors in $\mathbb{R}^2$.

***1.1.6.** Prove the last four parts of Theorem 1.1.3.

• **1.1.7.** Given n point masses m_i at the points with position vectors $\boldsymbol{r}_i$ in either two or three dimensions, their center of mass is defined as the point with position vector

$$\boldsymbol{r} = \frac{1}{M}\sum_{i=1}^{n} m_i\boldsymbol{r}_i, \text{ where } M = \sum_{i=1}^{n} m_i \text{ is the total mass.}$$

(If the masses are equal, the center of mass is called the points' centroid.) If three mass points are given with $m_1 = 2$, $m_2 = 3$, $m_3 = 5$, $\boldsymbol{r}_1 = (2, -1, 4)$, $\boldsymbol{r}_2 = (1, 5, -6)$, and $\boldsymbol{r}_3 = (-2, -5, 4)$, then find $\boldsymbol{r}$.

1.1.8. Show that the definition of the center of mass, given in Exercise 1.1.7, does not depend on the choice of the point O; that is, if the origin of a new coordinate system is denoted O' and the new position vectors $\boldsymbol{r}_i'$, then $\boldsymbol{r}' = \frac{1}{M}\sum_{i=1}^{n} m_i\boldsymbol{r}_i'$ gives the position vector of the center of mass in the new system.

• **1.1.9.** Let $\boldsymbol{a}$, $\boldsymbol{b}$, $\boldsymbol{c}$ be the position vectors of the vertices of a triangle. The point given by $\boldsymbol{p} = \frac{1}{3}(\boldsymbol{a} + \boldsymbol{b} + \boldsymbol{c})$ is the triangle's centroid. Show that it lies one third of the way from the midpoint of any side to the opposite vertex on the line joining these points. (Such a line is called a *median* of the triangle.) Draw an illustration.

1.1.10. Let $\boldsymbol{a}$, $\boldsymbol{b}$, $\boldsymbol{c}$, $\boldsymbol{d}$ be the position vectors of the vertices of a tetrahedron. The point given by $\boldsymbol{p} = \frac{1}{4}(\boldsymbol{a} + \boldsymbol{b} + \boldsymbol{c} + \boldsymbol{d})$ is the tetrahedron's centroid. Show that it lies one fourth of the way from the centroid of any face to the opposite vertex on the line joining these points. (Such a line is called a median of the tetrahedron.) Draw an illustration. Notice that the centroid divides each median in the ratio 1 to 3, in contrast to the ratio 1 to 2 for triangles. Now one vertex balances three vertices instead of two.

1.1.11. Show that for any tetrahedron a line joining the midpoints of opposite edges passes through the tetrahedron's centroid (defined in problem 1.1.10). Thus all three such lines meet in the centroid.

MATLAB Exercises 1.1

In MATLAB, vectors mean coordinate vectors of any length and are denoted by names of up to 19 characters. Extra characters beyond 19 are ignored. There are several ways of entering them. In the following very simple exercises we explore these and various arithmetic operations with vectors. (In the exercises we use boldface characters for vectors and keywords; in MATLAB you have to ignore this convention.)

ML1.1.1. Enter $\boldsymbol{u} = [1\ {-3}\ 6\ 0]$. (Be sure to leave spaces between the numbers.) Enter $\boldsymbol{v} = [1, -2, 2, 5]$. Describe and explain the results of the commands

a. $\boldsymbol{u} + \boldsymbol{v}$

b. $2\boldsymbol{u}$

c. $2 * \boldsymbol{u}$

d. $\boldsymbol{s} = \boldsymbol{u}/2$

e. $\boldsymbol{u}/\boldsymbol{v}$

f. $\boldsymbol{u}./\boldsymbol{v}$

g. $\boldsymbol{u}(2)$

h. $\boldsymbol{t} = \boldsymbol{u}/\boldsymbol{u}(2)$

i. **format rat**; $\boldsymbol{s}$

j. $\boldsymbol{t}$

k. **format short**; $\boldsymbol{s}$

l. $\boldsymbol{s} = 3:8$

m. $\boldsymbol{t} = 1:0.2:2.8$

n. **length**$(\boldsymbol{t})$

ML1.1.2. Let $\boldsymbol{r}_1 = (23, -31, 0)$, $\boldsymbol{r}_2 = (35, 14, -72)$, $\boldsymbol{r}_3 = (52, -25, 44)$, and $\boldsymbol{r}_4 = (12, 52, 24)$ be the position vectors of the vertices of a tetrahedron. Use MATLAB to compute the coordinates of its centroid. (See Exercise 1.1.10.)

ML1.1.3. Redo the computations of Exercise 1.1.7 with MATLAB.

1.2. LENGTH AND DOT PRODUCT OF VECTORS IN $\mathbb{R}^n$

In two and three dimensions the vectors we have discussed have an additional property not covered by Definition 1.1.1, namely length. This concept is not necessarily defined in general vector spaces such as those mentioned at the end of Section 1.1.

If we depict a vector $(x, y) \in \mathbb{R}^2$ by an arrow $\boldsymbol{p}$ anywhere in the xy system, then the Theorem of Pythagoras tells us that the length of $\boldsymbol{p}$, denoted by $|\boldsymbol{p}|$, is given by $|\boldsymbol{p}| = \sqrt{x^2 + y^2}$, and so we define $|(x, y)|$ as this square root.

In $\mathbb{R}^3$ we can deduce a similar formula as follows. For the quantities shown in Figure 1.14, by two applications of the Theorem of Pythagoras we obtain: $d^2 = x^2 + y^2$ and $|\boldsymbol{p}|^2 = d^2 + z^2$, and so $|\boldsymbol{p}|^2 = x^2 + y^2 + z^2$. Thus we define $|(x, y, z)|^2 = x^2 + y^2 + z^2$.

FIGURE 1.14

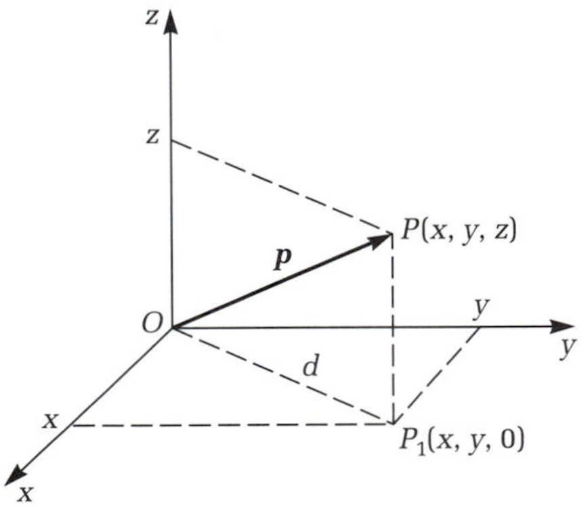

These formulas suggest the following generalization for any positive integer n:

DEFINITION 1.2.1. We define the *length* or *absolute value* of any vector $(x_1, x_2, \ldots, x_n)$ in $\mathbb{R}^n$ by

$$|(x_1, x_2, \ldots, x_n)| = \sqrt{x_1^2 + x_2^2 + \cdots + x_n^2}. \tag{1.2.1}$$

This length has some basic properties, summarized below.

THEOREM 1.2.1. For any vectors $\boldsymbol{p}$, $\boldsymbol{q}$ in $\mathbb{R}^n$ and any scalar c we have

1. $|\boldsymbol{p}| \geq 0$, with equality holding only for $\boldsymbol{p} = \boldsymbol{0}$.
2. $|c\boldsymbol{p}| = |c||\boldsymbol{p}|$, and
3. $|\boldsymbol{p} + \boldsymbol{q}| \leq |\boldsymbol{p}| + |\boldsymbol{q}|$. (Triangle Inequality)

The proofs of the first two parts are straightforward and are therefore omitted, and the proof of the third part (whose name is explained by Figure 1.8) is left to Exercise 1.2.13 on page 23. The three properties of the absolute-value function listed above hold for many other functions as well, and any function satisfying them is called a *norm* on $\mathbb{R}^n$. Those vector spaces in which a norm is defined are called *normed vector spaces*.

The absolute value $|\boldsymbol{p}|$ can be used to associate with any nonzero vector $\boldsymbol{p} \in \mathbb{R}^n$ a vector of length 1, called a *unit vector*, pointing in the same direction, namely $\boldsymbol{u}_p = \dfrac{\boldsymbol{p}}{|\boldsymbol{p}|}$. That $|\boldsymbol{u}_p| = 1$ can be seen from Part 2 of Theorem 1.2.1 with $c = \dfrac{1}{|\boldsymbol{p}|}$.

DEFINITION 1.2.2. The *distance* between two points P, Q with position vectors $\boldsymbol{p}, \boldsymbol{q} \in \mathbb{R}^n$ is defined as the length of $\overrightarrow{PQ} = \boldsymbol{p} - \boldsymbol{q}$; that is, as $|\boldsymbol{p} - \boldsymbol{q}|$.

If we want to define multiplication of vectors in $\mathbb{R}^n$, the most natural idea is to multiply them componentwise. However, another procedure makes a more useful definition, as suggested by the following applications.

Suppose we have n commodities with unit prices $(p_1, p_2, \ldots, p_n)$ and we want to buy the quantities $(q_1, q_2, \ldots, q_n)$ of each. The total amount we have to pay is then given by $p_1q_1 + p_2q_2 + \cdots + p_nq_n$. In probability theory the same expression gives the expected value of a random variable, with the q_i denoting the possible values and the p_i their probabilities. In physics the x-coordinate of the center of mass of point-masses p_i having x-coordinates q_i is given by the same formula divided by the total mass. Later we shall see that the formula in physics that gives the work done by a force moving an object also reduces to the same kind of expression, and in geometry too we shall put it to good use in several ways. Consequently the following definition has been adopted.

DEFINITION 1.2.3. For any vectors $\boldsymbol{p} = (p_1, p_2, \ldots, p_n)$ and $\boldsymbol{q} = (q_1, q_2, \ldots, q_n)$ in $\mathbb{R}^n$, their *scalar* or *dot product* is

$$\boldsymbol{p} \cdot \boldsymbol{q} = p_1q_1 + p_2q_2 + \cdots + p_nq_n. \tag{1.2.2}$$

EXAMPLE 1.2.1. Let $\boldsymbol{p} = (1, 2, 3)$ and $\boldsymbol{q} = (1, 4, -3)$. Then

$$\boldsymbol{p} \cdot \boldsymbol{q} = 1 \cdot 1 + 2 \cdot 4 + 3 \cdot (-3) = 0. \tag{1.2.3}$$

<<>>

In addition to illustrating the computation of such products, this example shows that the scalar product of two vectors can well be zero even if the factors are not. This will turn out to be a very important and useful property.

As we can see, this kind of product results in a scalar, which explains its first name, as opposed to other products to be defined later. As for the second name, we generally denote this product by a dot.

The scalar product has some simple properties:

THEOREM 1.2.2. For any vectors $\boldsymbol{p}$, $\boldsymbol{q}$ and $\boldsymbol{r}$ in $\mathbb{R}^n$ and any scalar c we have

1. $\boldsymbol{p} \cdot \boldsymbol{q} = \boldsymbol{q} \cdot \boldsymbol{p}$
2. $\boldsymbol{p} \cdot (\boldsymbol{q} + \boldsymbol{r}) = \boldsymbol{p} \cdot \boldsymbol{q} + \boldsymbol{p} \cdot \boldsymbol{r}$
3. $c(\boldsymbol{p} \cdot \boldsymbol{q}) = (c\boldsymbol{p}) \cdot \boldsymbol{q} = \boldsymbol{p} \cdot (c\boldsymbol{q})$
4. $\boldsymbol{p} \cdot \boldsymbol{p} = |\boldsymbol{p}|^2$

This theorem can be proved simply by substituting the values of the dot products from their definition, and is left to the reader. Also, as usual, these properties can be combined and extended to several vectors and scalars and to subtraction in an obvious manner. Furthermore, $\boldsymbol{p} \cdot \boldsymbol{p}$ will sometimes be written as $\boldsymbol{p}^2$.

Let us mention that any product (which may be defined by some equation other than Equation (1.2.2); see for instance Exercise 1.2.19) satisfying the first three properties of Theorem 1.2.2 is called an *inner product*,[4] and the vector spaces on which one is defined are called *inner product spaces*. The fourth property in Theorem 1.2.2 can then be used to define a corresponding norm.

Let us return to the case of zero products and prove the following fact:

LEMMA 1.2.1. Let $\boldsymbol{p}$ and $\boldsymbol{q}$ be any vectors in $\mathbb{R}^2$ or $\mathbb{R}^3$. Then $\boldsymbol{p}$ is orthogonal[5] to $\boldsymbol{q}$ if and only if $\boldsymbol{p} \cdot \boldsymbol{q} = 0$.

[4] Hermann Grassmann (1809–1877), who invented the dot product (together with most of vector algebra), gave this name to it to distinguish it from his exterior product, which we shall not discuss. He was led to these names by some geometrical considerations.

[5] "Orthogonal" is synonymous with "perpendicular" but is, for some reason, preferred in this context.

Proof Assume first $\boldsymbol{p} \perp \boldsymbol{q}$. Then by the Theorem of Pythagoras

$$|\boldsymbol{p} - \boldsymbol{q}|^2 = |\boldsymbol{p}|^2 + |\boldsymbol{q}|^2 \tag{1.2.4}$$

(see Figure 1.15). Using Part 4 of Theorem 1.2.2 and the abbreviation $\boldsymbol{p}^2 = \boldsymbol{p} \cdot \boldsymbol{p}$, we can rewrite this expression as

FIGURE 1.15

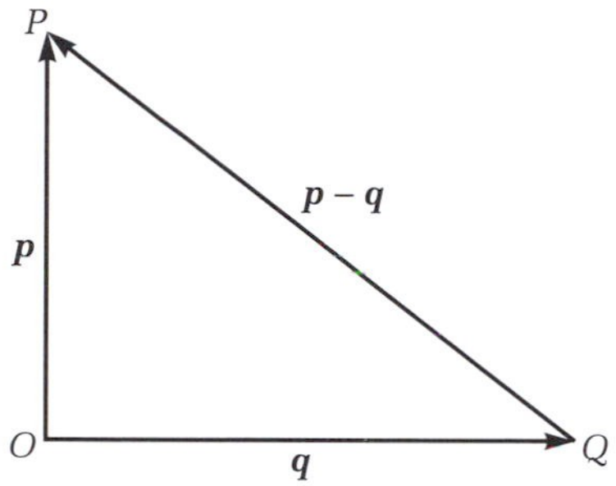

$$(\boldsymbol{p} - \boldsymbol{q})^2 = \boldsymbol{p}^2 + \boldsymbol{q}^2 \tag{1.2.5}$$

and as

$$\boldsymbol{p}^2 - 2\boldsymbol{p} \cdot \boldsymbol{q} + \boldsymbol{q}^2 = \boldsymbol{p}^2 + \boldsymbol{q}^2, \tag{1.2.6}$$

from which $\boldsymbol{p} \cdot \boldsymbol{q} = 0$ follows at once.

Conversely, if $\boldsymbol{p} \cdot \boldsymbol{q} = 0$, then Equation (1.2.6) holds, which implies Equation (1.2.4). But the converse of the Pythagorean Theorem says that Equation (1.2.4) implies that the OPQ triangle is a right triangle with $\boldsymbol{p} \perp \boldsymbol{q}$, provided $\boldsymbol{p} \neq \boldsymbol{0}$ and $\boldsymbol{q} \neq \boldsymbol{0}$ hold. To avoid these exceptions it is generally agreed to call the zero vector orthogonal to every vector, and then we can conclude $\boldsymbol{p} \perp \boldsymbol{q}$ whether $\boldsymbol{p}$ and $\boldsymbol{q}$ are $\boldsymbol{0}$ or not.

<<>>

Whatever the angle between $\boldsymbol{p}$ and $\boldsymbol{q}$, we have this very important characterization of the scalar product:

THEOREM 1.2.3. Let $\boldsymbol{p}$ and $\boldsymbol{q}$ be any vectors in $\mathbb{R}^2$ or $\mathbb{R}^3$. Then

$$\boldsymbol{p} \cdot \boldsymbol{q} = |\boldsymbol{p}||\boldsymbol{q}| \cos \theta, \tag{1.2.7}$$

where $\theta \in [0, \pi]$ is the angle between $\boldsymbol{p}$ and $\boldsymbol{q}$.

Proof Given the vectors $\boldsymbol{p}$ and $\boldsymbol{q}$ drawn from the point O, let us assume first, as shown in Figure 1.16, that $0 \leq \theta \leq \pi/2$ holds. Drop a perpendicular from P onto the line of $\boldsymbol{q}$. Name the point of intersection P_1 and denote the vector $\overrightarrow{OP_1}$ also $\boldsymbol{p}_1$. The latter is called the *orthogonal projection* of $\boldsymbol{p}$ onto the line of $\boldsymbol{q}$. Let us write $\boldsymbol{p}_2 = \boldsymbol{p} - \boldsymbol{p}_1$. Then we have $\boldsymbol{p}_2 \perp \boldsymbol{q}$, and we can write

$$\boldsymbol{p}_1 = c\boldsymbol{u}_q, \tag{1.2.8}$$

where $\boldsymbol{u}_q = \boldsymbol{q}/|\boldsymbol{q}|$ is the unit vector in the direction of $\boldsymbol{q}$, and c is the length of $\boldsymbol{p}_1$. Thus $\boldsymbol{p}$ is decomposed as

$$\boldsymbol{p} = \boldsymbol{p}_1 + \boldsymbol{p}_2 \tag{1.2.9}$$

into two components, parallel and orthogonal to $\boldsymbol{q}$ respectively. If we take the dot product of both sides of Equation (1.2.9) with $\boldsymbol{q}$, then, since $\boldsymbol{p}_2 \cdot \boldsymbol{q} = 0$, we get

$$\boldsymbol{p} \cdot \boldsymbol{q} = \boldsymbol{p}_1 \cdot \boldsymbol{q}. \tag{1.2.10}$$

FIGURE 1.16

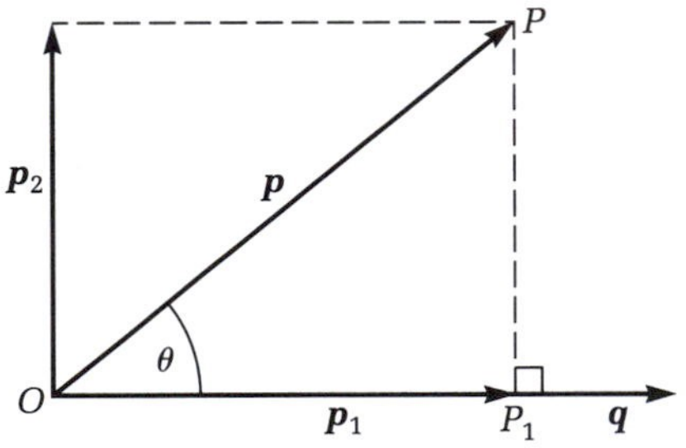

Substituting $\boldsymbol{p}_1$ into this expression from Equation (1.2.8) and using $\boldsymbol{q} \cdot \boldsymbol{q} = |\boldsymbol{q}|^2$, we find

$$\boldsymbol{p} \cdot \boldsymbol{q} = c\boldsymbol{u}_q \cdot \boldsymbol{q} = c\frac{\boldsymbol{q}}{|\boldsymbol{q}|} \cdot \boldsymbol{q} = c\frac{\boldsymbol{q} \cdot \boldsymbol{q}}{|\boldsymbol{q}|} = c\frac{|\boldsymbol{q}|^2}{|\boldsymbol{q}|} = c|\boldsymbol{q}|. \tag{1.2.11}$$

On the other hand, the definition of $\cos\theta$ applied to the OP_1P triangle shows that

$$c = |\boldsymbol{p}| \cos\theta \tag{1.2.12}$$

and substituting this expression into Equation (1.2.11) we obtain

$$\boldsymbol{p} \cdot \boldsymbol{q} = |\boldsymbol{p}||\boldsymbol{q}| \cos\theta, \tag{1.2.13}$$

as was to be shown.

We still have to see what happens if $\pi/2 < \theta \le \pi$. Then, as Figure 1.17 shows, $\boldsymbol{p}_1$ points in the direction opposite that of $\boldsymbol{q}$ and so we can still write $\boldsymbol{p}_1 = c\boldsymbol{u}_q$, although with a negative valued c. But $\cos\theta$ is also negative in this case and therefore Equation (1.2.12) remains valid. Since the sign of c is the only thing changed in the argument above, we arrive at the same conclusion as in the first case.

FIGURE 1.17

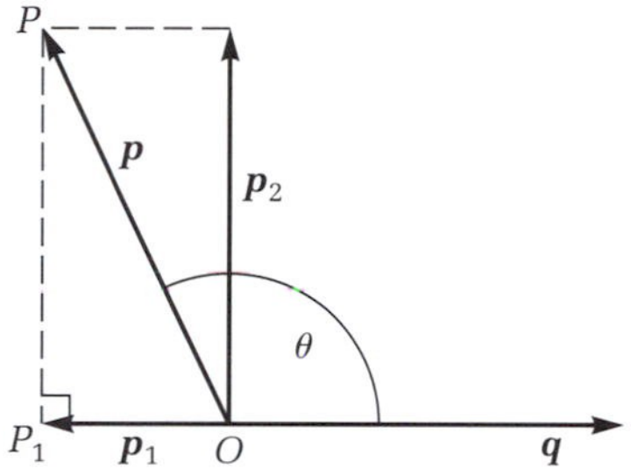

<<>>

The proof above has an important byproduct, which is worth stating separately:

COROLLARY 1.2.1. The orthogonal projection of $\boldsymbol{p}$ onto $\boldsymbol{q}$ is the vector

$$\boldsymbol{p}_1 = |\boldsymbol{p}|\cos\theta\,\frac{\boldsymbol{q}}{|\boldsymbol{q}|} = \frac{\boldsymbol{p}\cdot\boldsymbol{q}}{|\boldsymbol{q}|^2}\boldsymbol{q} \tag{1.2.14}$$

and its length is given by

$$|\boldsymbol{p}_1| = |\boldsymbol{p}||\cos\theta| = \frac{|\boldsymbol{p}\cdot\boldsymbol{q}|}{|\boldsymbol{q}|}. \tag{1.2.15}$$

One of the principal uses of Theorem 1.2.3 is computing the angle between two vectors given in coordinate form:

EXAMPLE 1.2.2. Let $\boldsymbol{p} = (1, 2, 3)$ and $\boldsymbol{q} = (1, -2, 2)$. Then

$$\cos\theta = \frac{\boldsymbol{p}\cdot\boldsymbol{q}}{|\boldsymbol{p}||\boldsymbol{q}|} = \frac{1\cdot 1 + 2\cdot(-2) + 3\cdot 2}{\sqrt{1^2+2^2+3^2}\cdot\sqrt{1^2+(-2)^2+2^2}} \approx .2673 \tag{1.2.16}$$

and $\theta \approx 74.5°$.

<<>>

EXAMPLE 1.2.3. Let us consider the same vectors $\boldsymbol{p}$, $\boldsymbol{q}$ as in Example 1.2.2, and decompose $\boldsymbol{p}$ into the sum of two vectors $\boldsymbol{p}_1$ and $\boldsymbol{p}_2$, parallel and orthogonal to $\boldsymbol{q}$ respectively, as in the proof of Theorem 1.2.3. Then, from Corollary 1.2.1,

$$\boldsymbol{p}_1 = \frac{\boldsymbol{p}\cdot\boldsymbol{q}}{|\boldsymbol{q}|^2}\boldsymbol{q} = \frac{1\cdot 1 + 2\cdot(-2) + 3\cdot 2}{1^2 + (-2)^2 + 2^2}(1, -2, 2) = \frac{1}{3}(1, -2, 2) \quad (1.2.17)$$

and

$$\boldsymbol{p}_2 = \boldsymbol{p} - \boldsymbol{p}_1 = (1, 2, 3) - \frac{1}{3}(1, -2, 2) = \frac{1}{3}(2, 8, 7). \quad (1.2.18)$$

We can easily check that $\boldsymbol{p}_2$ is orthogonal to $\boldsymbol{q}$ by computing their dot product:

$$\boldsymbol{p}_2 \cdot \boldsymbol{q} = \frac{1}{3}(2, 8, 7)\cdot(1, -2, 2) = \frac{1}{3}(2 - 16 + 14) = 0. \quad (1.2.19)$$

<<>>

Let us note that for $n > 3$, Equation (1.2.7) of Theorem 1.2.3 can be used as the *definition* of the angle between vectors in $\mathbb{R}^n$, because Cauchy's inequality (see Exercise 1.2.12) ensures that $|\cos\theta|$ computed this way will not exceed 1.

Also, given the properties of the dot product expressed in Theorem 1.2.2, Theorem 1.2.3 is equivalent to the Law of Cosines: In Figure 1.18 we have $(\boldsymbol{p} - \boldsymbol{q})^2 = \boldsymbol{p}^2 + \boldsymbol{q}^2 - 2\boldsymbol{p}\cdot\boldsymbol{q}$. By using Theorem 1.2.3, this equation can be rewritten as $|\boldsymbol{p} - \boldsymbol{q}|^2 = |\boldsymbol{p}|^2 + |\boldsymbol{q}|^2 - 2|\boldsymbol{p}||\boldsymbol{q}|\cos\theta$, and this is exactly the Law of Cosines for the OPQ triangle in either $\mathbb{R}^2$ or $\mathbb{R}^3$.

FIGURE 1.18

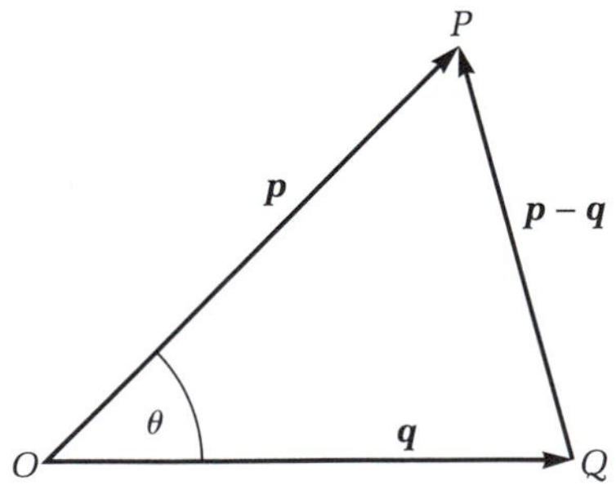

FIGURE 1.19

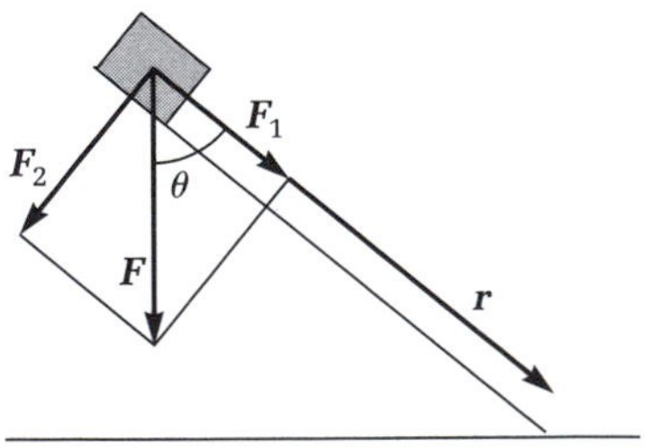

In first discussing the dot product we mentioned that in physics it has another very important use in defining work. Theorem 1.2.3 enables us to describe this application in more detail. Indeed, if $\boldsymbol{F}$ is a force acting on some object and $\boldsymbol{r}$ is the object's displacement caused by $\boldsymbol{F}$, then the corresponding work W is given by $|\boldsymbol{r}|$ times the magnitude $|\boldsymbol{F}| \cos \theta$ of the orthogonal projection of $\boldsymbol{F}$ onto the line of motion; that is, $W = \boldsymbol{F} \cdot \boldsymbol{r}$. Note that $\boldsymbol{F}$ does not have to point in the same direction as $\boldsymbol{r}$. For example, if $\boldsymbol{F}$ denotes the force of gravity moving something down an incline as shown in Figure 1.19, then $\boldsymbol{F}$ can be decomposed into the sum of two forces: $\boldsymbol{F}_1$ along $\boldsymbol{r}$ and $\boldsymbol{F}_2$ orthogonal to $\boldsymbol{r}$. The force $\boldsymbol{F}_2$ does not cause any motion; it just presses the object to the slope. The force $\boldsymbol{F}_1$, on the other hand, is the sole cause of the motion and the work W is proportional to its magnitude $|\boldsymbol{F}| \cos \theta$.

At this point we should mention certain unit vectors that are often used to make formulas simpler. These are the *coordinate unit vectors* or *standard vectors*

$$\boldsymbol{i} = (1, 0), \ \ \boldsymbol{j} = (0, 1) \text{ in } \mathbb{R}^2;$$

$$\boldsymbol{i} = (1, 0, 0), \ \ \boldsymbol{j} = (0, 1, 0), \ \ \boldsymbol{k} = (0, 0, 1) \text{ in } \mathbb{R}^3, \text{ and}$$

$$\boldsymbol{e}_1 = (1, 0, 0, \ldots, 0), \ \ \boldsymbol{e}_2 = (0, 1, 0, \ldots, 0), \ldots, \ \ \boldsymbol{e}_n = (0, 0, \ldots, 0, 1) \text{ in } \mathbb{R}^n.$$

Every vector in these spaces can be decomposed into components along these unit vectors:

$$(x, y) = x\boldsymbol{i} + y\boldsymbol{j} \tag{1.2.20}$$

$$(x, y, z) = x\boldsymbol{i} + y\boldsymbol{j} + z\boldsymbol{k} \tag{1.2.21}$$

$$(x_1, x_2, \ldots, x_n) = x_1\boldsymbol{e}_1 + x_2\boldsymbol{e}_2 + \cdots + x_n\boldsymbol{e}_n. \tag{1.2.22}$$

Writing $\boldsymbol{r} = (x, y, z)$, we can easily see that in $\mathbb{R}^3$

$$\boldsymbol{r} \cdot \boldsymbol{i} = x, \ \ \boldsymbol{r} \cdot \boldsymbol{j} = y, \ \ \boldsymbol{r} \cdot \boldsymbol{k} = z \tag{1.2.23}$$

hold, but just the first two of these equations hold in $\mathbb{R}^2$; and for $\boldsymbol{x} = (x_1, x_2, \ldots, x_n)$ in $\mathbb{R}^n$ we have

$$\boldsymbol{x} \cdot \boldsymbol{e}_i = x_i \ \text{ for } i = 1, 2, \ldots, n. \tag{1.2.24}$$

Exercises 1.2

1.2.1. Let $\boldsymbol{p} = (5, 5)$ and $\boldsymbol{q} = (1, -7)$.

a. Determine $\boldsymbol{p} + \boldsymbol{q}$ and $\boldsymbol{p} - \boldsymbol{q}$.

b. Represent $\boldsymbol{p}$, $\boldsymbol{q}$, $\boldsymbol{p} + \boldsymbol{q}$, and $\boldsymbol{p} - \boldsymbol{q}$ by arrows in a parallelogram.

c. Compute $|\boldsymbol{p}|$, $|\boldsymbol{q}|$, $|\boldsymbol{p} + \boldsymbol{q}|$, and $|\boldsymbol{p} - \boldsymbol{q}|$.

d. Is $|\boldsymbol{p} + \boldsymbol{q}|^2 = |\boldsymbol{p}|^2 + |\boldsymbol{q}|^2$?

1.2.2. Let $\boldsymbol{p} = (2, -2, 1)$ and $\boldsymbol{q} = (2, 3, 2)$. Show that $|\boldsymbol{p} + \boldsymbol{q}|^2 = |\boldsymbol{p}|^2 + |\boldsymbol{q}|^2$ and $|\boldsymbol{p} - \boldsymbol{q}|^2 = |\boldsymbol{p}|^2 + |\boldsymbol{q}|^2$. Interpret geometrically!

1.2.3. Let P, Q, and R be the vertices of a triangle in $\mathbb{R}^2$ or $\mathbb{R}^3$. Use vectors to show that the line segment joining the midpoints of any two sides of the triangle is parallel to and one-half the length of the third side. (Note: two vectors are parallel if and only if one is a scalar multiple of the other.)

• **1.2.4.** Find the angle between the vectors $\boldsymbol{p} = (-2, 4)$ and $\boldsymbol{q} = (3, -5)$.

• **1.2.5.** Find the angle between the vectors $\boldsymbol{p} = (1, -2, 4)$ and $\boldsymbol{q} = (3, 5, 2)$.

1.2.6. Find the angles between various diagonals of the unit cube (defined in Exercise 1.1.4), and between its edges and diagonals.

1.2.7. The line segments joining the centers of the faces of the unit cube form a regular octahedron. Find the angles between its various edges, and try to draw it.

• **1.2.8.** Consider a triangle in the xy plane with vertices $A = (1, 3)$, $B = (2, 4)$, and $C = (4, -1)$. Find (a) the orthogonal projection of $\overrightarrow{AB}$ onto the line of BC, (b) the distance of A from that line, and (c) the area of the triangle.

1.2.9. Decompose the vector $\boldsymbol{p} = (2, -3, 1)$ into components parallel and perpendicular to the vector $\boldsymbol{q} = (12, 3, 4)$.

1.2.10. Prove the so-called parallelogram law for the norm:

$$|\boldsymbol{p} + \boldsymbol{q}|^2 + |\boldsymbol{p} - \boldsymbol{q}|^2 = 2|\boldsymbol{p}|^2 + 2|\boldsymbol{q}|^2$$

for any vectors in $\mathbb{R}^n$. Interpret geometrically!

1.2.11. Using dot products, prove the Theorem of Thales: If we take a point P on a circle and form a triangle by joining it to the opposite ends of an arbitrary diameter, then the angle at P is a right angle. (See Figure 1.20.)

FIGURE 1.20

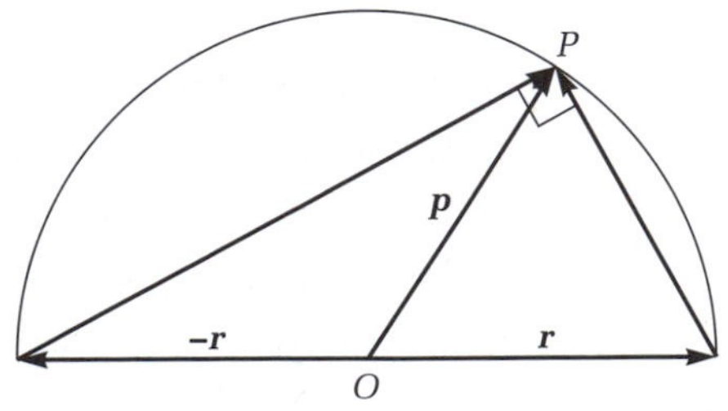

***1.2.12.** Prove Cauchy's[6] Inequality: $|\boldsymbol{p} \cdot \boldsymbol{q}| \le |\boldsymbol{p}||\boldsymbol{q}|$ for any vectors $\boldsymbol{p}$, $\boldsymbol{q}$ in $\mathbb{R}^n$, with equality holding if and only if $\boldsymbol{p}$ is parallel to $\boldsymbol{q}$. (In two or three dimensions this inequality is trivial, since it follows from Theorem 1.1.3 and the fact that $|\cos\theta| \le 1$. For $n > 3$, here is a hint: $(\boldsymbol{p} - \lambda\boldsymbol{q})^2 \ge 0$ for any scalar λ. Expand the left-hand side to obtain a quadratic function of λ. The graph of this function is a parabola above the λ-axis. Find the λ-value of the lowest point in terms of $\boldsymbol{p}$ and $\boldsymbol{q}$, substitute it into the inequality, and simplify.)

***1.2.13.** Using the result of Exercise 1.2.12, prove the Triangle Inequality (Part 3 of Theorem 1.2.1).

***1.2.14.** **a.** Prove the inequality $||\boldsymbol{p}| - |\boldsymbol{q}|| \le |\boldsymbol{p} - \boldsymbol{q}|$ for any vectors in $\mathbb{R}^n$.

b. When do we have equality in Part (a)? Explain!

1.2.15. Let $\boldsymbol{p}$ be any vector in $\mathbb{R}^2$ and $\boldsymbol{u}_p$ the unit vector in its direction. Show that

a. the vector $\boldsymbol{p}$ can be written $\boldsymbol{p} = |\boldsymbol{p}|\,(\cos\phi, \sin\phi)$, where ϕ is the angle from the positive x-axis to $\boldsymbol{p}$;

b. $\boldsymbol{u}_p = (\cos\phi, \sin\phi)$.

1.2.16. Let $\boldsymbol{p}$ be any vector in $\mathbb{R}^3$ and $\boldsymbol{u}_p$ the unit vector in its direction. Show that

a. the components $\boldsymbol{u}_p \cdot \boldsymbol{i}$, $\boldsymbol{u}_p \cdot \boldsymbol{j}$, $\boldsymbol{u}_p \cdot \boldsymbol{k}$ of $\boldsymbol{u}_p$ equal the cosines of the angles α_1, α_2, α_3 between $\boldsymbol{p}$ and the positive coordinate axes (these are called the *direction cosines* of $\boldsymbol{p}$);

b. $\cos^2\alpha_1 + \cos^2\alpha_2 + \cos^2\alpha_3 = 1$ (What formula in $\mathbb{R}^2$ does this one correspond to?);

c. $\boldsymbol{p} = |\boldsymbol{p}|\,(\cos\alpha_1, \cos\alpha_2, \cos\alpha_3)$.

• **1.2.17.** Find the direction cosines (see Exercise 1.2.16) of $\boldsymbol{p} = (3, -4, 12)$, and the angles α_1, α_2, α_3.

1.2.18. Prove the formula $\cos(\alpha - \beta) = \cos\alpha\cos\beta + \sin\alpha\sin\beta$ by considering the scalar product of two unit vectors $\boldsymbol{e}_a = (\cos\alpha, \sin\alpha)$ and $\boldsymbol{e}_b = (\cos\beta, \sin\beta)$.

***1.2.19.** Show that in $\mathbb{R}^2$ an inner product may be defined by $\boldsymbol{p} \cdot \boldsymbol{q} = 2p_1q_1 + p_2q_2$; that is, this product also satisfies the first three properties of Theorem 1.2.2. What is the geometrical meaning of this product?

[6]Named after Augustin Louis Cauchy (1789–1857) and also sometimes after V. I. Bunyaklovsky (1804–1889) and H. A. Schwarz (1843–1921), who generalized it to integrals.

***1.2.20.** Consider an *oblique coordinate system* in the plane with axes labeled ξ and η, as shown in Figure 1.21. Given a vector $\boldsymbol{p}$, let p_1 and p_2 denote the *orthogonal* scalar components of $\boldsymbol{p}$; that is, the signed lengths of the orthogonal projections of $\boldsymbol{p}$ onto the axes, and let p^1 and p^2 denote the *parallel* scalar components of $\boldsymbol{p}$. (Notice that in p^1 and p^2 the 1 and 2 are superscripts, not exponents!)

a. Show that if $\boldsymbol{p} \cdot \boldsymbol{q} = |\boldsymbol{p}||\boldsymbol{q}| \cos\theta$ as usual, then $\boldsymbol{p} \cdot \boldsymbol{q} = p_1 q^1 + p_2 q^2 = p^1 q_1 + p^2 q_2$.

b. Express $\boldsymbol{p} \cdot \boldsymbol{q}$ in the form $\sum_{i,j} g_{ij} p^i q^j$; that is, find appropriate constants g_{ij}.

(In differential geometry and in the theory of relativity such coordinates are very important. The quantities p^1 and p^2 are called the contravariant components and p_1 and p_2 the covariant components of $\boldsymbol{p}$, because of their behavior under coordinate transformations. In Cartesian coordinate systems they coincide.)

FIGURE 1.21

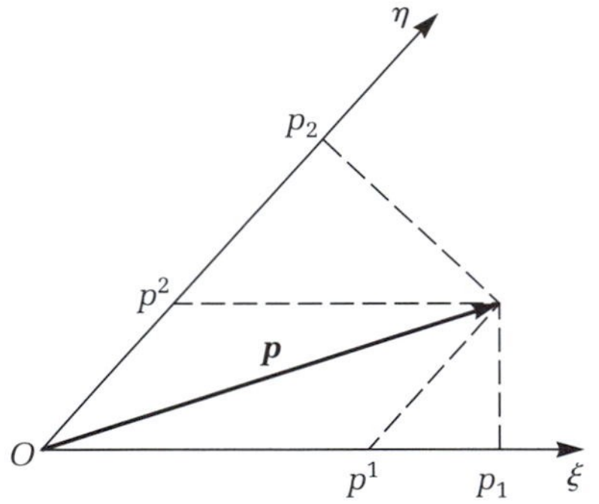

MATLAB Exercises 1.2

In MATLAB the functions **norm**($\boldsymbol{u}$) and **dot**($\boldsymbol{u}$,$\boldsymbol{v}$) return the length and dot product of vectors, respectively. The command **rand**(1,n) generates an n-vector with random components uniformly distributed between 0 and 1.

ML1.2.1. **a.** For $n = 2, 3, 4, 10, 20$, find the cosine of and the angle $\theta(n)$ between $\boldsymbol{u} = [1, 1, \ldots, 1]$ and $\boldsymbol{v} = [1, 0, 0, \ldots, 0]$. (To enter $\boldsymbol{u}$ and $\boldsymbol{v}$ use, for each n, $\boldsymbol{u}$ = **ones**(1,n), $\boldsymbol{v}$ = **zeros**(1,n), $\boldsymbol{v}(1) = 1$. The MATLAB function **acos**(x) gives the inverse cosine in radians.)

b. Make a conjecture for the value of $\lim_{n\to\infty} \theta(n)$ and prove it.

ML1.2.2. **a.** For n = 10, 50, 100, 500, find the angle $\theta(n)$ between $\boldsymbol{u}$ = **rand**$(1,n) - 1/2$ and $\boldsymbol{v}$ = **rand**$(1,n) - 1/2$. Use the up-arrow key to repeat each step several times.

b. What do you observe? Can you give a heuristic explanation?

ML1.2.3. Use MATLAB to decompose the vector $\boldsymbol{u} = [1, 1, 1, 1, 1]$ into the sum of a vector parallel to $\boldsymbol{v} = [1, 2, 3, 4, 5]$ and orthogonal to it.

1.3. LINES AND PLANES

We now have the necessary machinery for developing equations for lines and planes in $\mathbb{R}^n$ and for computing corresponding intersections, distances, and angles.

We start with lines. Let $\boldsymbol{p}_0 = (x_0, y_0, z_0)$ and $\boldsymbol{v} = (v_1, v_2, v_3) \neq \boldsymbol{0}$ be arbitrary vectors of $\mathbb{R}^3$; that is, triples of numbers. We consider $\boldsymbol{p}_0$ as the position vector, starting at O, of a point P_0 in three-dimensional space, but place the representative arrow of $\boldsymbol{v}$ conveniently at P_0. Let L denote the line drawn through the point P_0 along $\boldsymbol{v}$. (See Figure 1.22). Then obviously the position vector $\boldsymbol{p} = (x, y, z)$ of any point P on L can be written $\boldsymbol{p}_0 + t\boldsymbol{v}$ for some appropriate number t. Conversely, $\boldsymbol{p}_0 + t\boldsymbol{v}$ is, for every number t, the position vector of a point P on L. Thus

$$\boldsymbol{p} = \boldsymbol{p}_0 + t\boldsymbol{v} \tag{1.3.1}$$

is the desired equation of the line L.

FIGURE 1.22

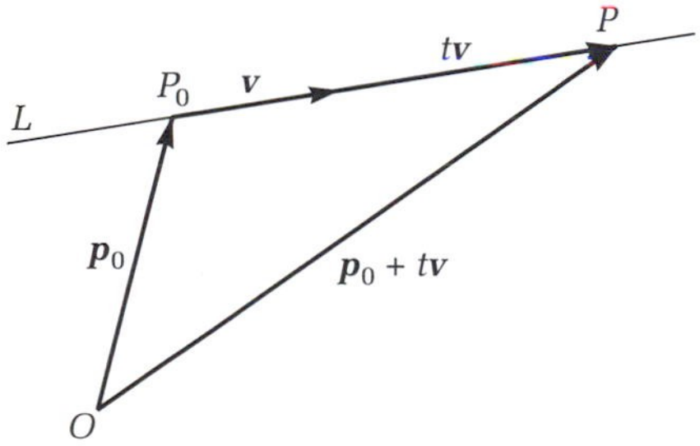

The variable t is called a *parameter* and Equation (1.3.1) a *parametric vector equation* of the line L. It describes the line L *with a scale superimposed on it*. This scale has $t = 0$ at P_0 and $t = 1$ at the point with position vector $\boldsymbol{p}_0 + \boldsymbol{v}$. Clearly, the same line has many parametric representations, since there are many ways of putting a scale on it. For example, replacing t by $2t$ in Equation (1.3.1) we get a different parametric equation of the same line, in which only the scale has been changed to one with intervals of doubled lengths. The point P_0 can also be changed to any other point of the line; this change would just move the zero point of the scale, and would result in a different equation, but one that still describes the same line. (See, for instance, Exercise 1.3.9.)

Also, we may think of t as denoting time, and then Equation (1.3.1) describes not only the line L but also the motion of a point along L, which is at the position P at time t and at P_0 at time 0. In this interpretation the vector $\boldsymbol{v}$ stands for the velocity of the moving point, but in general it is called a *direction vector* of L.

Because equality of vectors means equality of components, Equation (1.3.1) is equivalent to the three *parametric scalar equations*

$$x = x_0 + tv_1, \quad y = y_0 + tv_2, \quad z = z_0 + tv_3. \tag{1.3.2}$$

The situation is entirely analogous in $\mathbb{R}^n$. In fact, Equation (1.3.1) remains unchanged, but with the vectors reinterpreted as lying in $\mathbb{R}^n$.

If none of the components of $\boldsymbol{v}$ is zero, then we can solve each of the equations (1.3.2) for t and we obtain nonparametric equations for L:

$$\frac{x - x_0}{v_1} = \frac{y - y_0}{v_2} = \frac{z - z_0}{v_3}. \tag{1.3.3}$$

Notice that a line in three dimensions is given by the two nonparametric equations (1.3.3) rather than by just one equation as in two dimensions. As we will see shortly, the explanation for this difference is that each of the two equations describes a plane and the line is then represented as the intersection of these planes.

EXAMPLE 1.3.1. Let us find equations for the line L that passes through the points $A(2, -3, 5)$ and $B(6, 1, -8)$. We may take either one of the given points as P_0, and the vector $\overrightarrow{AB}$ as $\boldsymbol{v}$. We put $\boldsymbol{p}_0 = (2, -3, 5)$ and $\boldsymbol{v} = (6 - 2, 1 + 3, -8 - 5) = (4, 4, -13)$, and so we obtain the parametric equation

$$\boldsymbol{p} = (2, -3, 5) + t(4, 4, -13) \tag{1.3.4}$$

or

$$(x, y, z) = (2 + 4t, -3 + 4t, 5 - 13t) \tag{1.3.5}$$

for the line L. The corresponding scalar equations are

$$x = 2 + 4t, \quad y = -3 + 4t, \quad z = 5 - 13t, \tag{1.3.6}$$

and eliminating t leads to the nonparametric equations

$$\frac{x - 2}{4} = \frac{y + 3}{4} = \frac{z - 5}{-13}. \tag{1.3.7}$$

<<>>

If a component of $\boldsymbol{v}$ is zero, then the corresponding equation is already in nonparametric form, and the line is parallel to one or two of the coordinate planes, as in Example 1.3.2.

EXAMPLE 1.3.2. Let us find equations for the line L that passes through the point $A(2, -3, 5)$ parallel to the z-axis. In this case only the z-coordinate varies and the parametric vector equation can immediately be written as

$$\boldsymbol{p} = (2, -3, 5) + t(0, 0, 1). \tag{1.3.8}$$

In components, this is

$$x = 2, \quad y = -3, \quad z = 5 + t, \tag{1.3.9}$$

and the nonparametric equations are just the first two of these.

<<>>

EXAMPLE 1.3.3. Let us find the intersection of the two lines L_1 and L_2 given by

$$\boldsymbol{p} = (4, 3, 9) + s(2, -3, 7) \text{ and } \boldsymbol{p} = (3, 2, 0) + t(-1, 4, 2), \tag{1.3.10}$$

if there is one. (Notice that we used two different parameters s and t, for using only one would have meant looking not just for the point of intersection but also for two moving points to be there at the same time. See Exercise 1.3.9.) The obvious way to attack this problem is to equate the corresponding scalar components of the two expressions for $\boldsymbol{p}$ and solve the resulting equations for s and t. Thus we decompose the vector equation

$$(4, 3, 9) + s(2, -3, 7) = (3, 2, 0) + t(-1, 4, 2) \tag{1.3.11}$$

as

$$4 + 2s = 3 - t, \quad 3 - 3s = 2 + 4t, \quad 9 + 7s = 2t. \tag{1.3.12}$$

We can easily solve these equations to obtain $s = -1$ and $t = 1$. (Notice that in general we cannot expect to have solutions for two unknowns in three equations, which corresponds to the fact that in three dimensions most lines avoid each other. More about this subject in Chapter 2.) Substituting these parameter values into either of Equations (1.3.10), we obtain the position vector of the point of intersection as

$$\boldsymbol{p} = (2, 6, 2). \tag{1.3.13}$$

<<>>

EXAMPLE 1.3.4. Find the distance of the point $A(9, 13, -1)$ from the line L given by $\boldsymbol{p} = (-1, -2, 4) + t(3, 1, -5)$. First, pick any point Q on L, say $Q = (-1, -2, 4)$. Second, decompose $\boldsymbol{r} = \overrightarrow{QA}$ into two components $\boldsymbol{r}_1$ and $\boldsymbol{r}_2$, respectively, parallel and orthogonal to $\boldsymbol{v} = (3, 1, -5)$. Then $\boldsymbol{r} = (10, 15, -5)$ and $\boldsymbol{r}_1 = \frac{\boldsymbol{r} \cdot \boldsymbol{v}}{\boldsymbol{v}^2}\boldsymbol{v} = \frac{70}{35}(3, 1, -5) = (6, 2, -10)$. Consequently, $\boldsymbol{r}_2 = \boldsymbol{r} - \boldsymbol{r}_1 = (4, 13, 5)$ and the required distance is obtained as $|\boldsymbol{r}_2| = \sqrt{210}$.

<<>>

Let us now consider planes. To obtain parametric equations we may proceed very much as for lines, but with *two* vectors in place of $\boldsymbol{v}$ and *two* parameters s and t instead of one. Thus, let $\boldsymbol{p}_0 \in \mathbb{R}^3$ be regarded as the radius vector of a fixed point P_0 of the plane S we want to describe, P a variable point, and $\boldsymbol{u}$ and $\boldsymbol{v}$ two nonparallel, nonzero vectors of $\mathbb{R}^3$ with their representative arrows drawn in S, as shown in Figure 1.23.

FIGURE 1.23

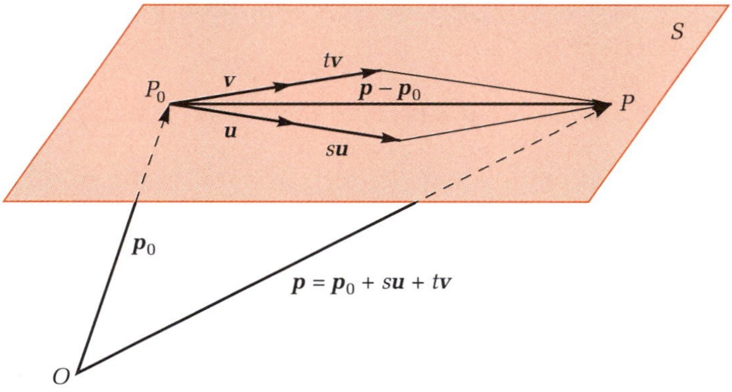

Then $\boldsymbol{p}$ can be expressed as

$$\boldsymbol{p} = \boldsymbol{p}_0 + s\boldsymbol{u} + t\boldsymbol{v}, \tag{1.3.14}$$

where s and t are appropriate real numbers. Conversely, any pair $s, t \in \mathbb{R}$ determines a point of S through Equation (1.3.14). This then is the parametric vector equation of S. In components, this vector equation becomes the set of three scalar equations

$$x = x_0 + su_1 + tv_1,\ \ y = y_0 + su_2 + tv_2,\ \ z = z_0 + su_3 + tv_3. \tag{1.3.15}$$

EXAMPLE 1.3.5. Let us write a parametric vector equation for the plane S containing the two intersecting lines of Example 1.3.3. We may take any point of either line to be P_0 and the same vectors for $\boldsymbol{u}$ and $\boldsymbol{v}$ as in Example 1.3.3. Then if we take $P_0 = (4, 3, 9)$, say, S will be described by the equation

$$\boldsymbol{p} = (4, 3, 9) + s(2, -3, 7) + t(-1, 4, 2). \qquad (1.3.16)$$

Let us write this expression out in components as in Equation (1.3.15) and eliminate s and t: Then

$$x = 4 + 2s - t, \;\; y = 3 - 3s + 4t, \;\; z = 9 + 7s + 2t. \qquad (1.3.17$$

Bring the constant terms to the left, multiply the first of these equations in turn by 4 and 2, and add the results to the second and third equations, respectively, to get

$$4(x - 4) + (y - 3) = 5s \text{ and } 2(x - 4) + (z - 9) = 11s. \qquad (1.3.18)$$

Now multiply the first of these equations by 11, the second one by (−5), and add the results. Then we obtain this nonparametric equation for S:

$$34(x - 4) + 11(y - 3) - 5(z - 9) = 0. \qquad (1.3.19)$$

<<>>

If we eliminate s and t from Equations (1.3.15) as in the example above, then in general we end up with an equation of the form

$$a(x - x_0) + b(y - y_0) + c(z - z_0) = 0, \qquad (1.3.20)$$

where a, b, c are appropriate numbers arising from the elimination process.[7] It is reasonable to ask what their geometric meaning may be. Now $x - x_0$, $y - y_0$, $z - z_0$ are the components of the vector $\boldsymbol{p} - \boldsymbol{p}_0$ and if we consider a, b, c to be the components of a vector $\boldsymbol{n}$, then Equation (1.3.20) may be written in vector form as

$$\boldsymbol{n} \cdot (\boldsymbol{p} - \boldsymbol{p}_0) = 0. \qquad (1.3.21)$$

This equation shows that the vector $\boldsymbol{n}$ is orthogonal to the variable vector $\boldsymbol{p} - \boldsymbol{p}_0$ lying in S (see Figure 1.24), and so it must be orthogonal to the plane S; that is, to every vector in S. Such a vector is called a *normal vector* of S, and one usually says it is *normal* to S rather than orthogonal to S.

[7] In Chapter 6 (page 212) we discuss a shortcut method for such eliminations in three dimensions, which involves the so-called cross product of vectors.

FIGURE 1.24

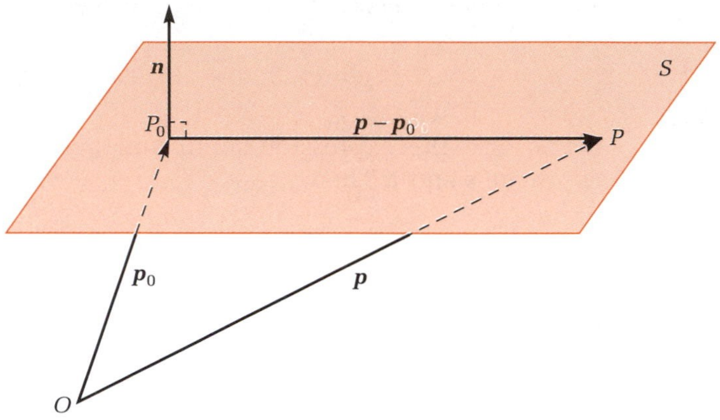

Equation (1.3.21) may be rewritten as

$$\boldsymbol{n} \cdot \boldsymbol{p} = \boldsymbol{n} \cdot \boldsymbol{p}_0 \tag{1.3.22}$$

and if we set $d = \boldsymbol{n} \cdot \boldsymbol{p}_0$ and write Equation (1.3.22) in components, then we get

$$ax + by + cz = d \tag{1.3.23}$$

as the simplest type of equation for a plane.

It is also natural to ask what the geometric meaning of d is. From its definition we see that $d = |\boldsymbol{n}||\boldsymbol{p}_0| \cos \theta$ (see Figure 1.25), and this is $|\boldsymbol{n}|$ times the projection of $\boldsymbol{p}_0$ onto the line of $\boldsymbol{n}$. The vector $\boldsymbol{p}_0$ joins the origin to S. Thus the projection of $\boldsymbol{p}_0$ onto the line of $\boldsymbol{n}$, which is perpendicular to S, gives the distance of O from S if θ is acute, and the negative of this distance if θ is obtuse. Thus d equals $|\boldsymbol{n}|$ times the distance of O from S if $\boldsymbol{n}$ points from O toward S and the negative of this distance otherwise.

FIGURE 1.25

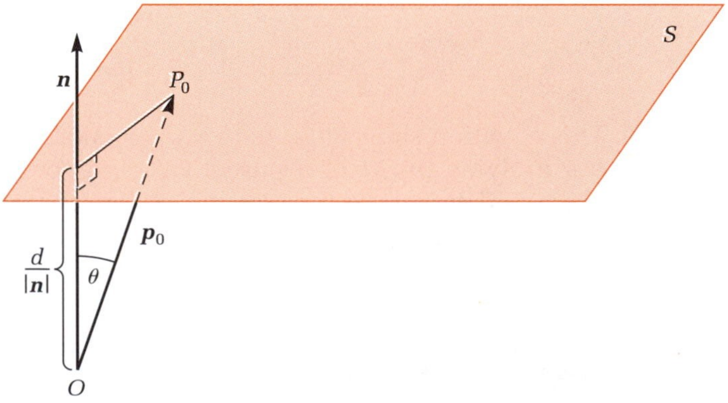

Although planes can be visualized only in three dimensions, the vector equations just deduced remain valid in $\mathbb{R}^n$ for $n > 3$ as well, with the obvious reinterpretation of the vectors. There is a difference, however, in the meaning of the parametric and the nonparametric equations, which did not occur in $\mathbb{R}^3$, namely that the parametric Equation (1.3.14) describes two-dimensional sets in $\mathbb{R}^n$, which may justifiably still be called planes, but the nonparametric Equation (1.3.21) or (1.3.22) describes $(n-1)$-dimensional sets that are usually called *hyperplanes*.

EXAMPLE 1.3.6. Given a plane S with equation $4x - 4y + 7z = 1$ and a point $A(5, 2, -3)$, find the distance D of A from S. We may solve this problem by picking any point P of S, say $P(2, 0, -1)$, and projecting the vector $\overrightarrow{AP} = (-3, -2, 2)$ onto $\boldsymbol{n}$. From the equation of S we can read off $\boldsymbol{n} = (4, -4, 7)$. Thus

$$D = \frac{|\overrightarrow{AP} \cdot \boldsymbol{n}|}{|\boldsymbol{n}|} = \frac{|-12 + 8 + 14|}{\sqrt{16 + 16 + 49}} = \frac{10}{9}. \tag{1.3.24}$$

<<>>

EXAMPLE 1.3.7. Find the perpendicular distance D between the lines L_1 and L_2 given by the equations

$$x = 1 + 4s, \;\; y = s, \;\; z = -2 + 3s, \tag{1.3.25}$$

and

$$x = 2 + 2t, \;\; y = -1 + t, \;\; z = 0. \tag{1.3.26}$$

We can solve this problem by first finding a vector $\boldsymbol{n}$ that is orthogonal to the direction vectors $\boldsymbol{u} = (4, 1, 3)$ and $\boldsymbol{v} = (2, 1, 0)$ of the given lines and second, by projecting the vector joining an arbitrary point P of L_1 to an arbitrary point Q of L_2 onto $\boldsymbol{n}$. To find an appropriate $\boldsymbol{n}$ we may consider a plane S that contains the vectors $\boldsymbol{u}$ and $\boldsymbol{v}$ through any point P_0 and find a normal vector of S as in Example 1.3.5. Whether L_1 and L_2 intersect or not, it is most convenient to choose $P_0 = O$ and the parametric equations

$$\boldsymbol{p} = s\boldsymbol{u} + t\boldsymbol{v}; \tag{1.3.27}$$

that is,

$$x = 4s + 2t, \;\; y = s + t, \;\; z = 3s \tag{1.3.28}$$

for the plane S. Eliminating the parameters yields

$$3x - 6y - 2z = 0, \tag{1.3.29}$$

and so we obtain $\boldsymbol{n} = (3, -6, -2)$. We may choose $P = (1, 0, -2)$ and $Q = (2, -1, 0)$ and then $\overrightarrow{PQ} = (1, -1, 2)$. The required distance can therefore be computed as

$$D = \frac{|\overrightarrow{PQ} \cdot \boldsymbol{n}|}{|\boldsymbol{n}|} = \frac{|3 + 6 - 4|}{\sqrt{9 + 36 + 4}} = \frac{5}{7}. \tag{1.3.30}$$

<<>>

Exercises 1.3

In the first eight exercises, find equations for the indicated lines in both parametric and nonparametric forms.

- **1.3.1.** Through $P_0(1, -2, 4)$ and along $\boldsymbol{v} = (2, 3, -5)$.
- **1.3.2.** Through $P_0(1, -2, 4)$ and parallel to the y-axis.
- **1.3.3.** Through $P_0(7, -2, 5)$ and $P_1(5, 6, -3)$.

1.3.4. Through $P_0(1, -2, 4)$ and $P_1(1, 6, -3)$.

1.3.5. Through $P_0(1, -2, 4)$ and $P_1(1, -2, -3)$.

- **1.3.6.** Through $P_0(5, 4, -8)$ and normal to the plane given by $3x - 4y + 3z = 7$.
- **1.3.7.** Through $P_0(5, 4, -8)$ and parallel both to the plane given by $3x - 4y + 3z = 7$ and to the xy-plane. Hint: The direction vector of the line sought must be orthogonal to the normal vectors of the two planes.

1.3.8. Through $P_0(1, -2, 4)$ and parallel to the planes given by $3x - 4y + 3z = 7$ and $-x + 3y + 4z = 8$.

1.3.9. **a.** Plot the lines $\boldsymbol{p} = (4, 3) + t(2, -3)$ and $\boldsymbol{p} = (3, 2) + t(-1, 4)$ in $\mathbb{R}^2$, indicating the points with $t = 0, \pm 1$ on each.

b. Explain why there is no common t value for the point of intersection.

c. Change the parametrization of each line (that is, write new equations for them, employing a new parameter) so that the new common parameter, say s, is 0 for both lines at the point of intersection.

1.3.10. Show that in $\mathbb{R}^n$, for any n, $\boldsymbol{p} = t\boldsymbol{a} + (1 - t)\boldsymbol{b}$ is a parametric vector equation of the line through the two points with position vectors $\boldsymbol{a}$, $\boldsymbol{b}$. (The numbers t and $1 - t$ are called *barycentric coordinates of* P, since this formula gives the center of mass of two point masses of size t and $1 - t$ at A and B respectively (cf. Exercise 1.1.7.)) What is the *geometric* meaning of t and $1 - t$?

***1.3.11.** Show that in $\mathbb{R}^n$, for any n, $\boldsymbol{p} = r\boldsymbol{a} + s\boldsymbol{b} + t\boldsymbol{c}$, with $0 \le r, s, t$; $r + s + t = 1$, represents the points P of a triangle with vertices given by the position vectors $\boldsymbol{a}$, $\boldsymbol{b}$, $\boldsymbol{c}$. (The numbers r, s, t are again called barycentric coordinates of P, for reason like the one in the previous exercise.) *Hint:* Generalize the solution of Exercise 1.1.9.

In the next eight exercises, find equations for the indicated planes in both parametric and nonparametric forms.

• **1.3.12.** Through $P_0(1, -2, 4)$ and containing the line given by $\boldsymbol{p} = (3, -2, 1) + t(2, 1, -3)$.

1.3.13. Through O and containing the line given by $\boldsymbol{p} = (3, -2, 1) + t(2, 1, -3)$.

1.3.14. Through O and orthogonal to the line given by $\boldsymbol{p} = (3, -2, 1) + t(2, 1, -3)$.

1.3.15. Through $P_0(5, 4, -8)$ and orthogonal to the line given by $\boldsymbol{p} = (3, -2, 1) + t(2, 1, -3)$.

1.3.16. Through $P_0(1, -2, 4)$ and parallel to the plane given by $3x - 4y + 3z = 7$.

1.3.17. Through $P_0(5, 4, -8)$ and parallel to the plane given by $7x + y + 2z = 8$.

1.3.18. Through the points O, $P_1(1, 6, -3)$ and $P_2(7, -2, 5)$.

1.3.19. Through the points $P_0(5, 4, -8)$, $P_1(1, 6, -3)$, and $P_2(7, -2, 5)$.

In the next six exercises find the points of intersection.

• **1.3.20.** Of the two lines $\boldsymbol{p} = (5, 1, 1) + s(-2, 1, 6)$ and $\boldsymbol{p} = (3, -2, 1) + t(2, 1, -3)$.

• **1.3.21.** Of the two lines $\boldsymbol{p} = (-5, 4, -1) + s(2, 1, -7)$ and $\boldsymbol{p} = (9, -9, -2) + t(2, -4, 5)$.

• **1.3.22.** Of the line $\boldsymbol{p} = (5, 1, 1) + s(-2, 1, 6)$ and the plane $7x + y + 2z = 8$.

1.3.23. Of the line $\boldsymbol{p} = (3, -2, 6) + s(-3, 5, 7)$ and the plane $3x + 2y - 2z = 3$.

1.3.24. Of the line $\boldsymbol{p} = (3, -2, 6) + s(-3, 5, 7)$ and the plane $\boldsymbol{p} = (4, -2, 1) + s(-2, 1, 3) + t(1, 3, 2)$.

1.3.25. Of the line $\boldsymbol{p} = (3, 2, -4) + s(7, -5, 4)$ and the plane $\boldsymbol{p} = (0, -2, 1) + s(-3, 0, 3) + t(2, -3, 4)$.

In the next six exercises find the distances.

• **1.3.26.** Between the point $P_0(1, -2, 4)$ and the plane $3x + 2y - 2z = 3$.

• **1.3.27.** Between the point $P_0(3, 4, 0)$ and the plane $y - 2z = 5$.

1.3.28. Between the lines $\boldsymbol{p} = (3, -2, 6) + s(-3, 5, 7)$ and $\boldsymbol{p} = (5, 1, 1) + t(-2, 1, 6)$.

1.3.29. Between the lines $\boldsymbol{p} = (2, 1, 5) + s(-4, 0, 3)$ and $\boldsymbol{p} = (0, -2, 3) + t(5, 0, -2)$.

1.3.30. Between the point $P_0(3, 4, 0)$ and the line L: $\boldsymbol{p} = (3, -2, 6) + s(-3, 5, 7)$. (*Hint:* Pick an arbitrary point Q on L and decompose the vector $\overrightarrow{QP_0}$ into components parallel and perpendicular to L.)

1.3.31. Between the point $P_0(1, -2, 4)$ and the line $\boldsymbol{p} = (3, 2, -4) + s(7, -5, 4)$. (See the hint in Exercise 1.3.30.)

1.3.32. What is the geometric meaning of Equation (1.3.21) in $\mathbb{R}^2$? Make a drawing and explain.

1.3.33. Let the equation of a plane S be given in the form $\boldsymbol{n} \cdot \boldsymbol{p} = d$, with $|\boldsymbol{n}| = 1$. Let us define a function by $f(\boldsymbol{q}) = \boldsymbol{n} \cdot \boldsymbol{q} - d$, where $\boldsymbol{q}$ is the position vector of any point Q, whether in S or not. Show that the value of $f(\boldsymbol{q})$ equals the signed distance of Q from S, which is positive if $\boldsymbol{n}$ points from S toward Q, and negative if $\boldsymbol{n}$ points from Q toward S.

1.3.34. Redo Exercise 1.3.26 by using the result of Exercise 1.3.33.

1.3.35. Redo Exercise 1.3.27 by using the result of Exercise 1.3.33.

1.3.36. Let P denote a variable point on a line L given by $\boldsymbol{p} = \boldsymbol{p}_0 + t\boldsymbol{v}$, and let Q denote any point in space not on L. Show that $\overrightarrow{PQ}$ is orthogonal to $\boldsymbol{v}$ if and only if the distance between P and Q is minimized as a function of t.

***1.3.37.** Find an equation for the *normal transverse* L of the lines given in Exercise 1.3.29. (This term means the line connecting the given ones orthogonally.) *Hint:* First, find the direction vector $\boldsymbol{v}$ of L, then a plane S containing $\boldsymbol{v}$ and one of the given lines, and last, the point of intersection of S and the other line.

MATLAB Exercises 1.3

In these six exercises find the distances using MATLAB:

ML1.3.1. Between the point $P_0(1, -2, 4)$ and the plane $3x + 2y - 2z = 3$.

ML1.3.2. Between the point $P_0(1, -2, 4, 5)$ and the hyperplane $3x + 2y - 2z + w = 3$ in $\mathbb{R}^4$.

ML1.3.3. Between the point $P_0(3, 4, 0)$ and the line $\boldsymbol{p} = (3, -2, 6) + s(-3, 5, 7)$.

ML1.3.4. Between the point $P_0(3, 4, 0, 3, 4, 0)$ and the line $\boldsymbol{p} = (3, -2, 6, 3, -2, 6) + s(-3, 5, 7, -3, 5, 7)$ in $\mathbb{R}^6$.

ML1.3.5. Between the lines $\boldsymbol{p} = (3, -2, 6, 4) + s(-3, 5, 7, 1)$ and $\boldsymbol{p} = (5, 1, 1, 2) + t(-2, 1, 6, 2)$ in $\mathbb{R}^4$.

ML1.3.6. Between the lines $\boldsymbol{p} = (2, 1, 5, 2, 1, 5) + s(-4, 0, 3, -4, 0, 3)$ and $\boldsymbol{p} = (0, -2, 3, 0, -2, 3) + t(5, 0, -2, 5, 0, -2)$ in $\mathbb{R}^6$.

<<2>> SYSTEMS OF LINEAR EQUATIONS, MATRICES

2.1. GAUSSIAN ELIMINATION

Systems of linear equations arise in many applications and we have already encountered a few simple cases in Chapter 1. Other examples in which they occur are: least-squares fitting of lines to observed data, methods for the approximate solution of various differential equations, Kirchhoff's equations relating currents and potentials in electrical circuits, and many economic models. In most of these applications the number of equations and unknowns can be quite large, sometimes in the hundreds or thousands, and it is very important therefore to understand the structure of such systems and to apply systematic and efficient methods for their solution. Even more important is that, as we shall see, studying such systems leads to several new concepts and theories that are at the heart of linear algebra.

We begin with a simple example:

EXAMPLE 2.1.1. Let us solve the following system of three equations in three unknowns:

$$\begin{aligned} 2x + 3y - z &= 8 \\ 4x - 2y + z &= 5 \\ x + 5y - 2z &= 9. \end{aligned} \tag{2.1.1}$$

(Geometrically this problem amounts to finding the point of intersection of three planes.) We want to proceed as follows: multiply both sides of the first equation by 2 and subtract the result from the second equation to eliminate the $4x$, and subtract $1/2$ times the first equation from the third equation to eliminate the x. The system is then changed into the equivalent, new system:[1]

[1]Two systems are called *equivalent* if their solution sets are the same. All elimination steps in this section, like the ones above, are designed to produce equivalent but simpler systems.

$$\begin{aligned} 2x + 3y - z &= 8 \\ -8y + 3z &= -11 \\ \tfrac{7}{2}y - \tfrac{3}{2}z &= 5. \end{aligned} \tag{2.1.2}$$

As our next step we want to get rid of the $7y/2$ term in the last equation, again without affecting the solution set. We can achieve this elimination by multiplying the middle equation by $-7/16$ and subtracting the result from the last equation. Then we get:

$$\begin{aligned} 2x + 3y - z &= 8 \\ -8y + 3z &= -11 \\ \frac{-3}{16}z &= \frac{3}{16}. \end{aligned} \tag{2.1.3}$$

At this point we can easily find the solution by starting with the last equation and working our way back up: First, we find $z = -1$, and second, substituting this result into the middle equation we get $-8y - 3 = -11$, which yields $y = 1$. Last, we enter these values into the top equation and obtain $2x + 3 + 1 = 8$; thus $x = 2$.

<<>>

The example above exhibits many of the essential features of Gaussian elimination,[2] the foremost method for solving such systems. Before discussing these features, however, let us mention that the way the computations were presented was the way a computer would be programmed to do them. For people, slight variations are preferable: we would rather avoid fractions, and if we want to eliminate, say, x from an equation beginning with bx using an equation beginning with ax, with a and b nonzero integers, then we multiply one by a and the other by b to get abx in both. Also, we would sometimes add and sometimes subtract, depending on the signs of the terms involved, where computers would always subtract. Last, we might use the equations in a different sequence because we can, and computers cannot, easily discern an advantage in doing so, which for computers may not even be an advantage.[3] For example, right at the start of the example above we could have put the last equation on top because it begins with x rather than $2x$, and used that equation as we have used the one beginning with $2x$.

The essence of the method, whichever way the computations are arranged, is to subtract multiples of the first equation from the others so

[2]Named after Carl Friedrich Gauss (1777–1855). It is ironic that in spite of his many great achievements he is best remembered for this simple but widely used method and for the so-called Gaussian Distribution that was mistakenly attributed to him but had been discovered by Abraham de Moivre in the 1730s.

[3]Computer programs sometimes have to reorder the equations but for different reasons, namely to avoid division by zero and to minimize roundoff errors.

that the leftmost term in the first equation eliminates all the corresponding terms below it, then to use the leftmost term in the new second equation in the same way to eliminate the corresponding term (or terms if there are more equations) below that, and so on, down to the last equation. Next, we work our way up by solving the last equation first, then substituting its solution into the previous equation, solving that, and so on. The two phases of the method are called *forward elimination* and *back substitution*. As will be seen shortly, a few complications can and do frequently arise, which make the theory that follows even more interesting and necessary. First, however, we introduce a crucial notational simplification.

Notice that in the computations of Example 2.1.1 the variables x, y, z were not really used and were unnecessary until the very last step in writing down the solutions. All the computations were done on the coefficients only. In computer programs there is no convenient way (and no need, either) to enter the variables. The coefficients are usually arranged in a rectangular array enclosed in parentheses or brackets, called a *matrix* (plural: *matrices*) and designated by a capital letter, as in

$$A = \begin{bmatrix} 2 & 3 & -1 \\ 4 & -2 & 1 \\ 1 & 5 & -2 \end{bmatrix}. \tag{2.1.4}$$

This matrix contains the coefficients on the left side of System (2.1.1) in the same arrangement and is therefore referred to as the *coefficient matrix* or just the *matrix* of that system. We may include the numbers at the right sides of the equations as well:

$$B = \begin{bmatrix} 2 & 3 & -1 & 8 \\ 4 & -2 & 1 & 5 \\ 1 & 5 & -2 & 9 \end{bmatrix}. \tag{2.1.5}$$

This is called the *augmented matrix* of the system. It is often written with a vertical line before its last column as

$$B = \left[\begin{array}{ccc|c} 2 & 3 & -1 & 8 \\ 4 & -2 & 1 & 5 \\ 1 & 5 & -2 & 9 \end{array}\right]. \tag{2.1.6}$$

The matrix A is a 3×3 (read: "three by three") matrix and B is a 3×4 matrix. Similarly, if a matrix has m rows and n columns, we call it an $m \times n$ matrix.

The general form of an $m \times n$ matrix is

$$A = \begin{bmatrix} a_{11} & a_{12} & \cdots & a_{1n} \\ a_{21} & a_{22} & \cdots & a_{2n} \\ \cdots & & & \\ a_{m1} & a_{m2} & \cdots & a_{mn} \end{bmatrix}, \tag{2.1.7}$$

where the a_{11}, a_{12}, ..., a_{mn} (read "a sub one–one, a sub one–two," etc.) are arbitrary real numbers. They are called the entries of the matrix A, with a_{ij} denoting the entry at the intersection of the ith row and jth column. Thus in the double subscript ij the order is important. Furthermore, the matrix A is often denoted by $[a_{ij}]$ or (a_{ij}).

Two matrices are said to be equal if they have the same shape and their corresponding entries are equal.

A matrix consisting of a single row or column is called a *row vector* or a *column vector*, and if we want to emphasize the size n, a row n-vector or a column n-vector.

By definition, a system of m linear equations for n unknowns $x_1, x_2, \ldots, x_n$ has the general form

$$\begin{aligned} a_{11}x_1 + a_{12}x_2 + \cdots + a_{1n}x_n &= b_1 \\ a_{21}x_1 + a_{22}x_2 + \cdots + a_{2n}x_n &= b_2 \\ &\cdots \\ a_{m1}x_1 + a_{m2}x_2 + \cdots + a_{mn}x_n &= b_m \end{aligned} \tag{2.1.8}$$

with the coefficient matrix A given in Equation (2.1.7) having arbitrary entries and the b_i denoting arbitrary numbers as well. We shall frequently find it useful to collect the x_i and the b_i values into two *column* vectors and write such systems as

$$\begin{bmatrix} a_{11} & a_{12} & \cdots & a_{1n} \\ a_{21} & a_{22} & \cdots & a_{2n} \\ \cdots \\ a_{m1} & a_{m2} & \cdots & a_{mn} \end{bmatrix} \begin{bmatrix} x_1 \\ x_2 \\ \cdots \\ x_n \end{bmatrix} = \begin{bmatrix} b_1 \\ b_2 \\ \cdots \\ b_m \end{bmatrix}, \tag{2.1.9}$$

or abbreviated as

$$A\mathbf{x} = \mathbf{b}. \tag{2.1.10}$$

We consider the expression $A\mathbf{x}$ as a new kind of product. Let us state its definition explicitly:

DEFINITION 2.1.1.

For any $m \times n$ matrix A and any column n-vector $\mathbf{x}$, we define $A\mathbf{x}$ as the column m-vector given by

$$A\mathbf{x} = \begin{bmatrix} a_{11} & a_{12} & \ldots & a_{1n} \\ a_{21} & a_{22} & \ldots & a_{2n} \\ \cdots \\ a_{m1} & a_{m2} & \ldots & a_{mn} \end{bmatrix} \begin{bmatrix} x_1 \\ x_2 \\ \cdots \\ x_n \end{bmatrix} = \begin{bmatrix} a_{11}x_1 + a_{12}x_2 + \ldots + a_{1n}x_n \\ a_{21}x_1 + a_{22}x_2 + \ldots + a_{2n}x_n \\ \cdots \\ a_{m1}x_1 + a_{m2}x_2 + \ldots + a_{mn}x_n \end{bmatrix}. \tag{2.1.11}$$

Notice that the components of the column vector $\mathbf{x}$ show up across every row of $A\mathbf{x}$; they get "flipped."

This product is extended and discussed in detail in Section 2.2. Here we need it only for notational convenience, and in the proof of Theorem 2.1.2 we shall need its obvious properties (whose proof is left as Exercise 2.1.11):

$$A(\mathbf{x}+\mathbf{y}) = A\mathbf{x} + A\mathbf{y} \text{ and } A(c\mathbf{x}) = c(A\mathbf{x}). \tag{2.1.12}$$

The reason for using column vectors $\mathbf{x}$ and $\boldsymbol{b}$ is explained in Section 2.2, although for $\boldsymbol{b}$ at least, the choice is rather natural, for then the right sides of Equations (2.1.8) and (2.1.9) match. *Henceforth all vectors will be column vectors unless explicitly designated otherwise.*

EXAMPLE 2.1.2. Using matrix notation we would write the computations of Example 2.1.1 as

$$\left[\begin{array}{ccc|c} 2 & 3 & -1 & 8 \\ 4 & -2 & 1 & 5 \\ 1 & 5 & -2 & 9 \end{array}\right] \to \left[\begin{array}{ccc|c} 2 & 3 & -1 & 8 \\ 0 & -8 & 3 & -11 \\ 0 & 7/2 & -3/2 & 5 \end{array}\right] \to \left[\begin{array}{ccc|c} 2 & 3 & -1 & 8 \\ 0 & -8 & 3 & -11 \\ 0 & 0 & -3/16 & 3/16 \end{array}\right].$$

<<>>

Notice that the arrows between the matrices do not designate equality, they just indicate the flow of the computation. For two matrices to be equal, all the corresponding entries must be equal, and here they are clearly not equal.

Let us review the steps in Example 2.1.2. We copied the first row, then we took 4/2 times the entries of the first row in order to change the 2 into a 4, and subtracted those multiples from the corresponding entries of the second row. (We express this operation more briefly by saying that we subtracted 4/2 times the first row from the second row.) Then we took 1/2 times the entries of the first row to change the 2 into a 1 and subtracted them from the third row. In all this computation the entry 2 in the first row played a pivotal role and is therefore called the *pivot* for these operations. In general, a pivot is an entry whose multiples are used to obtain zeroes below it or we already have all zeroes there, and the first nonzero entry remaining in the last row after the reduction is also called a pivot. (The precise definition is given on page 44). Thus, in this calculation the pivots are the numbers 2, –8, –3/16.

The operations we used are called *elementary row operations*. There are three kinds:

1. Multiplication of a row by a nonzero number.
2. Exchange of two rows.
3. Subtraction of a multiple of a row from another row.

Each of these operations changes the augmented matrix of a system of linear equations into the augmented matrix of an equivalent system; that is, a system having the same set of solutions. For this reason, any two matrices obtainable from each other by a finite number of elementary row operations are said to be *row-equivalent*.

We have used only the third type of elementary operation so far. The first kind is not necessary for Gaussian elimination but will be used later in further reductions. The second kind must be used if we encounter a zero where we need a pivot, as in this example:

EXAMPLE 2.1.3. Let us solve this system of 4 equations in 3 unknowns:

$$\begin{aligned} x_1 + 2x_2 \phantom{{}- x_3} &= 2 \\ 3x_1 + 6x_2 - x_3 &= 8 \\ x_1 + 2x_2 + x_3 &= 0 \\ 2x_1 + 5x_2 - 2x_3 &= 9. \end{aligned} \tag{2.1.13}$$

We do this solution in matrix form as follows, indicating the row operations between the matrices in an obvious manner. (The rows may be considered to be vectors and so we designate them by boldface letters.)

$$\left[\begin{array}{ccc|c} 1 & 2 & 0 & 2 \\ 3 & 6 & -1 & 8 \\ 1 & 2 & 1 & 0 \\ 2 & 5 & -2 & 9 \end{array}\right] \begin{array}{l} \mathbf{r}_1 \to \mathbf{r}_1 \\ \mathbf{r}_2 \to \mathbf{r}_2 - 3\mathbf{r}_1 \\ \mathbf{r}_3 \to \mathbf{r}_3 - \mathbf{r}_1 \\ \mathbf{r}_4 \to \mathbf{r}_4 - 2\mathbf{r}_1 \end{array} \left[\begin{array}{ccc|c} 1 & 2 & 0 & 2 \\ 0 & 0 & -1 & 2 \\ 0 & 0 & 1 & -2 \\ 0 & 1 & -2 & 5 \end{array}\right]$$

$$\begin{array}{l} \mathbf{r}_1 \to \mathbf{r}_1 \\ \mathbf{r}_2 \to \mathbf{r}_4 \\ \mathbf{r}_3 \to \mathbf{r}_3 \\ \mathbf{r}_4 \to \mathbf{r}_2 \end{array} \left[\begin{array}{ccc|c} 1 & 2 & 0 & 2 \\ 0 & 1 & -2 & 5 \\ 0 & 0 & 1 & -2 \\ 0 & 0 & -1 & 2 \end{array}\right] \begin{array}{l} \mathbf{r}_1 \to \mathbf{r}_1 \\ \mathbf{r}_2 \to \mathbf{r}_2 \\ \mathbf{r}_3 \to \mathbf{r}_3 \\ \mathbf{r}_4 \to \mathbf{r}_4 + \mathbf{r}_3 \end{array} \left[\begin{array}{ccc|c} 1 & 2 & 0 & 2 \\ 0 & 1 & -2 & 5 \\ 0 & 0 & 1 & -2 \\ 0 & 0 & 0 & 0 \end{array}\right]. \tag{2.1.14}$$

The back-substitution phase should start with the third row because the fourth one just expresses the trivial equation $0 = 0$. The third row gives $x_3 = -2$, the second row corresponds to $x_2 - 2x_3 = 5$, and so $x_2 = 1$, and the first row yields $x_1 + 2x_2 = 2$, from which $x_1 = 0$.

<<>>

As the example above shows, the number of equations and the number of unknowns need not be the same. In this case the four equations described four planes in three-dimensional space, having a single point of intersection given by the *unique solution* we have found. Of course, in general four planes need not have a point of intersection in common or may have an entire line or plane as their intersection (in the latter case the four equations would describe the same plane). If there is no intersection, then the equations have no solution and are said to be *inconsistent* and the system is *overdetermined*. Inconsistency of the system can happen with just two or three planes as well, for instance if two of them are parallel, and in two dimensions with parallel lines. Thus, before attacking the general theory, we discuss examples of inconsistent systems and systems with infinitely many solutions, which are called *underdetermined*.

EXAMPLE 2.1.4. Consider the system given by the matrix

$$[A|\boldsymbol{b}] = \left[\begin{array}{ccc|c} 1 & 2 & 0 & 2 \\ 3 & 6 & -1 & 8 \\ 1 & 2 & 1 & 4 \end{array}\right]. \tag{2.1.15}$$

Subtracting $3\boldsymbol{r}_1$ from $\boldsymbol{r}_2$, and $\boldsymbol{r}_1$ from $\boldsymbol{r}_3$, we get

$$[A'|\boldsymbol{b}'] = \left[\begin{array}{ccc|c} 1 & 2 & 0 & 2 \\ 0 & 0 & -1 & 2 \\ 0 & 0 & 1 & 2 \end{array}\right]. \tag{2.1.16}$$

The last two rows of A' represent the contradictory equations $x_3 = -2$ and $x_3 = 2$. These two equations describe parallel planes. Thus A had to represent an inconsistent system.

Usually the row operations produce equations of new planes, two of which have turned out to be parallel in this case. The planes corresponding to the rows of A are, however, not parallel; only the three lines of intersection of pairs of them are (see Exercise 2.1.12), like the three parallel edges of a prism; and that is why there is no point of intersection common to all three planes.

We may carry the matrix reduction one step further and obtain by adding the second row to the third one:

$$[A''|\boldsymbol{b}''] = \left[\begin{array}{ccc|c} 1 & 2 & 0 & 2 \\ 0 & 0 & -1 & 2 \\ 0 & 0 & 0 & 4 \end{array}\right]. \tag{2.1.17}$$

This matrix provides the single self-contradictory equation $0 = 4$ from its last row. There is no geometrical interpretation for such an equation but algebraically it is the best way of establishing the inconsistency. Thus this is the typical pattern we shall obtain in the general case whenever there is no solution.

<<>>

Next we modify the matrix of the last example so that all three planes intersect in a single line:

EXAMPLE 2.1.5. Let

$$[A|\boldsymbol{b}] = \left[\begin{array}{ccc|c} 1 & 2 & 0 & 2 \\ 3 & 6 & -1 & 8 \\ 1 & 2 & 1 & 0 \end{array}\right]. \tag{2.1.18}$$

We can reduce this matrix to

$$[A'|\boldsymbol{b}'] = \left[\begin{array}{ccc|c} 1 & 2 & 0 & 2 \\ 0 & 0 & -1 & 2 \\ 0 & 0 & 0 & 0 \end{array}\right], \tag{2.1.19}$$

which represents just two planes, since the last equation has become the trivial identity $0 = 0$. Algebraically, the second row gives $x_3 = -2$, and the first row relates x_1 to x_2. We can choose an arbitrary value for either x_1 or x_2 and solve for the other. In some other examples, however, we have no choice; but even then *we can always solve for the variable corresponding to the pivot, which cannot be zero, and that is what we always do.* Thus, we set x_2 equal to a parameter t and solve for x_1, to obtain $x_1 = 2 - 2t$. We can write the solutions in vector form as (remember: the convention is to use column vectors)

$$\begin{bmatrix} x_1 \\ x_2 \\ x_3 \end{bmatrix} = \begin{bmatrix} 2 \\ 0 \\ -2 \end{bmatrix} + t \begin{bmatrix} -2 \\ 1 \\ 0 \end{bmatrix}. \tag{2.1.20}$$

This is the parametric vector equation of the line of intersection of the three planes defined by the rows of $[A|\boldsymbol{b}]$. The coordinates of each of its points comprise one of the infinitely many solutions of the system.

<<>>

EXAMPLE 2.1.6. Let us solve the system

$$\begin{aligned} 2x_1 + 3x_2 - 2x_3 + 4x_4 &= 4 \\ -6x_1 - 8x_2 + 6x_3 - 2x_4 &= 1 \\ 4x_1 + 4x_2 - 4x_3 - x_4 &= -7. \end{aligned} \tag{2.1.21}$$

These equations represent three hyperplanes in four dimensions.[4] We can proceed as in the preceding examples:

$$\left[\begin{array}{cccc|c} 2 & 3 & -2 & 4 & 4 \\ -6 & -8 & 6 & -2 & 1 \\ 4 & 4 & -4 & -1 & -7 \end{array}\right] \begin{array}{l} \boldsymbol{r}_1 \to \boldsymbol{r}_1 \\ \boldsymbol{r}_2 \to \boldsymbol{r}_2 + 3\boldsymbol{r}_1 \\ \boldsymbol{r}_3 \to \boldsymbol{r}_3 - 2\boldsymbol{r}_1 \end{array} \left[\begin{array}{cccc|c} 2 & 3 & -2 & 4 & 4 \\ 0 & 1 & 0 & 10 & 13 \\ 0 & -2 & 0 & -9 & -15 \end{array}\right]$$

$$\begin{array}{l} \boldsymbol{r}_1 \to \boldsymbol{r}_1 \\ \boldsymbol{r}_2 \to \boldsymbol{r}_2 \\ \boldsymbol{r}_3 \to \boldsymbol{r}_3 + 2\boldsymbol{r}_2 \end{array} \left[\begin{array}{cccc|c} 2 & 3 & -2 & 4 & 4 \\ 0 & 1 & 0 & 10 & 13 \\ 0 & 0 & 0 & 11 & 11 \end{array}\right]. \tag{2.1.22}$$

The variables that have pivots as coefficients, x_1, x_2, x_4 in this case, are called *basic variables*. They can be solved for in terms of the other, so-called *free variables* that correspond to the pivot-free columns. The free variables are usually replaced by parameters, but this is just a formality to show that they can be chosen freely.

Thus we set $x_3 = t$, and find the solutions again as the points of a line given by

[4]A hyperplane in $\mathbb{R}^4$ is a copy of $\mathbb{R}^3$, just as a plane in $\mathbb{R}^3$ is a copy of $\mathbb{R}^2$.

$$\begin{bmatrix} x_1 \\ x_2 \\ x_3 \\ x_4 \end{bmatrix} = \begin{bmatrix} -9/2 \\ 3 \\ 0 \\ 1 \end{bmatrix} + t\begin{bmatrix} 1 \\ 0 \\ 1 \\ 0 \end{bmatrix}. \tag{2.1.23}$$

<<>>

EXAMPLE 2.1.7. Consider the system

$$\begin{aligned} 2x_1 + 3x_2 - 2x_3 + 4x_4 &= 2 \\ -6x_1 - 9x_2 + 7x_3 - 8x_4 &= -3 \\ 4x_1 + 6x_2 - x_3 + 20x_4 &= 13. \end{aligned} \tag{2.1.24}$$

We solve this system as follows:

$$\left[\begin{array}{rrrr|r} 2 & 3 & -2 & 4 & 2 \\ -6 & -9 & 7 & -8 & -3 \\ 4 & 6 & -1 & 20 & 13 \end{array}\right] \begin{array}{l} \boldsymbol{r}_1 \to \boldsymbol{r}_1 \\ \boldsymbol{r}_2 \to \boldsymbol{r}_2 + 3\boldsymbol{r}_1 \\ \boldsymbol{r}_3 \to \boldsymbol{r}_3 - 2\boldsymbol{r}_1 \end{array} \left[\begin{array}{rrrr|r} 2 & 3 & -2 & 4 & 2 \\ 0 & 0 & 1 & 4 & 3 \\ 0 & 0 & 3 & 12 & 9 \end{array}\right]$$

$$\begin{array}{l} \boldsymbol{r}_1 \to \boldsymbol{r}_1 \\ \boldsymbol{r}_2 \to \boldsymbol{r}_2 \\ \boldsymbol{r}_3 \to \boldsymbol{r}_3 - 3\boldsymbol{r}_2 \end{array} \left[\begin{array}{rrrr|r} 2 & 3 & -2 & 4 & 2 \\ 0 & 0 & 1 & 4 & 3 \\ 0 & 0 & 0 & 0 & 0 \end{array}\right]. \tag{2.1.25}$$

Because the pivots are in columns 1 and 3, the basic variables are x_1 and x_3 and the free variables x_2 and x_4. Thus we use two parameters and set $x_2 = s$ and $x_4 = t$. Then the second row leads to $x_3 = 3 - 4t$ and the first row to $2x_1 + 3s - 2(3 - 4t) + 4t = 2$; that is, to $2x_1 = 8 - 3s - 12t$. Putting all these results together we obtain the two-parameter set of solutions

$$\begin{bmatrix} x_1 \\ x_2 \\ x_3 \\ x_4 \end{bmatrix} = \begin{bmatrix} 4 \\ 0 \\ 3 \\ 0 \end{bmatrix} + s\begin{bmatrix} -3/2 \\ 1 \\ 0 \\ 0 \end{bmatrix} + t\begin{bmatrix} -6 \\ 0 \\ -4 \\ 1 \end{bmatrix}, \tag{2.1.26}$$

which is also a parametric vector equation of a plane in $\mathbb{R}^4$.

<<>>

We are now at a point where we can summarize the lessons from our examples: Given m equations for n unknowns, we consider their augmented matrix:

$$[A|\boldsymbol{b}] = \left[\begin{array}{cccc|c} a_{11} & a_{12} & \dots & a_{1n} & b_1 \\ a_{21} & a_{22} & \dots & a_{2n} & b_2 \\ \dots & & & & \dots \\ a_{m1} & a_{m2} & \dots & a_{mn} & b_m \end{array}\right], \tag{2.1.27}$$

and reduce it using elementary row operations according to the following algorithm:

1. Search the first column from the top down for the first nonzero coefficient, then, if necessary, the second column, then the third column, and so on. The entry thus found is called the *current pivot*.
2. Put the row containing the current pivot on top (unless it is already there).
3. Subtract appropriate multiples of the first row from each of the other rows to obtain all zeroes below the current pivot in its column.
4. Repeat the preceding steps on the submatrix[5] consisting of all those elements of $[A|\mathbf{b}]$ which lie lower than and to the right of the last pivot.
5. Stop when no further pivot can be found.

These steps constitute the *forward-elimination* phase of Gaussian elimination (the second phase will be discussed shortly), and they lead to a matrix of the form described below:

DEFINITION 2.1.2.

A matrix is said to be in *echelon form*[6] or an echelon matrix if it has a staircase-like pattern characterized by these properties:

a. The all-zero rows (if any) are at the bottom.
b. Calling the leftmost nonzero entry of each nonzero row a leading entry, we have the leading entry in each lower row to the right of the leading entry in every higher row.

These properties imply that in an echelon matrix U all the entries in a column below a leading entry are zero. If U arises from the reduction of a matrix A by the forward-elimination algorithm above, then the pivots of A become the leading entries of U. Also, if we were to apply the algorithm to an echelon matrix, then it would not be changed and we would find that its leading entries are its pivots.

Notice that although a given matrix is row equivalent to many different echelon matrices (just multiply any nonzero row of one of them by two, for example), the algorithm above leads to a single well-defined one in each case. Furthermore, it can be proved that the number and locations, though not the values, of the pivots are unique for all echelon matrices obtainable from the same A. Consequently, the results of Theorem 2.1.1 below, even though they depend on the pivots, are valid unambiguously.

Here is a possible $m \times (n + 1)$ echelon matrix obtainable from the matrix $[A|\mathbf{b}]$ above:

[5] A submatrix of a given matrix A is a matrix obtained by deleting any rows and/or columns of A.

[6] "Echelon" in French means "rung of a ladder," and in English it is used for some ladder-like military formations and rankings.

$$[U|\boldsymbol{c}] = \left[\begin{array}{ccccccc|c} p_1 & * & * & * & \cdots & * & * & c_1 \\ 0 & p_2 & * & * & \cdots & * & * & c_2 \\ 0 & 0 & 0 & p_3 & \cdots & * & * & c_3 \\ \cdots & & & & & & & \\ 0 & 0 & 0 & 0 & \cdots & p_r & * & c_r \\ 0 & 0 & 0 & 0 & \cdots & 0 & 0 & c_{r+1} \\ 0 & 0 & 0 & 0 & \cdots & 0 & 0 & 0 \end{array}\right]. \tag{2.1.28}$$

The first n columns constitute the echelon matrix U obtained from A and the last column is the corresponding reduction of $\boldsymbol{b}$. The p_i denote the pivots, and the entries denoted by $*$ and by c_i specify numbers that may or may not be zero. The number r is very important, since it determines the character of the solutions, and has a special name:

DEFINITION 2.1.3.

The number r of nonzero rows of an echelon matrix U obtained by the forward-elimination phase of the Gaussian elimination algorithm from a matrix A is called the *rank* of A.[7,8]

We can now describe the back-substitution phase of Gaussian elimination:

6. If $r < m$ and $c_{r+1} \neq 0$ hold, then the row containing c_{r+1} corresponds to the self-contradictory equation $0 = c_{r+1}$, and so the system has no solutions. (This case occurs in Example 2.1.4, where $m = 3$, $r = 2$ and $c_{r+1} = c_3 = 4$.)
7. If $r = m$ or $c_{r+1} = 0$, then, for every i such that the ith column contains no pivot, the variable x_i is a free variable and we set it equal to a parameter s_i. (In Example 2.1.6, for instance, $r = m = 3$ and x_3 is free. In Example 2.1.7 we have $m = 3$, $r = 2$, and $c_{r+1} = c_3 = 0$ and the free variables are x_2 and x_4.) If $r = n$, then there are no free variables. (In Example 2.1.2, for instance, $r = m = n = 3$.)
8. In any of the cases in Part 7 we solve for the basic variables x_i corresponding to the pivots p_i, starting with x_r in the rth row and working our way up row by row.

The Gaussian elimination algorithm proves this theorem:

THEOREM 2.1.1.

Consider the $m \times n$ system

$$A\mathbf{x} = \boldsymbol{b}. \tag{2.1.29}$$

[7]Of course, r is also the rank of U, since the algorithm applied to U would leave U unchanged.
[8]Some books call this quantity the *row-rank* of A until they define the column-rank and show that the two are equal.

Suppose the matrix $[A|\boldsymbol{b}]$ is reduced by the algorithm above to the echelon matrix $[U|\boldsymbol{c}]$ with r nonzero rows in the part U corresponding to A.

If $r = m$; that is, if U has no zero rows, then the system (2.1.29) is consistent. If $r < m$, then the system is consistent if and only if $c_{r+1} = 0$.

For a consistent system

a. there is a unique solution if and only if there are no free variables; that is, if $r = n$;
b. if $r < n$, then there is an $(n - r)$-parameter infinite set of solutions of the form

$$\mathbf{x} = \boldsymbol{u}_0 + \sum_{i=1}^{n-r} s_i \boldsymbol{u}_i. \tag{2.1.30}$$

We may state the uniqueness condition $r = n$ in another way by saying that the pivots are the diagonal entries $u_{11}, u_{22}, \ldots, u_{nn}$ of U, that is, that U has the form

$$U = \begin{bmatrix} p_1 & * & * & \cdots & * \\ 0 & p_2 & * & \cdots & * \\ 0 & 0 & p_3 & \cdots & * \\ \cdots & & & & \\ 0 & 0 & 0 & \cdots & p_r \\ 0 & 0 & 0 & \cdots & 0 \\ \cdots & & & & \\ 0 & 0 & 0 & \cdots & 0 \end{bmatrix}. \tag{2.1.31}$$

Although pivots are never zero, a matrix of this form, with $p_i = 0$ allowed, is referred to as an *upper triangular matrix*.

Notice that for any $m \times n$ matrix we have $0 \le r \le \min(m, n)$, because r equals the number of pivots and there can be only one pivot in each row and in each column. If, for a matrix A, $r = \min(m, n)$ holds, then A is said to have *full rank*, and otherwise it is said to be *rank deficient*.

In case the solutions of (2.1.29) are given by Equation (2.1.30), the latter is called *the general solution* of the system, as opposed to a *particular solution* obtained by substituting particular values for the parameters into Equation (2.1.30).

It is customary and very useful to distinguish two types of linear systems depending on the value of $\boldsymbol{b}$:

DEFINITION 2.1.4.

A system of linear equations as given in (2.1.29) is called *homogeneous* if $\boldsymbol{b} = \boldsymbol{0}$, and *inhomogeneous* otherwise.

We may restate part of Theorem 2.1.1 for homogeneous systems as follows:

COROLLARY 2.1.1. A homogeneous system

$$A\mathbf{x} = \mathbf{0} \tag{2.1.32}$$

is always consistent. It always has the so-called *trivial solution* $\mathbf{x} = \mathbf{0}$. If $r = n$, then it has only this solution; and if $m < n$ or, more generally, if $r < n$ holds, then it has nontrivial solutions as well.

There is an important relationship between the solutions of corresponding homogeneous and inhomogeneous systems, the analog of which is indispensable for solving many differential equations:

THEOREM 2.1.2. If $\mathbf{x} = \mathbf{x}_b$ is any particular solution of the inhomogeneous equation

$$A\mathbf{x} = \mathbf{b}, \tag{2.1.33}$$

then

$$\mathbf{x} = \mathbf{x}_b + \mathbf{u} \tag{2.1.34}$$

is its general solution if and only if

$$\mathbf{u} = \sum_{i=1}^{n-r} s_i \mathbf{u}_i. \tag{2.1.35}$$

is the general solution of the corresponding homogeneous equation

$$A\mathbf{u} = \mathbf{0}. \tag{2.1.36}$$

Proof Assume first that (2.1.35) is the general solution of (2.1.36). (Certainly, the Gaussian elimination algorithm would give it in this form.) Then applying A to (2.1.34) we get

$$A\mathbf{x} = A(\mathbf{x}_b + \mathbf{u}) = A\mathbf{x}_b + A\mathbf{u} = \mathbf{b} + \mathbf{0} = \mathbf{b}. \tag{2.1.37}$$

Thus, any solution of the homogeneous equation (2.1.36) leads to a solution of the form (2.1.34) of the inhomogeneous equation (2.1.33).

Conversely, assume that (2.1.34) is a solution of the inhomogeneous equation (2.1.33). Then

$$A\mathbf{u} = A(\mathbf{x} - \mathbf{x}_b) = A\mathbf{x} - A\mathbf{x}_b = \mathbf{b} - \mathbf{b} = \mathbf{0}. \tag{2.1.38}$$

This equation shows that the $\mathbf{u}$ given by (2.1.35) is indeed a solution of (2.1.36), or, in other words, that any solution of the inhomogeneous equation (2.1.33) leads to a solution of the form (2.1.35) of the homogeneous equation (2.1.36).

<<>>

This theorem establishes a one-to-one pairing of the solutions of the two equations (2.1.33) and (2.1.36). Geometrically this result means that the solutions of $A\boldsymbol{u} = \boldsymbol{0}$ are the position vectors of the points of the hyperplane through the origin given by (2.1.35), and the solutions of $A\mathbf{x} = \boldsymbol{b}$ are those of a parallel hyperplane obtained from the first one by shifting it by the vector $\mathbf{x}_b$. (See Figure 2.1.) Note that we could have shifted by the coordinate vector of any other point of the second hyperplane, that is, by any other particular solution $\mathbf{x}_b'$ of (2.1.33) (see Figure 2.2), and we would have obtained the same new hyperplane.

FIGURE 2.1

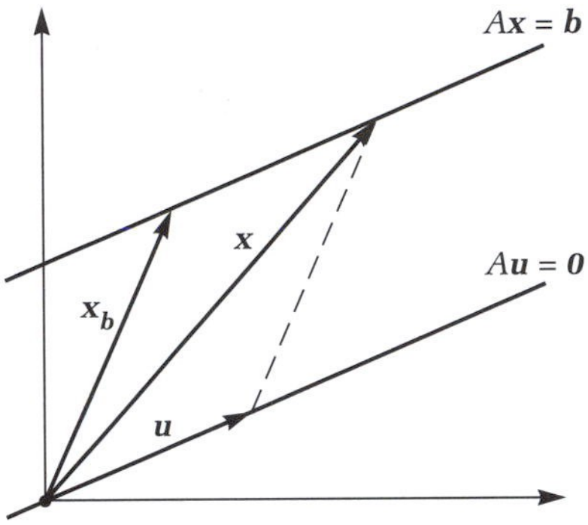

$\mathbf{x} = \mathbf{x}_b + \boldsymbol{u}$

FIGURE 2.2

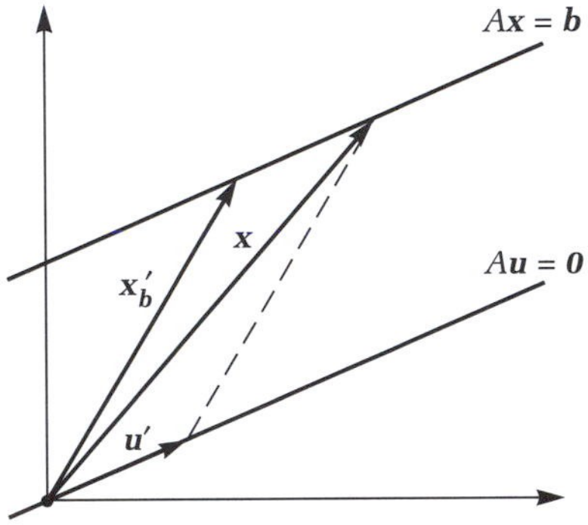

$\mathbf{x} = \mathbf{x}_b' + \boldsymbol{u}'$

Sometimes the forward elimination procedure is carried further so as to obtain leading entries in the echelon matrix that equal 1, and to obtain 0 entries in the basic columns not just below but also above the pivots. This method is called *Gauss-Jordan elimination* and the final matrix a *reduced echelon matrix* or a *row-reduced echelon matrix*. We give one example of this method:

EXAMPLE 2.1.8. Let us continue the reduction of Example 2.1.7 on page 44, starting with the echelon matrix obtained in the forward elimination phase:

$$\left[\begin{array}{cccc|c} 2 & 3 & -2 & 4 & 2 \\ 0 & 0 & 1 & 4 & 3 \\ 0 & 0 & 0 & 0 & 0 \end{array}\right] \begin{array}{l} \boldsymbol{r}_1 \to \boldsymbol{r}_1/2 \\ \boldsymbol{r}_2 \to \boldsymbol{r}_2 \\ \boldsymbol{r}_3 \to \boldsymbol{r}_3 \end{array} \left[\begin{array}{cccc|c} 1 & \frac{3}{2} & -1 & 2 & 1 \\ 0 & 0 & 1 & 4 & 3 \\ 0 & 0 & 0 & 0 & 0 \end{array}\right]$$

$$\begin{array}{l} \boldsymbol{r}_1 \to \boldsymbol{r}_1 + \boldsymbol{r}_2 \\ \boldsymbol{r}_2 \to \boldsymbol{r}_2 \\ \boldsymbol{r}_3 \to \boldsymbol{r}_3 \end{array} \left[\begin{array}{cccc|c} 1 & \frac{3}{2} & 0 & 6 & 4 \\ 0 & 0 & 1 & 4 & 3 \\ 0 & 0 & 0 & 0 & 0 \end{array}\right]. \tag{2.1.39}$$

From here on we proceed exactly as in the Gaussian elimination algorithm: we assign parameters s and t to the free variables x_2 and x_4, and solve for the basic variables x_1 and x_3. The latter step is now trivial, since all the work has already been done. The equations corresponding to the final matrix are

$$x_1 + \frac{3}{2}s + 6t = 4 \quad \text{and} \quad x_3 + 4t = 3. \tag{2.1.40}$$

Thus we find the same general solution as before:

$$\begin{aligned} x_1 &= 4 - \frac{3}{2}s - 6t \\ x_2 &= \phantom{4 - \frac{3}{2}}s \\ x_3 &= 3 \phantom{- \frac{3}{2}s} - 4t \\ x_4 &= \phantom{4 - \frac{3}{2}s - 6} t. \end{aligned} \tag{2.1.41}$$

We can write this solution in vector form as

$$\begin{bmatrix} x_1 \\ x_2 \\ x_3 \\ x_4 \end{bmatrix} = \begin{bmatrix} 4 \\ 0 \\ 3 \\ 0 \end{bmatrix} + s\begin{bmatrix} -3/2 \\ 1 \\ 0 \\ 0 \end{bmatrix} + t\begin{bmatrix} -6 \\ 0 \\ -4 \\ 1 \end{bmatrix}. \tag{2.1.42}$$

Notice how the numbers in the first and third rows of this solution correspond to the entries of the last matrix in (2.1.39), which is in reduced echelon form.

<<>>

As can be seen from the example above, the reduced echelon form combines the results of both the forward-elimination and back-substitution phases of Gaussian elimination, and the general solution can simply be read from it. In general, if $[R|\boldsymbol{c}]$ is the reduced echelon matrix corresponding to the system $A\mathbf{x} = \boldsymbol{b}$, then we assign parameters s_j to the free variables x_j; and the ith row of the corresponding system $R\mathbf{x} = \boldsymbol{c}$ becomes

$$x_k + \sum_{j>k} r_{ij}s_j = c_i, \tag{2.1.43}$$

where x_k is a basic variable and i is the index of the row whose leading entry is in the kth place. Thus the general solution is given by

$$\begin{aligned} x_j &= s_j && \text{if } x_j \text{ is free, and} \\ x_k &= c_i - \sum_{j>k} r_{ij}s_j && \text{if } x_k \text{ is basic.} \end{aligned} \tag{2.1.44}$$

Gauss-Jordan elimination is rarely used for solving systems because a variant of Gaussian elimination, which we shall study in Section 8.1, is frequently more efficient; but Gauss-Jordan elimination is the preferred method for inverting matrices, as we shall see in Section 2.3. Also, it is sometimes helpful that the reduced echelon form of a matrix is unique (we do not prove this fact here), and that the solution of any system is immediately visible in it.

Exercises 2.1

Find all solutions of the systems in the first four exercises by Gaussian elimination.

• **2.1.1.**
$$\begin{aligned} 2x_1 + 3x_2 - 3x_3 &= 0 \\ x_1 + 5x_2 + 2x_3 &= 1 \\ -4x_1 + 6x_3 &= 2 \end{aligned}$$

2.1.2.
$$\begin{aligned} 2x_1 + 2x_2 - 3x_3 &= 0 \\ x_1 + 5x_2 + 2x_3 &= 0 \\ -4x_1 + 6x_3 &= 0 \end{aligned}$$

• **2.1.3.**
$$\begin{aligned} 2x_1 + 2x_2 - 3x_3 &= 0 \\ x_1 + 5x_2 + 2x_3 &= 1 \end{aligned}$$

• **2.1.4.** $2x_1 + 2x_2 - 3x_3 = 0$

In the next six exercises, find all solutions of the systems given by their augmented matrices.

• **2.1.5.**
$$\left[\begin{array}{rrr|r} 1 & 0 & -1 & 0 \\ -2 & 3 & -1 & 0 \\ -6 & 6 & 0 & 0 \end{array}\right]$$

2.1.6.
$$\left[\begin{array}{rrr|r} 1 & 0 & -1 & 1 \\ -2 & 3 & -1 & 0 \\ -6 & 6 & 0 & 0 \end{array}\right]$$

2.1.7.
$$\left[\begin{array}{rrrr|r} 1 & 4 & 9 & 2 & 0 \\ 2 & 2 & 6 & -3 & 0 \\ 2 & 7 & 16 & 3 & 0 \end{array}\right]$$

2.1.8.
$$\left[\begin{array}{rrr|r} 2 & 4 & 1 & 7 \\ 0 & 1 & 3 & 7 \\ 3 & 3 & -1 & 9 \\ 1 & 2 & 3 & 11 \end{array}\right]$$

2.1.9.
$$\left[\begin{array}{rrrrr|r} 3 & -6 & -1 & 1 & 5 & 0 \\ -1 & 2 & 2 & 3 & 3 & 0 \\ 4 & -8 & -3 & -2 & 1 & 0 \end{array}\right]$$

2.1.10.
$$\left[\begin{array}{rrrr|r} 3 & -6 & -1 & 1 & 5 \\ -1 & 2 & 2 & 3 & 3 \\ 4 & -8 & -3 & -2 & 1 \end{array}\right]$$

2.1.11. Prove Equations (2.1.12) on page 39.

2.1.12. Show that the three planes defined by the rows of the matrix in Equation (2.1.15) on page 41 have parallel lines of intersection.

• **2.1.13.** List all possible forms of 2×2 echelon matrices in a manner similar to that of Equation (2.1.31), with p_i for the pivots and $*$ for the entries that may or may not be zero.

2.1.14. List all possible forms of 3×3 echelon matrices. (*Hint:* There are eight distinct forms.)

2.1.15. List all possible forms of 2×2 reduced echelon matrices.

2.1.16. List all possible forms of 3×3 reduced echelon matrices.

2.1.17. Solve Exercise 2.1.5 by Gauss-Jordan elimination.

2.1.18. Solve Exercise 2.1.6 by Gauss-Jordan elimination.

2.1.19. Solve Exercise 2.1.9 by Gauss-Jordan elimination.

2.1.20. Solve Exercise 2.1.10 by Gauss-Jordan elimination.

In the next four exercises, find conditions on the vector $\boldsymbol{b}$ that would make the equation $A\mathbf{x} = \boldsymbol{b}$ consistent for the given matrix A. (*Hint:* Reduce the augmented matrix using undetermined components b_i of $\boldsymbol{b}$, until the A in it is in echelon form, and set $c_{r+1} = c_{r+2} = \cdots = 0$.)

• **2.1.21.** $A = \begin{bmatrix} 1 & 0 & -1 \\ -2 & 3 & -1 \\ -6 & 6 & 0 \end{bmatrix}$

2.1.22. $A = \begin{bmatrix} 1 & -2 \\ 2 & -4 \\ -6 & 12 \end{bmatrix}$

2.1.23. $A = \begin{bmatrix} 1 & 2 & -6 \\ -2 & -4 & 12 \end{bmatrix}$

• **2.1.24.** $A = \begin{bmatrix} 1 & 0 & -1 \\ -2 & 3 & -1 \\ 3 & -3 & 0 \\ 2 & 0 & -2 \end{bmatrix}$

In each of the next two exercises, find two particular solutions $\mathbf{x}_b$ and $\mathbf{x}_b'$ of the given system and the general solution $\boldsymbol{u}$ of the corresponding homogeneous system. Write the general solution of the former as $\mathbf{x}_b + \boldsymbol{u}$ and also as $\mathbf{x}_b' + \boldsymbol{u}$, and show that the two forms are equivalent, that is, that the set of vectors of the form $\mathbf{x}_b + \boldsymbol{u}$ is identical with the set of vectors of the form $\mathbf{x}_b' + \boldsymbol{u}$.

2.1.25. $2x_1 + 3x_2 - 1x_3 = 4$
$3x_1 + 5x_2 + 2x_3 = 1$

2.1.26. $2x_1 + 2x_2 - 3x_3 - 2x_4 = 0$
$5x_1 + 5x_2 + 2x_3 + 5x_4 = 1$

2.1.27. Show that the system $A\mathbf{x} = \boldsymbol{b}$ is consistent if and only if A and $[A \,|\, \boldsymbol{b}]$ have the same rank.

MATLAB Exercises 2.1

In MATLAB, linear systems are entered in matrix form. We can enter a matrix by writing its entries between brackets, row by row from left to right, top to bottom, separating row entries by spaces or commas, and rows by semicolons. For example, the command $A = [2, 3; 1, -2]$ would produce the matrix

$$A = \begin{bmatrix} 2 & 3 \\ 1 & -2 \end{bmatrix}.$$

(The size of a matrix is automatic; no size declaration is needed or possible, unlike in most other computer languages.) The entry a_{ij} of the matrix A is denoted by $A(i,j)$ in MATLAB, the ith row by $A(i,:)$ and the jth column by $A(:,j)$.

The vector $\boldsymbol{b}$ must be entered as a column vector. This form can be achieved either by separating entries by semicolons or by writing a prime after the closing bracket, as in $\boldsymbol{b} = [1\ 2]'$. This operation would result in the column vector

$$\boldsymbol{b} = \begin{bmatrix} 1 \\ 2 \end{bmatrix}.$$

The augmented matrix can be formed by the command $[A\ \ \boldsymbol{b}]$. Sometimes we may wish to name it as, say, $A\_\boldsymbol{b} = [A\ \ \boldsymbol{b}]$ or simply as $C = [A\ \ \boldsymbol{b}]$. The command **rref**$(C)$ returns the reduced-echelon form of C.

The command $\mathbf{x} = A\backslash\boldsymbol{b}$ always returns a solution of the system $A\mathbf{x} = \boldsymbol{b}$. This is the unique solution if there is only one; it is a particular solution with as many zeroes as possible for components of $\mathbf{x}$ with the lowest subscripts; and it is the least-squares "solution" (discussed in Section 5.1) if the system is inconsistent. This is the most efficient method for finding a solution and is the one you should use whenever possible. On the other hand, to find the *general solution* of an underdetermined system this method does not work, and you should use **rref**$([A\ \ \boldsymbol{b}])$ to obtain the reduced echelon matrix, and proceed as in Example 2.1.8 or Equations (2.1.44).

ML2.1.1. **a.** Write MATLAB commands to implement elementary row operations on a matrix A.

b. Use these commands to reduce the matrix

$$A = \begin{bmatrix} 3 & -6 & -1 & 1 & 5 & 2 \\ -1 & 2 & 2 & 3 & 3 & 6 \\ 4 & -8 & -3 & -2 & 1 & 0 \end{bmatrix}$$

to reduced echelon form and compare your result to **rref**(A).

c. Write MATLAB commands to compute a matrix B with the same rows as the matrix A, but the first two switched.

d. Compare **rref**(B) with **rref**(A). Explain your result.

ML2.1.2. Find the general solution of $A\boldsymbol{x} = \boldsymbol{0}$ for

$$A = \begin{bmatrix} -1 & -2 & -1 & -1 & 1 \\ -1 & -2 & 0 & 3 & -1 \\ 1 & 2 & 1 & 1 & 1 \\ 0 & 0 & 2 & 8 & 2 \end{bmatrix}.$$

ML2.1.3. Let A be the same matrix as in Exercise ML2.1.2. and let

$$\boldsymbol{b} = \begin{bmatrix} 9 \\ 1 \\ -5 \\ -4 \end{bmatrix}.$$

a. Find the general solution of $A\boldsymbol{x} = \boldsymbol{b}$ using **rref**([A $\boldsymbol{b}$]).

b. Verify that $\boldsymbol{x} = A \backslash \boldsymbol{b}$ is indeed a particular solution by computing $A\boldsymbol{x}$ from it.

c. Find the parameter values in the general solution obtained in Part (a) that give the particular solution of Part (b).

d. To verify the result of Theorem 2.1.2 for this case, show that the general solution of Part (a) equals $\boldsymbol{x} = A \backslash \boldsymbol{b}$ plus the general solution to the homogeneous equation found in Exercise ML2.1.2.

ML2.1.4. Let A and $\boldsymbol{b}$ be the same as in Exercise ML2.1.3. The command $\boldsymbol{x}$ = **pinv**(A) $*$ $\boldsymbol{b}$ gives another particular solution of $A\boldsymbol{x} = \boldsymbol{b}$. (This result will be explained in Section 5.1.) Verify Theorem 2.1.2 for this particular solution, as in Part (d) of Exercise ML2.1.3.

ML2.1.5. Let A be the same matrix as in Exercise ML2.1.3 and let

$$\boldsymbol{b} = \begin{bmatrix} 1 \\ 2 \\ 3 \\ 4 \end{bmatrix}.$$

Compute $\boldsymbol{x} = A \backslash \boldsymbol{b}$ and substitute this vector into $A\boldsymbol{x}$. Explain how your result is possible. (*Hint:* Look at **rref**([A $\boldsymbol{b}$]).)

2.2. THE ALGEBRA OF MATRICES

Just as for vectors, we can define algebraic operations for matrices, and these operations will vastly extend their utility.

To motivate the forthcoming definitions it is helpful to consider matrices as functions or mappings. Thus if A is an $m \times n$ matrix, the equation $\boldsymbol{v} = A\boldsymbol{u}$ may be regarded as a mapping of any $\boldsymbol{u} \in \mathbb{R}^n$ to some $\boldsymbol{v} \in \mathbb{R}^m$. This is also reflected in the terminology: We frequently read $A\boldsymbol{u}$ as A being *applied* to $\boldsymbol{u}$ instead of A *times* $\boldsymbol{u}$. If $m = n$, we may consider this as a transformation of the vectors of $\mathbb{R}^n$ to corresponding vectors in the same space. Here is an illustration:

EXAMPLE 2.2.1. The matrix

$$R_\theta = \begin{bmatrix} \cos\theta & -\sin\theta \\ \sin\theta & \cos\theta \end{bmatrix} \tag{2.2.1}$$

represents the rotation of $\mathbb{R}^2$ around O by the angle θ, as can be seen in the following way. (See Figure 2.3.)

FIGURE 2.3

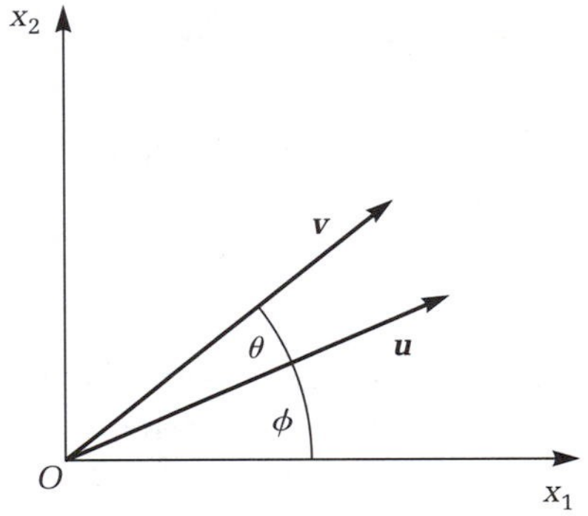

Let

$$\boldsymbol{u} = \begin{bmatrix} |\boldsymbol{u}|\cos\phi \\ |\boldsymbol{u}|\sin\phi \end{bmatrix} \tag{2.2.2}$$

be any vector in $\mathbb{R}^2$ (see Exercise 1.2.15 on page 23). Then, by Definition 2.1.1,

$$\boldsymbol{v} = R_\theta\boldsymbol{u} = \begin{bmatrix} \cos\theta & -\sin\theta \\ \sin\theta & \cos\theta \end{bmatrix}\begin{bmatrix} |\boldsymbol{u}|\cos\phi \\ |\boldsymbol{u}|\sin\phi \end{bmatrix}$$
$$= |\boldsymbol{u}|\begin{bmatrix} \cos\theta\cos\phi - \sin\theta\sin\phi \\ \sin\theta\cos\phi + \cos\theta\sin\phi \end{bmatrix} \tag{2.2.3}$$

and so

$$\mathbf{v} = |\mathbf{u}| \begin{bmatrix} \cos(\phi + \theta) \\ \sin(\phi + \theta) \end{bmatrix}. \tag{2.2.4}$$

This is indeed a vector of the same length as $\mathbf{u}$ and it encloses the angle $\phi + \theta$ with the x_1-axis.

<<>>

Such transformations are discussed in detail in Chapter 4. Here we just present the concept briefly to lay the groundwork for the following definitions. These are analogous to the familiar definitions for functions of real variables, where, given f and g, their sum $f + g$ is defined as the function such that $(f + g)(x) = f(x) + g(x)$ for every x, and cf such that $(cf)(x) = cf(x)$ for every x.

DEFINITION 2.2.1.

Let A and B be two $m \times n$ matrices and for any $\mathbf{x} \in \mathbb{R}^n$ let

$$\mathbf{u} = A\mathbf{x} \quad \text{and} \quad \mathbf{v} = B\mathbf{x}. \tag{2.2.5}$$

We define $A + B$ as the matrix that maps $\mathbf{x}$ to $\mathbf{u} + \mathbf{v}$ or, in other words, as the matrix for which we have

$$(A + B)\mathbf{x} = A\mathbf{x} + B\mathbf{x} \quad \text{for all } \mathbf{x} \in \mathbb{R}^n. \tag{2.2.6}$$

Similarly, for any scalar c, the matrix cA is defined as the matrix that maps $\mathbf{x}$ to $c\mathbf{u}$; that is, the matrix for which

$$(cA)\mathbf{x} = c(A\mathbf{x}) \quad \text{for all } \mathbf{x} \in \mathbb{R}^n. \tag{2.2.7}$$

The existence of the matrices $A + B$ and cA will be proved by Theorem 2.2.1 below, where they will be computed explicitly.

Definition 2.2.1 can be paraphrased as requiring that the order of the operations be reversible: on the right-hand side of Equation (2.2.6) we first *apply* A and B separately to $\mathbf{x}$ and then *add*, and on the left we first *add* A to B and then *apply* the sum to $\mathbf{x}$. Similarly, on the right side of Equation (2.2.7) we first apply A to $\mathbf{x}$ and then multiply by c, while on the left this order is reversed: A is first multiplied by c and then cA is applied to $\mathbf{x}$. We may regard Equation (2.2.6) as a new distributive rule and Equation (2.2.7) as a new "associative rule."

To write all these equations in components, let us first recall from Definition 2.1.1 that the product $A\mathbf{x}$ is a column m-vector whose ith component is given, for any i, by

$$(A\mathbf{x})_i = a_{i1}x_1 + a_{i2}x_2 + \cdots a_{in}x_n, \tag{2.2.8}$$

which can be abbreviated as

$$(A\mathbf{x})_i = \sum_j a_{ij}x_j. \tag{2.2.9}$$

Applying the same principle to $A + B$ we have

$$[(A + B)\mathbf{x}]_i = \sum_j (A + B)_{ij}x_j \tag{2.2.10}$$

and applying it to B we have

$$(B\mathbf{x})_i = \sum_j b_{ij}x_j. \tag{2.2.11}$$

Thus,

$$\begin{aligned}(A\mathbf{x} + B\mathbf{x})_i &= (A\mathbf{x})_i + (B\mathbf{x})_i \\ &= \sum_j a_{ij}x_j + \sum_j b_{ij}x_j = \sum_j (a_{ij} + b_{ij})x_j.\end{aligned} \tag{2.2.12}$$

Comparing Equations (2.2.6), (2.2.10), and (2.2.12) we obtain the first assertion below. The second one can be obtained similarly and is left as an exercise.

THEOREM 2.2.1. For any two matrices of the same shape we have

$$(A + B)_{ij} = a_{ij} + b_{ij} \text{ for all } i, j, \tag{2.2.13}$$

and for any scalar and any matrix

$$(cA)_{ij} = ca_{ij} \text{ for all } i, j. \tag{2.2.14}$$

This theorem can be paraphrased as: any entry of a sum of matrices equals the sum of the corresponding entries of the summands; and we multiply a matrix A by a scalar c by multiplying every entry by c. Notice again, as for vectors, the reversal of operations: "Entry of a sum = sum of entries" and "entry of (c times A) = c times entry of A."

Let us emphasize that only matrices of the same shape can be added to each other and that the sum has the same shape, in which case we call them *conformable for addition*, while for matrices of differing shapes there is no reasonable way of defining a sum.

We can also define multiplication of matrices in certain cases and this will prove to be an enormously fruitful operation. We want the product of two matrices to represent the performance of two mappings in succession. Let B map any $\mathbf{x} \in \mathbb{R}^n$ to a vector $\mathbf{y} = B\mathbf{x}$ of $\mathbb{R}^p$ and A map any $\mathbf{y} \in \mathbb{R}^p$ to $\mathbf{z} = A\mathbf{y}$ of $\mathbb{R}^m$. Then A must be an $m \times p$ matrix and B a $p \times n$ matrix. These mappings and Definition 2.2.2 are illustrated symbolically in Figure 2.4:

FIGURE 2.4

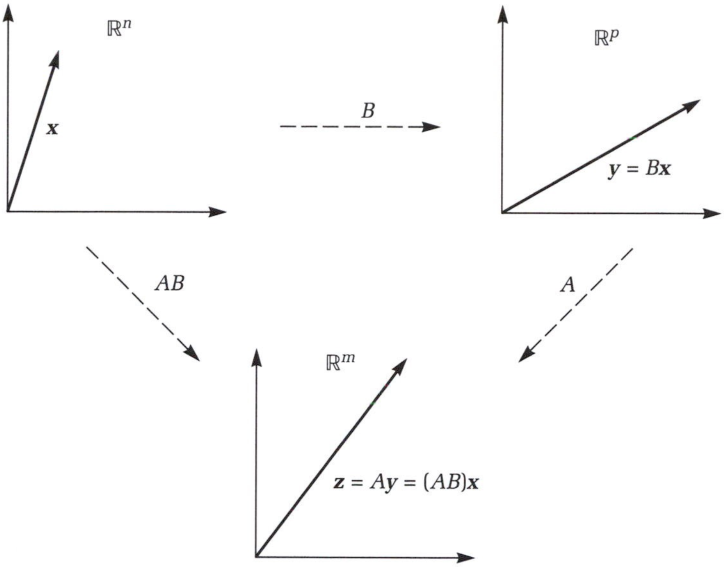

DEFINITION 2.2.2.

Let A be an $m \times p$ matrix and B a $p \times n$ matrix. For any $\mathbf{x} \in \mathbb{R}^n$ let the vector $\mathbf{y} = B\mathbf{x}$ be a corresponding element of $\mathbb{R}^p$ and, for any such $\mathbf{y}$, let the vector $\mathbf{z} = A\mathbf{y}$ be a corresponding element of $\mathbb{R}^m$. We define the product AB as the $m \times n$ matrix of the corresponding direct mapping of $\mathbf{x}$ to $\mathbf{z}$, that is, by the formula

$$(AB)\mathbf{x} = A(B\mathbf{x}) \text{ for all } \mathbf{x} \in \mathbb{R}^n. \tag{2.2.15}$$

That such a matrix always exists will be proved by Theorem 2.2.2, where it will be computed explicitly.

Let us emphasize that only for matrices A and B such that the number of columns of A (the p in the definition) equals the number of rows of B can the product AB be formed, in which case we call them *conformable for multiplication*.

Notice that this definition can also be viewed as a new associative law or as a reversal of the order of the two multiplications (but not of the factors).

EXAMPLE 2.2.2. Let

$$A = \begin{bmatrix} 3 & 5 \\ 1 & 2 \\ 2 & 4 \end{bmatrix} \text{ and } B = \begin{bmatrix} 2 & 1 \\ 0 & 3 \end{bmatrix}. \tag{2.2.16}$$

Then, applying Equation (2.2.11), for any $\mathbf{x}$ we have

$$B\mathbf{x} = \begin{bmatrix} 2 & 1 \\ 0 & 3 \end{bmatrix} \begin{bmatrix} x_1 \\ x_2 \end{bmatrix} = \begin{bmatrix} 2x_1 + x_2 \\ 3x_2 \end{bmatrix} \tag{2.2.17}$$

and similarly

$$A(B\mathbf{x}) = \begin{bmatrix} 3 & 5 \\ 1 & 2 \\ 2 & 4 \end{bmatrix} \begin{bmatrix} 2x_1 + x_2 \\ 3x_2 \end{bmatrix} = \begin{bmatrix} 6x_1 + 18x_2 \\ 2x_1 + \ 7x_2 \\ 4x_1 + 14x_2 \end{bmatrix} = \begin{bmatrix} 6 & 18 \\ 2 & 7 \\ 4 & 14 \end{bmatrix} \begin{bmatrix} x_1 \\ x_2 \end{bmatrix}. \tag{2.2.18}$$

Thus

$$AB = \begin{bmatrix} 6 & 18 \\ 2 & 7 \\ 4 & 14 \end{bmatrix}. \tag{2.2.19}$$

<<>>

From the definition we can easily deduce the following rule that gives the components of AB and shows that the vector $\mathbf{x}$ can be dispensed with in their computation:

THEOREM 2.2.2. Let A be an $m \times p$ matrix and B a $p \times n$ matrix. Then the product AB is an $m \times n$ matrix whose components are given by the formula

$$(AB)_{ik} = \sum_{j=1}^{p} a_{ij} b_{jk} \quad \text{for } i = 1, \ldots, m \text{ and } k = 1, \ldots, n. \tag{2.2.20}$$

Proof The equation $\boldsymbol{y} = B\mathbf{x}$ can be written in components as

$$y_j = (B\mathbf{x})_j = \sum_{k=1}^{n} b_{jk} x_k. \tag{2.2.21}$$

Also,

$$(A\boldsymbol{y})_i = \sum_{j=1}^{p} a_{ij} y_j. \tag{2.2.22}$$

Substituting from Equation (2.2.21) into (2.2.22) we get

$$(A(B\mathbf{x}))_i = \sum_{j=1}^{p} a_{ij} \left(\sum_{k=1}^{n} b_{jk} x_k \right) = \sum_{k=1}^{n} \left(\sum_{j=1}^{p} a_{ij} b_{jk} \right) x_k. \tag{2.2.23}$$

On the other hand we have

$$((AB)\mathbf{x})_i = \sum_{k=1}^{n} (AB)_{ik} x_k. \tag{2.2.24}$$

In view of Equation (2.2.15) the left-hand sides of (2.2.23) and (2.2.24) must be equal, and because the vector $\mathbf{x}$ can be chosen arbitrarily, the coefficients of x_k must agree on the right-hand sides. This result is the statement of the theorem.

<<>>

The special case of Theorem 2.2.2, in which $m = n = 1$, which is also a special case of the definition of $A\mathbf{x}$ on page 38, is worth stating separately:

COROLLARY 2.2.1. If A is a $1 \times p$ matrix; that is, a row p-vector

$$\boldsymbol{a} = (a_1, a_2, \ldots, a_p) \tag{2.2.25}$$

and B a $p \times 1$ matrix; that is, a column p-vector

$$\boldsymbol{b} = \begin{bmatrix} b_1 \\ b_2 \\ \cdots \\ b_p \end{bmatrix}, \tag{2.2.26}$$

then their matrix product $\boldsymbol{ab}$ is the same as their dot product as vectors, namely

$$\boldsymbol{ab} = \sum_{j=1}^{p} a_j b_j. \tag{2.2.27}$$

It is very important to observe that the matrix product is *not commutative.* This fact could be seen by direct computations, but it also follows from the product's definition as two mappings in succession, since mappings are generally not commutative. The latter is true even for transformations in the same space. Consider, for instance, the effect of a north–south stretch followed by a 90-degree rotation on a car facing north, and of the same operations performed in reverse order. In the first case we end up with a longer car facing west, and in the second with a wider car facing west.

For the two vectors in Corollary 2.2.1, the product $\boldsymbol{ba}$ is very different from $\boldsymbol{ab}$. If the column vector comes first, they do not even have to be equal in length. Changing $\boldsymbol{b}$ to a column m-vector and $\boldsymbol{a}$ to a row n-vector we get, by Theorem 2.2.2 with $p = 1$,

$$\boldsymbol{ba} = \begin{bmatrix} b_1 \\ b_2 \\ \cdots \\ b_m \end{bmatrix} (a_1, a_2, \ldots, a_n) = \begin{bmatrix} b_1a_1 & b_1a_2 & \cdots & b_1a_n \\ b_2a_1 & b_2a_2 & \cdots & b_2a_n \\ \cdots & & & \\ b_ma_1 & b_ma_2 & \cdots & b_ma_n \end{bmatrix}. \tag{2.2.28}$$

This matrix is called the *outer product* of the two vectors, in contrast to the much more important inner product given by Equation (2.2.27), presumably because the outer product is in the space of $m \times n$ matrices, which contains the spaces $\mathbb{R}^m$ and $\mathbb{R}^n$ of the factors, and those spaces, in turn, contain the space $\mathbb{R}^1$ of the inner product.

Even if the product AB is defined, often the product BA is not. For example, if A is 2×3, say, and B is 3×1, then AB is, by Definition 2.2.2, a 2×1 matrix, but BA is not defined because the inside numbers 1 and 2 in 3×1 and 2×3 do not match as required by Definition 2.2.2.

The interpretation of the product in Corollary 2.2.1 as a dot product suggests that the formula of Theorem 2.2.2 can be interpreted similarly:

COROLLARY 2.2.2. Let A be an $m \times p$ matrix and B a $p \times n$ matrix and let us denote the ith row of A by $\boldsymbol{a}^i$ and the kth column of B by $\boldsymbol{b}_k$; that is, let[9]

$$\boldsymbol{a}^i = (a_{i1}, a_{i2}, \ldots, a_{ip}) \tag{2.2.29}$$

and

$$\boldsymbol{b}_k = \begin{bmatrix} b_{1k} \\ b_{2k} \\ \cdots \\ b_{pk} \end{bmatrix}. \tag{2.2.30}$$

Then we have

$$(AB)_{ik} = \boldsymbol{a}^i \boldsymbol{b}_k \quad \text{for } i = 1, \ldots, m \text{ and } k = 1, \ldots, n. \tag{2.2.31}$$

Using this result to write out the entire product matrix, we obtain

$$AB = \begin{bmatrix} \boldsymbol{a}^1 \\ \boldsymbol{a}^2 \\ \cdots \\ \boldsymbol{a}^m \end{bmatrix} (\boldsymbol{b}_1, \boldsymbol{b}_1, \ldots, \boldsymbol{b}_n) = \begin{bmatrix} \boldsymbol{a}^1\boldsymbol{b}_1 & \boldsymbol{a}^1\boldsymbol{b}_2 & \cdots & \boldsymbol{a}^1\boldsymbol{b}_n \\ \boldsymbol{a}^2\boldsymbol{b}_1 & \boldsymbol{a}^2\boldsymbol{b}_2 & \cdots & \boldsymbol{a}^2\boldsymbol{b}_n \\ \cdots & & & \\ \boldsymbol{a}^m\boldsymbol{b}_1 & \boldsymbol{a}^m\boldsymbol{b}_2 & \cdots & \boldsymbol{a}^m\boldsymbol{b}_n \end{bmatrix}. \tag{2.2.32}$$

The last formula is analogous to the outer product in Equation (2.2.28), but the entries on the right are inner products of vectors rather than ordinary products of numbers. Equation (2.2.31) may be paraphrased as saying that the entry in the ith row and kth column of AB equals the dot product of the ith row of A with the kth column of B. This observation is very helpful in evaluating matrix products, as will be seen below.

Let us also comment on the use of superscripts and subscripts. The notation we follow for row and column vectors is standard in multilinear algebra and will serve us well later, but we have stayed with the more elementary standard usage of just subscripts for matrix elements. Thus our notation is a mixture of two conventions. To be consistent we should have used a^i_j instead of a_{ij} to denote an entry of A, because then a^i_j could have

[9]The i here is a superscript to distinguish a row of a matrix from a column and must not be mistaken for an exponent.

been properly interpreted as the jth component of the ith row $\boldsymbol{a}^i$, and also as the ith component of the jth column $\boldsymbol{a}_j$. However, since here we need no such sophistication, we have adopted the simpler convention.

EXAMPLE 2.2.3. Let

$$A = \begin{bmatrix} 2 & 4 \\ 3 & 7 \end{bmatrix} \text{ and } B = \begin{bmatrix} 3 & -1 \\ 5 & 6 \end{bmatrix}. \tag{2.2.33}$$

Then

$$AB = \begin{bmatrix} (2\ \ 4)\begin{bmatrix} 3 \\ 5 \end{bmatrix} & (2\ \ 4)\begin{bmatrix} -1 \\ 6 \end{bmatrix} \\ (3\ \ 7)\begin{bmatrix} 3 \\ 5 \end{bmatrix} & (3\ \ 7)\begin{bmatrix} -1 \\ 6 \end{bmatrix} \end{bmatrix} \tag{2.2.34}$$

and so

$$AB = \begin{bmatrix} 2 \cdot 3 + 4 \cdot 5 & 2 \cdot (-1) + 4 \cdot 6 \\ 3 \cdot 3 + 7 \cdot 5 & 3 \cdot (-1) + 7 \cdot 6 \end{bmatrix} = \begin{bmatrix} 26 & 22 \\ 44 & 39 \end{bmatrix}. \tag{2.2.35}$$

For further reference, notice that we can factor out the column vectors $\begin{bmatrix} 3 \\ 5 \end{bmatrix}$ and $\begin{bmatrix} -1 \\ 6 \end{bmatrix}$ in the columns of AB as given in Equation (2.2.34), and write AB as

$$AB = \left[\begin{bmatrix} 2 & 4 \\ 3 & 7 \end{bmatrix}\begin{bmatrix} 3 \\ 5 \end{bmatrix} \quad \begin{bmatrix} 2 & 4 \\ 3 & 7 \end{bmatrix}\begin{bmatrix} -1 \\ 6 \end{bmatrix} \right] = \left[A\begin{bmatrix} 3 \\ 5 \end{bmatrix} \quad A\begin{bmatrix} -1 \\ 6 \end{bmatrix} \right]. \tag{2.2.36}$$

Thus, in the product AB the matrix A can be distributed over the columns of B. Similarly, we can factor out the row vectors $(2\ \ 4)$ and $(3\ \ 7)$ from the rows of AB as given in Equation (2.2.34) and write AB also as

$$AB = \begin{bmatrix} (2\ \ 4)B \\ (3\ \ 7)B \end{bmatrix}; \tag{2.2.37}$$

that is, with the matrix B distributed over the rows of A.

<<>>

EXAMPLE 2.2.4. Let

$$A = \begin{bmatrix} 2 & -2 & 4 \\ 1 & 3 & 5 \end{bmatrix} \text{ and } B = \begin{bmatrix} 2 & -1 \\ 4 & -2 \\ 6 & 3 \end{bmatrix}. \tag{2.2.38}$$

Then

$$AB = \begin{bmatrix} 2 \cdot 2 - 2 \cdot 4 + 4 \cdot 6 & 2 \cdot (-1) - 2 \cdot (-2) + 4 \cdot 3 \\ 1 \cdot 2 + 3 \cdot 4 + 5 \cdot 6 & 1 \cdot (-1) + 3 \cdot (-2) + 5 \cdot 3 \end{bmatrix} = \begin{bmatrix} 20 & 14 \\ 44 & 8 \end{bmatrix}. \tag{2.2.39}$$

<<>>

EXAMPLE 2.2.5. The matrices

$$R_{30} = \frac{1}{2}\begin{bmatrix} \sqrt{3} & -1 \\ 1 & \sqrt{3} \end{bmatrix} \text{ and } R_{60} = \frac{1}{2}\begin{bmatrix} 1 & -\sqrt{3} \\ \sqrt{3} & 1 \end{bmatrix} \tag{2.2.40}$$

represent rotations by 30 and 60 degrees respectively, according to Example 2.2.1. Their product

$$R_{30}R_{60} = \begin{bmatrix} 0 & -1 \\ 1 & 0 \end{bmatrix} = R_{90} \tag{2.2.41}$$

represents the rotation by 90 degrees, as it should.

<<>>

An interesting use of matrices and matrix operations is provided by the following example, typical of a large number of similar applications involving so-called *incidence* or *connection matrices*:

EXAMPLE 2.2.6. Suppose that an airline has nonstop flights between cities A, B, C, D, E as described by the matrix

$$M = \begin{bmatrix} 0 & 1 & 0 & 1 & 0 \\ 1 & 0 & 0 & 0 & 0 \\ 0 & 0 & 0 & 0 & 1 \\ 1 & 0 & 0 & 0 & 1 \\ 0 & 0 & 1 & 1 & 0 \end{bmatrix}. \tag{2.2.42}$$

Here the entry m_{ij} is 1 if there is a nonstop connection from city i to city j, and 0 if there is not, with the cities labeled $1, 2, \ldots, 5$ instead of A, $B, \ldots, E$.

Then the entries of the matrix

$$M^2 = MM = \begin{bmatrix} 2 & 0 & 0 & 0 & 1 \\ 0 & 1 & 0 & 1 & 0 \\ 0 & 0 & 1 & 1 & 0 \\ 0 & 1 & 1 & 2 & 0 \\ 1 & 0 & 0 & 0 & 2 \end{bmatrix} \tag{2.2.43}$$

show the one-stop connections. Why? If we consider the entry

$$(M^2)_{ik} = \sum_{j=1}^{5} m_{ij}m_{jk} \tag{2.2.44}$$

of M^2, then the jth term equals 1 in this sum if and only if $m_{ij} = 1$ and $m_{jk} = 1$; that is, if we have a nonstop flight from i to j and another from j to k. If there are two such j values, then the sum will be equal to 2, showing that there are two choices for one-stop flights from i to k. Thus, for instance, $(M^2)_{11} = 2$ shows that there are two one-stop routes from A to A: Indeed, from A one can fly to B or D and back.

The entries of the matrix

$$M + M^2 = \begin{bmatrix} 2 & 1 & 0 & 1 & 1 \\ 1 & 1 & 0 & 1 & 0 \\ 0 & 0 & 1 & 1 & 1 \\ 1 & 1 & 1 & 2 & 1 \\ 1 & 0 & 1 & 1 & 2 \end{bmatrix} \tag{2.2.45}$$

show the number of ways of reaching one city from another with one-leg or two-leg flights. In particular the zero entries show, for instance, that B and E are not so connected. Evaluating $(M^3)_{25} = (M^3)_{52}$, we would similarly find that even those cities can be reached from each other with three-leg flights.

What are the vectors on which these matrices act; that is, what meaning can we give to an equation like $\boldsymbol{y} = M\boldsymbol{x}$? The answer is that if the components of $\boldsymbol{x}$ are restricted to just 0 and 1, then $\boldsymbol{x}$ may be regarded as representing a set of cities and $\boldsymbol{y}$ the set that can be reached nonstop from $\boldsymbol{x}$. Thus, for instance,

$$\boldsymbol{x} = \begin{bmatrix} 1 \\ 1 \\ 0 \\ 0 \\ 0 \end{bmatrix} \tag{2.2.46}$$

represents the set $\{A, B\}$, and then

$$\boldsymbol{y} = M\boldsymbol{x} = \begin{bmatrix} 1 \\ 1 \\ 0 \\ 1 \\ 0 \end{bmatrix} \tag{2.2.47}$$

represents the set $\{A, B, D\}$ that can be reached nonstop from $\{A, B\}$. (Again, if a number greater than 1 were to show up in $\boldsymbol{y}$, it would indicate that the corresponding city can be reached in more than one way.)

<<>>

As we have seen, a matrix can be regarded as a row vector of its columns and also as a column vector of its rows. Making full use of this choice we can rewrite the product of matrices in two more ways, corresponding to the particular cases shown in Equations (2.2.36) and (2.2.37). We obtain these formulas by factoring out the $\boldsymbol{a}^i$ coefficients in the rows of the matrix on the right of Equation (2.2.32) and by factoring out the $\boldsymbol{b}_j$ coefficients in the columns. Thus we get

COROLLARY 2.2.3. Let A and B be as in Corollary 2.2.2. With the same notations for the rows and columns as there, we have

$$AB = A(\boldsymbol{b}_1\ \boldsymbol{b}_2\ \cdots\ \boldsymbol{b}_n) = (A\boldsymbol{b}_1\ A\boldsymbol{b}_2\ \cdots\ A\boldsymbol{b}_n) \tag{2.2.48}$$

and

$$AB = \begin{bmatrix} \boldsymbol{a}^1 \\ \boldsymbol{a}^2 \\ \cdots \\ \boldsymbol{a}^m \end{bmatrix} B = \begin{bmatrix} \boldsymbol{a}^1 B \\ \boldsymbol{a}^2 B \\ \cdots \\ \boldsymbol{a}^m B \end{bmatrix}. \tag{2.2.49}$$

Although the matrix product is not commutative, it still has the most important properties expected of a product, namely associativity and distributivity:

THEOREM 2.2.3. Let A, B, and C be arbitrary matrices for which the expressions below all make sense. Then we have the associative law

$$A(BC) = (AB)C \tag{2.2.50}$$

and the distributive law

$$A(B + C) = AB + AC. \tag{2.2.51}$$

Proof Let A, B, and C be $m \times p$, $p \times q$, and $q \times n$ matrices respectively. Then we may evaluate the left side of Equation (2.2.50) using Equations (2.2.48) and (2.2.15) as follows:

$$\begin{aligned} A(BC) &= A(B(\boldsymbol{c}_1\ \boldsymbol{c}_2 \cdots \boldsymbol{c}_n)) = A(B\boldsymbol{c}_1\ B\boldsymbol{c}_2 \cdots B\boldsymbol{c}_n) = (A(B\boldsymbol{c}_1) \cdots A(B\boldsymbol{c}_n)) \\ &= ((AB)\boldsymbol{c}_1 \cdots (AB)\boldsymbol{c}_n) = (AB)(\boldsymbol{c}_1 \cdots \boldsymbol{c}_n) = (AB)C. \end{aligned} \tag{2.2.52}$$

We leave the proof of the distributive law to the reader.

<<>>

Once we have defined addition and multiplication of matrices, it is natural to ask what matrices take the place of the numbers 0 and 1, which play such an important role in the algebra of numbers. Zero is easy: we take any matrix with all entries equal to zero to be a zero matrix and, denoting it by O regardless of its shape, we have for any A of the same shape

$$A + O = A, \tag{2.2.53}$$

and whenever the product is defined

$$AO = O \text{ and } OA = O. \tag{2.2.54}$$

Note that the zero matrices on either side of each equal sign in (2.2.54) may be of different sizes although they are usually denoted by the same letter O.

Though a little less straightforward, it is still easy to see how to find analogs of 1. For any n, the $n \times n$ matrix

$$I = \begin{bmatrix} 1 & 0 & \dots & 0 \\ 0 & 1 & \dots & 0 \\ \dots & & & \\ 0 & 0 & \dots & 1 \end{bmatrix}, \tag{2.2.55}$$

with 1's along its "main diagonal" and zeroes everywhere else, has the properties

$$AI = A \text{ and } IA = A \tag{2.2.56}$$

whenever the products are defined. This fact can be verified by direct computation in any one of the product's forms, with A in the general form (a_{ij}). We may do it as follows: We write $I = (\delta_{ij})$, where

$$\delta_{ij} = \begin{cases} 1 \text{ if } i = j \\ 0 \text{ if } i \neq j \end{cases} \tag{2.2.57}$$

is called Kronecker's delta function and is the standard notation for the entries of the matrix I. With this notation, for A and I sized $m \times n$ and $n \times n$ respectively, Theorem 2.2.2 gives

$$(AI)_{ik} = \sum_{j=1}^{n} a_{ij}\delta_{jk} = a_{ik} \text{ for } i = 1, \dots, m \text{ and } k = 1, \dots, n, \tag{2.2.58}$$

because, by the definition of δ_{jk}, in the sum all terms are zero except the one with $j = k$, and that one gives a_{ik}. This result is, of course, equivalent to $AI = A$. We leave the proof of the other equation of (2.2.56) to the reader.

For any n the matrix I is called the *unit matrix* or the *identity matrix* of order n. We usually dispense with any indication of its order unless it is important and would be unclear from the context. In such cases we write it as I_n. Notice that the columns of I are the standard vectors $\mathbf{e}_i$ (regarded as column vectors, of course), that is,

$$I = (\mathbf{e}_1 \ \mathbf{e}_2 \ \dots \ \mathbf{e}_n). \tag{2.2.59}$$

In Section 2.3 we shall see how the inverse of a matrix can be defined in some cases. In closing this section we present briefly the promised reason for using *column* vectors in the equation $A\mathbf{x} = \mathbf{b}$. In a nutshell, we followed this procedure because otherwise the formalism of this whole section would have broken down. We used the product $A\mathbf{x}$ in Definition 2.2.2 of the general matrix product AB, which led to the formula of Theorem 2.2.2 for the components $(AB)_{ik}$. If we want to multiply AB by a third matrix C, we have no problem repeating the preceding procedure, that is, form the products $(AB)_{ik}c_{kl}$ and sum over k. However, had we used a *row* vector $\mathbf{x}$ in the beginning, that would have led to the formula $(AB)_{ik} = \sum_{j=1}^{n} a_{ij}b_{kj}$ and then multiplying this result by c_{kl} or c_{lk}, and summing over k we would have had to use the first subscript of b for summation in this product, unlike in the first one. Thus the associative law could not be maintained and the nice formulas of Corollary 2.2.3 would also cease to hold. Once we decided to use *rows* of A to multiply $\mathbf{x}$ in the product $A\mathbf{x}$, then we had to make $\mathbf{x}$ a column vector to end up with a decent formalism.

Exercises 2.2

2.2.1. Let

$$A = \begin{bmatrix} 2 & 3 \\ 1 & -2 \end{bmatrix} \text{ and } B = \begin{bmatrix} 3 & -4 \\ 2 & 2 \end{bmatrix}.$$

Find the matrices (a) $C = 2A + 3B$, and (b) $D = 4A - 3B$.

2.2.2. Prove Equation (2.2.14) of Theorem 2.2.1.

In the next six exercises, find the products of the given matrices in both orders; that is, both AB and BA, if possible.

• **2.2.3.** $A = \begin{bmatrix} 1 & -2 & 3 \end{bmatrix}$ and $B = \begin{bmatrix} 3 \\ 2 \\ 1 \end{bmatrix}$.

2.2.4. $A = \begin{bmatrix} 2 & 3 & 5 \\ 1 & -2 & 3 \end{bmatrix}$ and $B = \begin{bmatrix} 3 & -4 \\ 2 & 2 \\ 1 & -3 \end{bmatrix}$.

• **2.2.5.** $A = \begin{bmatrix} 2 & 3 & 5 \\ 1 & -2 & 3 \\ 3 & -4 & 2 \end{bmatrix}$ and $B = \begin{bmatrix} 3 & -4 \\ 2 & 2 \\ 1 & -3 \end{bmatrix}$.

2.2.6. $A = \begin{bmatrix} 1 & -2 & 3 & -4 \end{bmatrix}$ and $B = \begin{bmatrix} 3 \\ 2 \\ 1 \end{bmatrix}$.

2.2.7. $A = \begin{bmatrix} 1 & -2 & 3 & -4 \end{bmatrix}$ and $B = \begin{bmatrix} 3 & -4 \\ 2 & 2 \\ 1 & -3 \\ -2 & 5 \end{bmatrix}$.

2.2.8. $A = \begin{bmatrix} 2 & 3 & 5 \\ 1 & -2 & 3 \end{bmatrix}$ and $B = \begin{bmatrix} 3 & -4 \\ 2 & 2 \\ 1 & -3 \\ -2 & 5 \end{bmatrix}$.

2.2.9. Verify the associative law for the product of the matrices

$$A = [1 \ \ -2],\ B = \begin{bmatrix} 3 & -4 \\ 2 & 2 \end{bmatrix},\ \text{and}\ C = \begin{bmatrix} 1 & -3 \\ 3 & 0 \end{bmatrix}.$$

2.2.10. With the notation of Example 2.2.1, prove that for any two rotation matrices R_α and R_β we have $R_\alpha R_\beta = \mathrm{R}_{\alpha+\beta}$.

• **2.2.11.** Find two nonzero 2×2 matrices A and B such that $AB = O$.

2.2.12. Show that the cancellation law does not hold for matrix products: find nonzero matrices A, B, C such that $AB = AC$ but $B \neq C$.

***2.2.13.** Let A be an $m \times p$ matrix and B a $p \times n$ matrix. Show that the product AB can also be written in the following alternative forms:

a. $AB = \boldsymbol{a}_1\boldsymbol{b}^1 + \boldsymbol{a}_2\boldsymbol{b}^2 + \cdots + \boldsymbol{a}_p\boldsymbol{b}^p$,

b. $AB = \left(\sum_{i=1}^{p} \boldsymbol{a}_i b_{i1}, \sum_{i=1}^{p} \boldsymbol{a}_i b_{i2}, \ldots, \sum_{i=1}^{p} \boldsymbol{a}_i b_{in} \right)$,

c. $AB = \begin{bmatrix} \sum_{j=1}^{p} a_{1j}\boldsymbol{b}^j \\ \sum_{j=1}^{p} a_{2j}\boldsymbol{b}^j \\ \vdots \\ \sum_{j=1}^{p} a_{mj}\boldsymbol{b}^j \end{bmatrix}$.

2.2.14. Let A be any $n \times n$ matrix. Its powers for any nonnegative exponent k are defined by induction as $A^0 = I$ and $A^k = AA^{k-1}$. Show that the rules $A^kA^l = A^{k+l}$ and $(A^k)^l = A^{kl}$ hold, just as for real numbers.

• **2.2.15.** Find a nonzero 2×2 matrix A such that $A^2 = O$.

2.2.16. Find a 3×3 matrix A such that $A^2 \neq O$ but $A^3 = O$.

2.2.17. Find the number of three-leg flights connecting B and D in Example 2.2.6 by evaluating $(M^3)_{24} = (M^3)_{42}$.

2.2.18. Prove Equation (2.2.51) of Theorem 2.2.3.

The next five exercises deal with *block multiplication*:

2.2.19. Show that

$$\left[\begin{array}{cc|cc} 1 & -2 & 1 & 0 \\ 3 & 4 & 0 & 1 \end{array}\right]\left[\begin{array}{cc} 0 & 0 \\ 0 & 0 \\ \hline 3 & 2 \\ 1 & -1 \end{array}\right] = \begin{bmatrix} 1 & -2 \\ 3 & 4 \end{bmatrix}\begin{bmatrix} 0 & 0 \\ 0 & 0 \end{bmatrix} + \begin{bmatrix} 1 & 0 \\ 0 & 1 \end{bmatrix}\begin{bmatrix} 3 & 2 \\ 1 & -1 \end{bmatrix} = \begin{bmatrix} 3 & 2 \\ 1 & -1 \end{bmatrix}$$

***2.2.20.** Show that if two conformable matrices of any size are partitioned, as in Exercise 2.2.19, so that the products make sense, then $[A \quad B]\begin{bmatrix} C \\ D \end{bmatrix} = [AC + BD]$.

***2.2.21.** Show that if two conformable matrices of any size are partitioned into four submatrices each, so that the products and sums make sense, then

$$\begin{bmatrix} A & B \\ C & D \end{bmatrix}\begin{bmatrix} E & F \\ G & H \end{bmatrix} = \begin{bmatrix} AE + BG & AF + BH \\ CE + DG & CF + DH \end{bmatrix}.$$

2.2.22. Compute the product by block multiplication, using the result of Exercise 2.2.21:

$$\left[\begin{array}{cc|cc} 1 & -2 & 1 & 0 \\ 3 & 4 & 0 & 1 \\ \hline -1 & 0 & 0 & 0 \\ 0 & -1 & 0 & 0 \end{array}\right]\left[\begin{array}{cc|cc} 1 & -2 & 1 & 0 \\ 2 & 0 & -3 & 1 \\ \hline 0 & 0 & 2 & 3 \\ 0 & 0 & 7 & 4 \end{array}\right].$$

2.2.23. Partition the first matrix in Exercise 2.2.22 as

$$\left[\begin{array}{cc|cc} 1 & -2 & 1 & 0 \\ 3 & 4 & 0 & 1 \\ -1 & 0 & 0 & 0 \\ \hline 0 & -1 & 0 & 0 \end{array}\right].$$

Find the appropriate corresponding partition of the second matrix, and evaluate the product by using these blocks.

MATLAB Exercises 2.2

In MATLAB the product of matrices is denoted by $*$, and a power like A^k by A^k; both the same as for numbers. The unit matrix of order n is denoted by **eye**(n), and the $m \times n$ zero matrix by **zeros**(m,n). The command **rand**(m,n) returns an $m \times n$ matrix with random entries uniformly distributed between 0 and 1. The command **round**(A) rounds each entry of A to the nearest integer.

ML2.2.1. As in Example 2.2.1, let $\boldsymbol{v}$ denote the vector obtained from the vector $\boldsymbol{u}$ by a rotation through an angle θ.

a. Compute $\boldsymbol{v}$ for $\boldsymbol{u} = \begin{bmatrix} 2 \\ 5 \end{bmatrix}$ and each of $\theta = 15°, 30°, 45°, 60°, 75°$, and $90°$.

(MATLAB will compute the trig functions if you use radians.)

b. Use MATLAB to verify that $R_{75°} = R_{25°} * R_{50°}$.

ML2.2.2. Let

$$M = \begin{bmatrix} 0 & 1 & 0 & 1 & 0 & 0 \\ 1 & 0 & 0 & 0 & 1 & 1 \\ 0 & 0 & 0 & 0 & 0 & 1 \\ 1 & 0 & 0 & 0 & 1 & 0 \\ 0 & 1 & 0 & 1 & 0 & 1 \\ 0 & 1 & 1 & 0 & 1 & 0 \end{bmatrix}$$

be the connection matrix of an airline network as in Example 2.2.6.

a. Which cities can be reached from A with exactly two stops?

b. Which cities can be reached from A with two stops or fewer?

c. What is the maximum number of stops needed to reach any city from any other?

ML2.2.3. Let $\boldsymbol{a}$ = 10 * **rand**(1,4) – 5 and $\boldsymbol{b}$ = 10 * **rand**(1,4) – 5.

a. Compute $C = \boldsymbol{a} * \boldsymbol{b}'$ and **rank**(C) for ten instances of such $\boldsymbol{a}$ and $\boldsymbol{b}$. (Use the up-arrow key.)

b. Make a conjecture about **rank**(C) in general.

c. Prove your conjecture.

ML2.2.4. Let A = 10 * **rand**(2,4) – 5 and B = 10 * **rand**(4,2) – 5.

a. Compute $C = A * B$, $D = B * A$, **rank**(C), and **rank**(D) for ten instances of such A and B.

b. Make a conjecture about **rank**(C) and **rank**(D) in general.

ML2.2.5. In MATLAB you can enter blocks in a matrix in the same way as you enter scalars. Use this method to solve (a) Exercise 2.2.19, and (b) Exercise 2.2.22.

2.3. THE ALGEBRA OF MATRICES, CONTINUED

Although for vectors it is impossible to define division, for matrices it is possible in some very important cases.

We may try to follow the same procedure as for numbers. The fraction b/a has been defined as the solution of the equation $ax = b$, or as b times $1/a$, where $1/a$ is the solution to $ax = 1$. For matrices we mimic the latter formula: we look for the solution of matrix equations of the form

$$AX = I, \qquad (2.3.1)$$

where I is the $n \times n$ unit matrix, A a given matrix, and X an unknown one. In terms of mappings, because I represents the identity mapping or no change, this equation means that if a mapping is given by the matrix A, we are looking for the matrix X of the (right) inverse mapping; that is, of the mapping that is undone if followed by the mapping A. (As it turns out, and as should be evident from the geometrical meaning, the order of the factors does not matter if A is a mapping from $\mathbb{R}^n$ to itself.)

By the definition of the product, A must then be of size $n \times p$ and X of size $p \times n$ for some p. Then Equation (2.3.1) corresponds to n^2 scalar equations for np unknowns. Thus, if $p < n$ holds, then we have fewer unknowns than equations and generally no solutions apart from exceptional cases. On the other hand, if $p > n$ holds, then we have more unknowns than equations, and so generally infinitely many solutions. Because we are interested in finding unique solutions, we restrict our attention to those cases in which $p = n$ holds, or in other words to $n \times n$ or *square* matrices A. In this case n is called the *order* of A. For such A, Equation (2.3.1) may be written as

$$A(\mathbf{x}_1\ \mathbf{x}_2\ \dots\ \mathbf{x}_n) = (\mathbf{e}_1\ \mathbf{e}_2\ \dots\ \mathbf{e}_n) \tag{2.3.2}$$

and by Equation (2.2.48) on page 64 we can decompose this result into n separate systems

$$A\mathbf{x}_1 = \mathbf{e}_1, A\mathbf{x}_2 = \mathbf{e}_2, \dots, A\mathbf{x}_n = \mathbf{e}_n \tag{2.3.3}$$

for the n unknown n-vectors $\mathbf{x}_1, \mathbf{x}_2, \dots, \mathbf{x}_n$.

Before proceeding further with the general theory let us consider an example:

EXAMPLE 2.3.1. Let

$$A = \begin{bmatrix} 1 & 2 \\ 3 & 4 \end{bmatrix} \tag{2.3.4}$$

and so let us solve

$$\begin{bmatrix} 1 & 2 \\ 3 & 4 \end{bmatrix}\begin{bmatrix} x_{11} & x_{12} \\ x_{21} & x_{22} \end{bmatrix} = \begin{bmatrix} 1 & 0 \\ 0 & 1 \end{bmatrix} \tag{2.3.5}$$

or equivalently, the separate systems

$$\begin{bmatrix} 1 & 2 \\ 3 & 4 \end{bmatrix}\begin{bmatrix} x_{11} \\ x_{21} \end{bmatrix} = \begin{bmatrix} 1 \\ 0 \end{bmatrix} \text{ and } \begin{bmatrix} 1 & 2 \\ 3 & 4 \end{bmatrix}\begin{bmatrix} x_{12} \\ x_{22} \end{bmatrix} = \begin{bmatrix} 0 \\ 1 \end{bmatrix}. \tag{2.3.6}$$

Subtracting 3 times the first row from the second in both systems, we get

$$\begin{bmatrix} 1 & 2 \\ 0 & -2 \end{bmatrix}\begin{bmatrix} x_{11} \\ x_{21} \end{bmatrix} = \begin{bmatrix} 1 \\ -3 \end{bmatrix} \text{ and } \begin{bmatrix} 1 & 2 \\ 0 & -2 \end{bmatrix}\begin{bmatrix} x_{12} \\ x_{22} \end{bmatrix} = \begin{bmatrix} 0 \\ 1 \end{bmatrix}. \tag{2.3.7}$$

Adding the second row to the first and dividing the second row by −2, again in both systems, we find

$$\begin{bmatrix} 1 & 0 \\ 0 & 1 \end{bmatrix}\begin{bmatrix} x_{11} \\ x_{21} \end{bmatrix} = \begin{bmatrix} -2 \\ 3/2 \end{bmatrix} \text{ and } \begin{bmatrix} 1 & 0 \\ 0 & 1 \end{bmatrix}\begin{bmatrix} x_{12} \\ x_{22} \end{bmatrix} = \begin{bmatrix} 1 \\ -1/2 \end{bmatrix}. \tag{2.2.8}$$

Thus we obtain

$$x_{11} = -2,\ x_{21} = 3/2,\ x_{12} = 1,\ x_{22} = -1/2 \tag{2.2.9}$$

or in matrix form,

$$X = \begin{bmatrix} -2 & 1 \\ 3/2 & -1/2 \end{bmatrix}. \tag{2.3.10}$$

It is easy to check that this X is a solution of $AX = I$, and in fact of $XA = I$, too. Furthermore, since the two systems given by Equations (2.3.6) have the same matrix A on their left sides, the row-reduction steps were exactly the same for both, and can therefore be combined into the reduction of a single augmented matrix with the two columns of I on the right; that is, of $[A|I]$ as follows:

$$\left[\begin{array}{cc|cc} 1 & 2 & 1 & 0 \\ 3 & 4 & 0 & 1 \end{array}\right] \to \left[\begin{array}{cc|cc} 1 & 2 & 1 & 0 \\ 0 & -2 & -3 & 1 \end{array}\right] \to$$
$$\left[\begin{array}{cc|cc} 1 & 0 & -2 & 1 \\ 0 & -2 & -3 & 1 \end{array}\right] \to \left[\begin{array}{cc|cc} 1 & 0 & -2 & 1 \\ 0 & 1 & 3/2 & -1/2 \end{array}\right]. \tag{2.3.11}$$

<<>>

We can generalize the results of this example in part as a definition and in part as a theorem:

DEFINITION 2.3.1.

A matrix A is called *invertible* if it is a square matrix and there exists a square matrix X of the same size such that $AX = I$ and $XA = I$ hold. Such an X, if one exists, is called the *inverse* of A and is denoted by A^{-1}.

THEOREM 2.3.1.

A square matrix A is invertible if and only if the augmented matrix $[A|I]$ can be reduced by elementary row operations to the form $[I|X]$, and in that case X is the unique inverse A^{-1} of A.

Proof Obviously the reduction of $[A|I]$, if it ends up in the form $[I|X]$, produces the unique solution X of $AX = I$. By reversing the elementary row operations we can undo this reduction, or equivalently, if X is changed back to I, then $[X|I]$ is reduced to $[I|A]$. Since the augmented matrix $[X|I]$ may be considered that of a system $XY = I$ for an unknown matrix Y, the reverse reduction ending in $[I|A]$ gives the solution $Y = A$ for this new system. Thus, if X solves $AX = I$, then $XA = I$ holds, too, and A is invertible with X as its inverse A^{-1}.

On the other hand, if $[A|I]$ cannot be reduced to the form $[I|X]$, then the system $AX = I$ has no solution for the following reason: In this case the reduction of A must produce a zero row at the bottom of any corresponding echelon matrix U; and suppose it is the ith row of A that becomes this last row of U. Then the submatrix $[A|\mathbf{e}_i]$ will get reduced to $[U|\mathbf{e}_n]$, since the zero entries in the last column cannot affect the single 1 of it, which ends up at the bottom. The matrix $[U|\mathbf{e}_n]$, however, represents an inconsistent system, because the last row of U is zero but the last component of $\mathbf{e}_n$ is not.

<<>>

EXAMPLE 2.3.2. Let us find the inverse of the matrix

$$A = \begin{bmatrix} 2 & 3 & 5 \\ 1 & -2 & 3 \\ 3 & -4 & 2 \end{bmatrix} \tag{2.3.12}$$

if it exists. We form the augmented matrix $[A|I]$ and reduce it as follows:

$$\left[\begin{array}{ccc|ccc} 2 & 3 & 5 & 1 & 0 & 0 \\ 1 & -2 & 3 & 0 & 1 & 0 \\ 3 & -4 & 2 & 0 & 0 & 1 \end{array}\right] \to \left[\begin{array}{ccc|ccc} 1 & -2 & 3 & 0 & 1 & 0 \\ 2 & 3 & 5 & 1 & 0 & 0 \\ 3 & -4 & 2 & 0 & 0 & 1 \end{array}\right] \to$$

$$\left[\begin{array}{ccc|ccc} 1 & -2 & 3 & 0 & 1 & 0 \\ 0 & 7 & -1 & 1 & -2 & 0 \\ 0 & 2 & -7 & 0 & -3 & 1 \end{array}\right] \to \left[\begin{array}{ccc|ccc} 1 & 0 & -4 & 0 & -2 & 1 \\ 0 & 7 & -1 & 1 & -2 & 0 \\ 0 & 2 & -7 & 0 & -3 & 1 \end{array}\right] \to$$

$$\left[\begin{array}{ccc|ccc} 1 & 0 & -4 & 0 & -2 & 1 \\ 0 & 1 & 20 & 1 & 7 & -3 \\ 0 & 2 & -7 & 0 & -3 & 1 \end{array}\right] \to \left[\begin{array}{ccc|ccc} 1 & 0 & -4 & 0 & -2 & 1 \\ 0 & 1 & 20 & 1 & 7 & -3 \\ 0 & 0 & -47 & -2 & -17 & 7 \end{array}\right] \to$$

$$\left[\begin{array}{ccc|ccc} 1 & 0 & -4 & 0 & -2 & 1 \\ 0 & 1 & 20 & 1 & 7 & -3 \\ 0 & 0 & 1 & 2/47 & 17/47 & -7/47 \end{array}\right] \to$$

$$\left[\begin{array}{ccc|ccc} 1 & 0 & 0 & 8/47 & -26/47 & 19/47 \\ 0 & 1 & 0 & 7/47 & -11/47 & -1/47 \\ 0 & 0 & 1 & 2/47 & 17/47 & -7/47 \end{array}\right]. \tag{2.3.13}$$

Thus we can read off the inverse of A as

$$A^{-1} = \frac{1}{47}\begin{bmatrix} 8 & -26 & 19 \\ 7 & -11 & -1 \\ 2 & 17 & -7 \end{bmatrix}. \tag{2.3.14}$$

It is easy to check that we do indeed have $AA^{-1} = A^{-1}A = I$.

<<>>

Just as for numbers $b/a = a^{-1}b$ is the solution of $ax = b$, similarly we have for matrix equations a simple consequence of Definition 2.3.1:

THEOREM 2.3.2. If A is an invertible $n \times n$ matrix and B an arbitrary $n \times p$ matrix, then the equation

$$AX = B \tag{2.3.15}$$

has the unique solution

$$X = A^{-1}B. \tag{2.3.16}$$

Proof That $X = A^{-1}B$ is a solution can be seen easily by substituting it into Equation (2.3.15):

$$A(A^{-1}B) = (AA^{-1})B = IB = B, \tag{2.3.17}$$

and that it is the only solution can be seen in this way: Assume that Y is another solution, so that

$$AY = B \tag{2.3.18}$$

holds. Multiplying both sides of this equation by A^{-1} we get

$$A^{-1}(AY) = A^{-1}B \tag{2.3.19}$$

and this equation reduces to

$$(A^{-1}A)Y = Y = A^{-1}B, \tag{2.3.20}$$

which shows that $Y = X$.

<<>>

In case $p = 1$ holds, Equation (2.3.15) becomes our old friend

$$A\mathbf{x} = \mathbf{b}, \tag{2.3.21}$$

where $\mathbf{x}$ and $\mathbf{b}$ are n-vectors. Thus Theorem 2.3.2 provides a new way of solving this equation. Unfortunately, this technique has little practical significance, since computing the inverse of A is generally more difficult than solving Equation (2.3.21) by Gaussian elimination. In some theoretical considerations, however, it is useful to know that the solution of Equation (2.3.21) can be written as

$$\mathbf{x} = A^{-1}\mathbf{b}. \tag{2.3.22}$$

Also, if we have several equations like (2.3.21) with the same left sides, then they can be combined into an equation of the form (2.3.15) with

$p > 1$ and profitably solved by computing the inverse of A and using Theorem 2.3.2.

EXAMPLE 2.3.3. Let us solve

$$\begin{bmatrix} 1 & 2 \\ 3 & 4 \end{bmatrix} X = \begin{bmatrix} 2 & 3 & -5 \\ 4 & -1 & 3 \end{bmatrix}. \qquad (2.3.23)$$

From Example 2.3.1 we know that

$$\begin{bmatrix} 1 & 2 \\ 3 & 4 \end{bmatrix}^{-1} = \begin{bmatrix} -2 & 1 \\ 3/2 & -1/2 \end{bmatrix}. \qquad (2.3.24)$$

Hence, by Theorem 2.3.2, we obtain

$$X = \begin{bmatrix} -2 & 1 \\ 3/2 & -1/2 \end{bmatrix} \begin{bmatrix} 2 & 3 & -5 \\ 4 & -1 & 3 \end{bmatrix} = \begin{bmatrix} 0 & -7 & 13 \\ 1 & 5 & -9 \end{bmatrix}. \qquad (2.3.25)$$

<<>>

As we have just seen, if A is invertible, then Equation (2.3.22) provides the solution of (2.3.21) for any choice of $\boldsymbol{b}$. It is then natural to ask whether the converse is true; that is, whether the existence of a solution of Equation (2.3.21) for *every* $\boldsymbol{b}$ implies the invertibility of A (we know that a *single* $\boldsymbol{b}$ is not enough: (2.3.21) may be solvable for some right-hand sides and not for others). The answer is yes:

THEOREM 2.3.3. An $n \times n$ matrix A is invertible if and only if $A\mathbf{x} = \boldsymbol{b}$ has a solution for every n-vector $\boldsymbol{b}$.

Proof Half of this statement has already been proved; we include it just for the sake of completeness. Thus let us assume that $A\mathbf{x} = \boldsymbol{b}$ has a solution for any n-vector $\boldsymbol{b}$. Then this condition is certainly true for each standard vector $\mathbf{e}_i$ in the role of $\boldsymbol{b}$; that is, each of the equations

$$A\mathbf{x}_1 = \mathbf{e}_1,\ A\mathbf{x}_2 = \mathbf{e}_2,\ \ldots,\ A\mathbf{x}_n = \mathbf{e}_n \qquad (2.3.26)$$

has a solution by assumption. These equations can, however, be combined into the single equation

$$AX = I, \qquad (2.3.27)$$

whose augmented matrix is $[A|I]$. From the proof of Theorem 2.3.1 we know that the solution to this equation, if one exists, must be $X = A^{-1}$, and because we have stipulated the existence of a solution, the invertibility of A follows.

<<>>

The condition of solvability of $A\mathbf{x} = \boldsymbol{b}$ for *every* possible right side can be replaced by the requirement of uniqueness of the solution for a *single* $\boldsymbol{b}$:

THEOREM 2.3.4. An $n \times n$ matrix A is invertible if and only if $A\mathbf{x} = \boldsymbol{b}$ has a unique solution for some n-vector $\boldsymbol{b}$ and then also for all n-vectors $\boldsymbol{b}$.

Proof If A is invertible, then, by Theorem 2.3.2, $\mathbf{x} = A^{-1}\boldsymbol{b}$ gives the unique solution of $A\mathbf{x} = \boldsymbol{b}$ for any $\boldsymbol{b}$. Conversely if, for some $\boldsymbol{b}$, $A\mathbf{x} = \boldsymbol{b}$ has a unique solution, then Theorem 2.1.1 (page 45) shows that the rank of A equals n, consequently that $AX = I$ too has a unique solution. Of course, this solution must be A^{-1}.

<<>>

The vector $\boldsymbol{b}$ in Theorem 2.3.4 may be taken to be the zero vector. This case is sufficiently important for special mention:

COROLLARY 2.3.1. An $n \times n$ matrix A is invertible if and only if $A\mathbf{x} = \boldsymbol{0}$ has only the trivial solution.

A square matrix that does not have the nice properties of the theorems above is called *singular*. Let us collect some equivalent characterizations of these that follow from our considerations up to now.

THEOREM 2.3.5. An $n \times n$ matrix A is singular if and only if it has any (and then all) of the following properties:

1. A is not invertible.
2. The rank of A is less than n.
3. A is not row equivalent to I.
4. $A\mathbf{x} = \boldsymbol{b}$ has no solution for some $\boldsymbol{b}$.
5. Even if $A\mathbf{x} = \boldsymbol{b}$ has a solution for a given $\boldsymbol{b}$, it is not unique.
6. The homogeneous equation $A\mathbf{x} = \boldsymbol{0}$ has nontrivial solutions.

For numbers, the product and the inverse are connected by the formula $(ab)^{-1} = a^{-1}b^{-1}$. For matrices, we have an analogous result but with the significant difference that the order of the factors on the right must be reversed:

THEOREM 2.3.6. If A and B are invertible matrices of the same size, then so too is AB and

$$(AB)^{-1} = B^{-1}A^{-1}. \tag{2.3.28}$$

Proof The proof is very simple: Repeated application of the associative law and the definition of I give

$$(AB)(B^{-1}A^{-1}) = ((AB)B^{-1})A^{-1} = (A(BB^{-1}))A^{-1}$$
$$= (AI)A^{-1} = AA^{-1} = I \tag{2.3.29}$$

and similarly in the reverse order

$$(B^{-1}A^{-1})(AB) = I. \tag{2.3.30}$$

<<>>

There exists another simple operation for matrices, one that has no analog with numbers. Although we do not need it until later, we present it here since it rounds out our discussion of the algebra of matrices.

DEFINITION 2.3.2.

Given any matrix A, we define its *transpose* A^T as the matrix obtained from A by interchanging its rows with its corresponding columns. In components:

$$(A^T)_{ij} = a_{ji} \tag{2.3.31}$$

for all values of i and j.

From this definition it easily follows that if A is $m \times n$, then A^T is $n \times m$, and in particular that the transpose of a column n-vector is a row n-vector and vice versa. This fact is often used for avoiding the inconvenient appearance of tall column vectors by writing them as transposed rows:

EXAMPLE 2.3.4.

$$(x_1, x_2, \ldots, x_n)^T = \begin{bmatrix} x_1 \\ x_2 \\ \ldots \\ x_n \end{bmatrix}. \tag{2.3.32}$$

<<>>

EXAMPLE 2.3.5. Let

$$A = \begin{bmatrix} 2 & 3 & -5 \\ 4 & -1 & 3 \end{bmatrix}. \tag{2.3.33}$$

Then

$$A^T = \begin{bmatrix} 2 & 4 \\ 3 & -1 \\ -5 & 3 \end{bmatrix}. \tag{2.3.34}$$

<<>>

The transpose has some useful properties:

THEOREM 2.3.7. If A and B are matrices such that their product is defined, then

$$(AB)^T = B^T A^T, \tag{2.3.35}$$

and if A is invertible, then so too is A^T and

$$(A^{-1})^T = (A^T)^{-1}. \tag{2.3.36}$$

We leave the proof to the reader.

Exercises 2.3

In the first six exercises, find the inverse matrix if possible.

• **2.3.1.** $A = \begin{bmatrix} 2 & 3 \\ 4 & -1 \end{bmatrix}$

2.3.2. $A = \begin{bmatrix} 5 & 2 \\ 3 & 4 \end{bmatrix}$

• **2.3.3.** $A = \begin{bmatrix} 2 & 3 & 5 \\ 4 & -1 & 1 \\ 3 & 2 & -2 \end{bmatrix}$

2.3.4. $A = \begin{bmatrix} 0 & -6 & 2 \\ 3 & -1 & 0 \\ 4 & 3 & -2 \end{bmatrix}$

2.3.5. $A = \begin{bmatrix} 2 & 3 & 5 \\ 4 & -1 & 3 \\ 3 & 2 & 5 \end{bmatrix}$

2.3.6. $A = \begin{bmatrix} 1 & -1 & 0 & 0 \\ 0 & 1 & -1 & 0 \\ 0 & 1 & 1 & 0 \\ 1 & 0 & -1 & 1 \end{bmatrix}$

2.3.7. Find two invertible 2×2 matrices A and B such that $A \neq -B$ and $A + B$ is not invertible.

2.3.8. **a.** Given the 2×3 matrix

$$A = \begin{bmatrix} 2 & 0 & 4 \\ 4 & -1 & 1 \end{bmatrix},$$

find a 3×2 matrix X by Gauss-Jordan elimination such that $AX = I$ holds. (Such a matrix is called a *right-inverse* of A.)

b. Can you find a 3×2 matrix Y such that $YA = I$ holds?

2.3.9. **a.** Given the 3×2 matrix

$$A = \begin{bmatrix} 2 & -1 \\ 4 & -1 \\ 2 & 2 \end{bmatrix}$$

find a 2×3 matrix X by Gauss-Jordan elimination such that $XA = I$ holds. (Such a matrix is called a *left-inverse* of A.)

b. Can you find a 2×3 matrix Y such that $AY = I$ holds?

***2.3.10.** **a.** Try to formulate a general rule, based on the results of the last two exercises, for the existence of a right-inverse and for the existence of a left-inverse of a 2×3 and of a 3×2 matrix.

b. Same as above for an $m \times n$ matrix.

c. When would the right-inverse and the left-inverse be unique?

***2.3.11.** Show that if a square matrix has a right-inverse X and a left-inverse Y, then $Y = X$ must hold. (*Hint:* Modify the second part of the proof of Theorem 2.3.2.)

2.3.12. The matrix

$$E = \begin{bmatrix} 1 & 0 & 0 \\ c & 1 & 0 \\ 0 & 0 & 1 \end{bmatrix}$$

is obtained from the unit matrix by the elementary row operation of adding c times its first row to its second row. Show that for any 3×3 matrix A the same elementary row operation performed on A results in the product matrix EA. Also, find E^{-1} and describe the elementary row operation it corresponds to. (A matrix that produces the same effect by multiplication as an elementary row operation, like this E and the matrices P in the next two exercises, is called an *elementary matrix*.)

2.3.13. Find a matrix P such that, for any 3×3 matrix A, PA equals the matrix obtained from A by multiplying its first row by any nonzero scalar c. (*Hint:* Try $A = I$ first.) Find P^{-1}.

• **2.3.14.** Find a matrix P such that, for any 3×3 matrix A, PA equals the matrix obtained from A by exchanging its first and third rows. (*Hint:* Try $A = I$ first.) Find P^{-1}.

2.3.15. If A is any invertible matrix and c any nonzero scalar, what is the inverse of cA? Prove your answer.

2.3.16. For any invertible matrix A and any positive integer n we define $A^{-n} = (A^{-1})^n$. Show that in this case we also have $A^{-n} = (A^n)^{-1}$ and $A^{-m}A^{-n} = A^{-m-n}$ if m is a positive integer as well.

2.3.17. A square matrix with a single 1 in each row and in each column and zeroes everywhere else is called a *permutation matrix*. List all six 3×3 permutation matrices P and their inverses and show that, for any $3 \times n$ matrix A, PA equals the matrix obtained from A by a permutation of its rows, which is the same as the permutation of the rows of I that results in P. What is BP if B is $n \times 3$?

2.3.18. State six conditions corresponding to those of Theorem 2.3.5 for a matrix to be nonsingular.

• **2.3.19.** Prove that if A is an invertible matrix then so too is A^{-1} and $(A^{-1})^{-1} = A$. (*Hint:* Imitate the proof of Theorem 2.3.2 for the equation $A^{-1}X = I$.)

• **2.3.20.** Prove that if A, B, C are invertible matrices of the same order, then so too is ABC; and $(ABC)^{-1} = C^{-1}B^{-1}A^{-1}$.

***2.3.21.** Prove Theorem 2.3.7.

MATLAB Exercises 2.3

In MATLAB the transpose of A is denoted by A'. The reduction of $[A|I]$ can be achieved by the command **rref**([A **eye**(n)]) or **rref**([A **eye**(**size**(A))]). A^{-1} can also be obtained by writing **inv**(A). These commands or the command **rank**(A) can be used to determine whether A is singular or not.

ML2.3.1. Let A = **round**(10 * **rand**(4)), B = **triu**(A), and C = **tril**(A).

a. Find the inverses of B and C in **format rat** by using **rref**, if they exist, and verify that they are indeed inverses.

b. Repeat Part (a) five times. (Use the up-arrow key.)

c. Do you see any pattern? Make a conjecture and prove it.

ML2.3.2. Let A = **round**(10 * **rand**(3,5)).

a. Find a solution for $AX = I$ by using **rref**, or show that no solution exists.

b. If you have found a solution, verify that it satisfies $AX = I$.

c. If there is a solution, compute $A \backslash$ **eye**(3) and check whether it is a solution.

d. If there is a solution of $AX = I$, try to find a solution for $YA = I$ by using **rref**. (*Hint:* Rewrite this as $A^TY^T = I$ first.) Draw a conclusion.

e. Repeat Parts (a)–(d) three times.

<<3>> VECTOR SPACES AND SUBSPACES

3.1. GENERAL VECTOR SPACES

In Section 1.1 we mentioned that various sets of functions have the same kind of structure as the Euclidean vector spaces we have studied so far. (By structure we mean the operations and their basic rules on the given sets, together with all their implications.) In subsequent sections we develop several concepts such as subspaces and linear independence that are common to all these spaces and are important for all of them. Thus, though our focus will remain $\mathbb{R}^n$, it is advantageous to study all vector spaces together before taking them up individually, and in this section we define general vector spaces, list some specific examples, and show that the algebraic rules in Section 1.1 hold for all of them.

DEFINITION 3.1.1.

A set V is called a (real) vector space and its elements vectors if V is not empty and with each $\boldsymbol{p}, \boldsymbol{q} \in V$ and each real number c a unique sum $\boldsymbol{p} + \boldsymbol{q} \in V$ and a unique product $c\boldsymbol{p} \in V$ are associated satisfying the eight rules below:[1]

1. $\boldsymbol{p} + \boldsymbol{q} = \boldsymbol{q} + \boldsymbol{p}$ (commutativity of addition)
2. $(\boldsymbol{p} + \boldsymbol{q}) + \boldsymbol{r} = \boldsymbol{p} + (\boldsymbol{q} + \boldsymbol{r})$ (associativity of addition)
3. There is a vector $\boldsymbol{0}$ such that $\boldsymbol{p} + \boldsymbol{0} = \boldsymbol{p}$ for every $\boldsymbol{p}$ (existence of zero vector)
4. For every vector $\boldsymbol{p}$ there is an associated vector $-\boldsymbol{p}$ such that $\boldsymbol{p} + (-\boldsymbol{p}) = \boldsymbol{0}$ (existence of additive inverse)
5. $1\boldsymbol{p} = \boldsymbol{p}$ (rule of multiplication by 1)
6. $a(b\boldsymbol{p}) = (ab)\boldsymbol{p}$ (associativity of multiplication by scalars)
7. $(a + b)\boldsymbol{p} = a\boldsymbol{p} + b\boldsymbol{p}$ (first distributive law)
8. $a(\boldsymbol{p} + \boldsymbol{q}) = a\boldsymbol{p} + a\boldsymbol{q}$ (second distributive law)

[1]Such rules in definitions are usually called *axioms*.

EXAMPLE 3.1.1. The Euclidean space $\mathbb{R}^n$, for any n, is a vector space by this new definition as well, for it was taken as the paradigm of all vector spaces: Except for Axiom 4, we just changed Theorem 1.1.1 into this definition, thus that theorem proves that seven of the axioms above hold for $\mathbb{R}^n$. If we use the definition $-\boldsymbol{p} = (-1)\boldsymbol{p}$ from Section 1.1, then Axiom 4 too can be shown to hold for $\mathbb{R}^n$ as follows: $\boldsymbol{p} + (-\boldsymbol{p}) = 1\boldsymbol{p} + (-1)\boldsymbol{p} = [1 + (-1)]\boldsymbol{p} = 0\boldsymbol{p} = 0(p_1, p_2, \ldots, p_n) = (0p_1, 0p_2, \ldots, 0p_n) = \boldsymbol{0}$.

<<>>

EXAMPLE 3.1.2. The set $\mathcal{M}_{m,n}$ of all $m \times n$ matrices, with the usual rules of addition and multiplication of matrices by scalars, has the structure of a vector space, which is basically the same as that of $\mathbb{R}^{mn}$, except for the insignificant detail of the components being arranged in a rectangular array, rather than in a column.

<<>>

In the examples below we exhibit various vector spaces of real-valued functions in which, for any f, g, the sum $f + g$ and the product cf are defined in the usual way by

$$(f + g)(x) = f(x) + g(x) \tag{3.1.1}$$

and

$$(cf)(x) = c \cdot f(x) \tag{3.1.2}$$

for all x where the right-hand sides are defined. Thus the domain of $f + g$ is the intersection of the domains of f and g, and the domain of cf is the same as that of f.

EXAMPLE 3.1.3. Let D be any set of real numbers. The set $\mathcal{F}(D)$ of all real-valued functions on D with the operations above is clearly a vector space.

<<>>

Definition 3.1.1 implicitly states that the sum of two vectors and any scalar multiple of a vector must also be vectors *in the same space*. These are conditions that must be checked to determine whether a given set is a vector space or not. Although in the foregoing examples these conditions were obviously true, in many others they may not be or the operations can lead out of the set we started with. For example, the sum of two numbers between 0 and 1 may well be more than 1. Thus the interval $(0, 1)$ with the usual operations is not a vector space. We have a special name for sets that have the properties above:

DEFINITION 3.1.2.

A set S is said to be *closed under addition* if for every pair of elements $\boldsymbol{p}, \boldsymbol{q} \in S$ a sum $\boldsymbol{p} + \boldsymbol{q}$ is defined and belongs to S. The set S is said to be *closed under multiplication by scalars* if for every scalar c and $\boldsymbol{p} \in S$ the product $c\boldsymbol{p}$ is defined and belongs to S.

EXAMPLE 3.1.4. Let $\mathcal{P}_n = \{\boldsymbol{p} = P_n : P_n(x) = p_0 + p_1x + \cdots + p_nx^n; p_0, p_1, \ldots, p_n \in \mathbb{R}\}$ be the set of single-variable polynomials of degree n or less and the zero polynomial,[2] with the above rules of addition and of multiplication by scalars, which in this case mean

$$(p_0 + p_1x + \cdots + p_nx^n) + (q_0 + q_1x + \cdots + q_nx^n)$$

$$= (p_0 + q_0) + (p_1 + q_1)x + \cdots + (p_n + q_n)x^n \tag{3.1.3}$$

and

$$c(p_0 + p_1 + x + \cdots + p_nx^n) = cp_0 + cp_1x + \cdots + cp_nx^n. \tag{3.1.4}$$

With these rules the set $\mathcal{P}_n$ becomes a vector space: Clearly $\mathcal{P}_n$ is closed under addition and multiplication by scalars, and the polynomials of $\mathcal{P}_n$ satisfy the axioms of the definition and can be regarded as vectors. Verification of this fact would be straightforward, and we omit it, except for exhibiting the zero vector as $\boldsymbol{0} = 0 + 0x + \cdots + 0x^n$, which may, of course, be identified with the number 0 as well.

<<>>

EXAMPLE 3.1.5. The set $C[a, b]$ of continuous functions on the interval $[a, b]$ is a vector space for the following reasons: It is closed under addition and under multiplication by scalars, because the sum of two functions continuous on an interval is also continuous there, and any scalar multiple of such a function is continuous there as well. The axioms could again be easily verified.

<<>>

EXAMPLE 3.1.6. Consider the set $D[a, b]$ of discontinuous functions on the interval $[a, b]$. This is *not* a vector space because it is not closed under addition: The sum of two discontinuous functions need not be discontinuous. For example, if the only discontinuity of f is a jump of size 1 at some point $c \in [a, b]$ and g's only discontinuity is a jump of size -1 at c, then

[2]By convention the zero polynomial has no degree, because otherwise the additive rule for degrees in the multiplication of polynomials could not be maintained.

$f + g$ has no discontinuity at all, since the two jumps cancel each other. This cancellation holds trivially for $g = -f$, since then $f + (-f) = 0$, and the zero function is *continuous*. Thus also Axiom 3 is violated, because there is no zero vector in this space; the only candidate is outside the space.

<<>>

Subtraction of vectors can be defined just as before:

DEFINITION 3.1.3.

For all vectors $\boldsymbol{p}$ and $\boldsymbol{q}$ in any vector space V, we define

$$\boldsymbol{p} - \boldsymbol{q} = \boldsymbol{p} + (-\boldsymbol{q}). \tag{3.1.5}$$

We have a list of further properties of vectors just as in $\mathbb{R}^n$:

THEOREM 3.1.1.

For all vectors $\boldsymbol{p}, \boldsymbol{q}, \mathbf{x}$ in any vector space V and all scalars c and d we have

1. $0\boldsymbol{p} = \boldsymbol{0}$,
2. $c\boldsymbol{0} = \boldsymbol{0}$,
3. $\boldsymbol{p} + \mathbf{x} = \boldsymbol{q}$ if and only if $\mathbf{x} = \boldsymbol{q} - \boldsymbol{p}$,
4. If $c\boldsymbol{p} = \boldsymbol{0}$ then either $c = 0$ or $\boldsymbol{p} = \boldsymbol{0}$ or both,
5. $-\boldsymbol{p} = (-1)\boldsymbol{p}$,
6. $(-c)\boldsymbol{p} = c(-\boldsymbol{p}) = -(c\boldsymbol{p})$,
7. $c(\boldsymbol{p} - \boldsymbol{q}) = c\boldsymbol{p} - c\boldsymbol{q}$,
8. $(c - d)\boldsymbol{p} = c\boldsymbol{p} - d\boldsymbol{p}$.

Proof By Axiom 5 we have $\boldsymbol{p} = 1\boldsymbol{p}$. By the definition of the number 0, Axiom 7 and Axiom 5 again, this statement can be changed to $\boldsymbol{p} = (1 + 0)\boldsymbol{p} = 1\boldsymbol{p} + 0\boldsymbol{p} = \boldsymbol{p} + 0\boldsymbol{p}$. Change the order of the terms on the right using Axiom 1 and add $-\boldsymbol{p}$ to both sides: Then we get $\boldsymbol{p} + (-\boldsymbol{p}) = (0\boldsymbol{p} + \boldsymbol{p}) + (-\boldsymbol{p}) = 0\boldsymbol{p} + [\boldsymbol{p} + (-\boldsymbol{p})]$, where in the last step we used Axiom 2. Applying Axiom 4 on both sides, we obtain $\boldsymbol{0} = 0\boldsymbol{p} + \boldsymbol{0}$. Axiom 3 reduces this result to $\boldsymbol{0} = 0\boldsymbol{p}$, as we set out to prove.

To prove Part 2, observe that $c\boldsymbol{0} = c(0\boldsymbol{p})$ by Part 1 and, by Axiom 6, the ordinary multiplication of numbers and Part 1 again, this equation becomes $(c0)\boldsymbol{p} = 0\boldsymbol{p} = \boldsymbol{0}$.

The proof of Part 3 runs as follows: Suppose first that $\mathbf{x} = \boldsymbol{q} - \boldsymbol{p}$. Then $\boldsymbol{p} + \mathbf{x} = \boldsymbol{p} + (\boldsymbol{q} - \boldsymbol{p}) = (\boldsymbol{q} - \boldsymbol{p}) + \boldsymbol{p} = [\boldsymbol{q} + (-\boldsymbol{p})] + \boldsymbol{p} = \boldsymbol{q} + [(-\boldsymbol{p}) + \boldsymbol{p}] = \boldsymbol{q} + [\boldsymbol{p} + (-\boldsymbol{p})] = \boldsymbol{q} + \boldsymbol{0} = \boldsymbol{q}$.

To prove the converse, assume that $\boldsymbol{p} + \mathbf{x} = \boldsymbol{q}$. Then, subtracting $\boldsymbol{p}$ from both sides gives $(\boldsymbol{p} + \mathbf{x}) - \boldsymbol{p} = \boldsymbol{q} - \boldsymbol{p}$, and the left-hand side reduces to $(\mathbf{x} + \boldsymbol{p}) - \boldsymbol{p} = \mathbf{x} + [\boldsymbol{p} + (-\boldsymbol{p})] = \mathbf{x} + \boldsymbol{0} = \mathbf{x}$.

Part 4 may be proved by showing that $c \neq 0$ and $\boldsymbol{p} \neq \boldsymbol{0}$ cannot hold simultaneously if $c\boldsymbol{p} = \boldsymbol{0}$. Thus, suppose that $c\boldsymbol{p} = \boldsymbol{0}$ and $c \neq 0$ hold. Then $\boldsymbol{p} = \frac{1}{c}(c\boldsymbol{p}) = \frac{1}{c}\boldsymbol{0} = \boldsymbol{0}$, and so $\boldsymbol{p} \neq \boldsymbol{0}$ cannot be true in this case.

To prove Part 5, consider $\boldsymbol{p} + (-1)\boldsymbol{p} = 1\boldsymbol{p} + (-1)\boldsymbol{p} = [1 + (-1)]\boldsymbol{p} = 0\boldsymbol{p} = \boldsymbol{0}$. Subtracting $\boldsymbol{p}$ on both sides, we get $(-1)\boldsymbol{p} = -\boldsymbol{p}$, as in the proof of Part 3.

The proofs of the remaining statements are straightforward, and are left as exercises.

<<>>

Exercises 3.1

In the first ten exercises, determine whether the given set describes a vector space or not. Explain!

- **3.1.1.** The set of all polynomials of degree two and the zero polynomial.
- **3.1.2.** The set of all solutions (x, y) of the equation $2x + 3y = 0$.
- **3.1.3.** The set of all solutions (x, y) of the equation $2x + 3y = 1$.
- **3.1.4.** The set of all twice differentiable functions f for which $f''(x) + 2f(x) = 0$ holds.
- **3.1.5.** The set of all twice differentiable functions f for which $f''(x) + 2f(x) = 1$ holds.
- **3.1.6.** The set $\mathcal{P}$ of all polynomials in a single variable x.

3.1.7. The set of all ordered pairs of real numbers with addition and multiplication by scalars defined by

$$(p_1, p_2) + (q_1, q_2) = (p_1 + q_2, p_2 + q_1)$$

and

$$c(p_1, p_2) = (cp_1, cp_2).$$

3.1.8. The set of all ordered pairs of real numbers with addition and multiplication by scalars defined by

$$(p_1, p_2) + (q_1, q_2) = (p_1 + q_2, 0)$$

and

$$c(p_1, p_2) = (cp_1, cp_2).$$

3.1.9. The set of all ordered pairs of real numbers with addition and multiplication by scalars defined by

$$(p_1, p_2) + (q_1, q_2) = (p_1 + q_1, 0)$$

and

$$c(p_1, p_2) = (cp_1, cp_2).$$

3.1.10. The set of all ordered pairs of real numbers with addition and multiplication by scalars defined by

$$(p_1, p_2) + (q_1, q_2) = (p_1 + q_1, p_2 + q_2)$$

and

$$c(p_1, p_2) = (|c|p_1, |c|p_2).$$

***3.1.11.** Prove the last three parts of Theorem 3.1.1.

***3.1.12.** Prove that in any vector space if $\boldsymbol{p} + \boldsymbol{x} = \boldsymbol{p}$ holds for all $\boldsymbol{p}$, then $\boldsymbol{x} = \boldsymbol{0}$ must hold.

***3.1.13.** Prove that in any vector space we have this cancellation rule: If, for some $\boldsymbol{p}$, $\boldsymbol{q}$, $\boldsymbol{r}$ the equation $\boldsymbol{p} + \boldsymbol{q} = \boldsymbol{p} + \boldsymbol{r}$ holds, then $\boldsymbol{q} = \boldsymbol{r}$ must hold.

***3.1.14.** Show that we could define vector spaces by just seven axioms instead of eight if we replaced Axioms 3 and 4 by the single axiom:

3′. There is a vector $\boldsymbol{0}$ such that $0\boldsymbol{p} = \boldsymbol{0}$ holds for all vectors $\boldsymbol{p}$.

In other words, prove that if we define the zero vector by Axiom 3′ instead of Axiom 3, then this, in conjunction with the other axioms, implies both the additive property of the zero vector expressed in Axiom 3 and the existence of additive inverses expressed in Axiom 4, with $(-1)\boldsymbol{p}$ in the role of $-\boldsymbol{p}$.

MATLAB Exercises 3.1

ML3.1.1. Let V denote the set of all ordered quintuples of 0's and 1's, with addition defined by the MATLAB command $\boldsymbol{p}|\boldsymbol{q}$ and multiplication by scalars $c\&\boldsymbol{p}$.

a. Generate such vectors by the command **round**(**rand**(1,5)) and use MATLAB to check whether each of the eight vector space axioms is satisfied for those vectors and selected scalars.

b. If you think this is a vector space, prove it, and if not, explain why.

c. Do you get a vector space if the scalars are also restricted to 0's and 1's and their addition and multiplication are also defined by $c|d$ and $c\&d$, respectively?

ML3.1.2. Let V denote the set of all ordered quintuples of real numbers and define addition of vectors by the MATLAB command **max**($\boldsymbol{p}$,$\boldsymbol{q}$) and multiplication by scalars by componentwise multiplication, as in $\mathbb{R}^5$.

a. Generate such vectors by the command **round**(10 * **rand**(1, 5) − 5) and use MATLAB to check whether each of the eight vector space axioms is satisfied for those vectors and selected scalars.

b. If you think this is a vector space, prove it, and if not, explain why.

3.2. SUBSPACES, SPAN, AND INDEPENDENCE OF VECTORS

In the solution of linear systems and in the parametric description of planes and hyperplanes we have encountered expressions like $s\boldsymbol{u} + t\boldsymbol{v}$, or more generally of the type

$$\sum_{i=1}^{n} s_i \boldsymbol{u}_i, \tag{3.2.1}$$

with n being any positive integer. Such expressions are called *linear combinations* of the vectors involved. In many applications we need to consider the set of all linear combinations of the given vectors as the coefficients vary. Such sets describe hyperplanes through the origin, and we shall explore various questions concerning the vectors $\boldsymbol{u}_i$ that "generate" them; such as how many $\boldsymbol{u}_i$ we need and how some can be added, omitted or changed. Thus, it is useful to begin by characterizing these sets in a general way without involving the $\boldsymbol{u}_i$ vectors:

DEFINITION 3.2.1. A nonempty subset U of a vector space X is called a *subspace of* X if it is a vector space with addition of vectors and multiplication by scalars being the same as in X.

In Section 3.1.1, we defined a vector space as a set U such that for any pair $\boldsymbol{p}$, $\boldsymbol{q}$ of vectors in U and any scalar c the sum $\boldsymbol{p} + \boldsymbol{q}$ and the product $c\boldsymbol{p}$ belong to U; that is, so that U is *closed* under these operations, and eight algebraic rules hold. Now, for a subset of a vector space, as U is here, the algebraic rules holding in X remain valid for the vectors of U because the operations are the same in U as in X. *Thus, to test whether a nonempty subset U of a vector space is a subspace, it is enough to test whether it is closed under addition and under multiplication by scalars.* Let us look at some examples:

EXAMPLE 3.2.1. Consider in the space $\mathbb{R}^3$ the set U of vectors[3] whose third coordinate is 0; that is, let $U = \{\boldsymbol{u} \in \mathbb{R}^3 \,|\, \boldsymbol{u} = (u_1, u_2, 0)^T\}$. It is easy to see that this set is nonempty, and is a subspace of $\mathbb{R}^3$, since if $\boldsymbol{u}, \boldsymbol{v} \in U$ and

[3] Remember the convention that all vectors are to be column vectors, but they may be written as row vectors transposed.

c is any scalar, then $\boldsymbol{u} + \boldsymbol{v} = (u_1, u_2, 0)^T + (v_1, v_2, 0)^T = (u_1 + v_1, u_2 + v_2, 0)^T$, and so the third component of $\boldsymbol{u} + \boldsymbol{v}$ being zero, $\boldsymbol{u} + \boldsymbol{v}$ also belongs to U and similarly $c\boldsymbol{u}$ does too.

<<>>

EXAMPLE 3.2.2. Consider in the space $\mathbb{R}^3$ the set U of vectors whose third coordinate is 1; that is, let $U = \{\boldsymbol{u} \in \mathbb{R}^3 \mid \boldsymbol{u} = (u_1, u_2, 1)^T\}$. To see that this set is not a subspace of $\mathbb{R}^3$, all we have to do is exhibit two vectors in U whose sum is not in U, or one vector $\boldsymbol{u}$ in U and a scalar c whose product $c\boldsymbol{u}$ is not in U. In this example any two vectors of U will add up to a vector with 2 for its third component, and so this sum vector will be outside of U. Alternatively, c times any vector $\boldsymbol{u}$ of U will result in a vector having c as its third component, and then $c\boldsymbol{u}$ is outside of U if $c \neq 1$ or 0.

<<>>

In general, to prove that a subset U of a vector space *is* a subspace we have to prove the closure properties for all $\boldsymbol{u}, \boldsymbol{v} \in U$ and all scalars c (making sure also that U is not empty), while to prove that U is *not* a subspace all we need is a single counterexample to either of the closure requirements.

Note that in any vector space X the set $\{\boldsymbol{0}\}$, consisting of the zero vector alone, is a subspace of X and so too is the whole space X. These are called the *trivial subspaces* of X, while all the others are its *nontrivial* or *proper subspaces.*

EXAMPLE 3.2.3. Let us find the smallest subspace U of $\mathbb{R}^3$ that contains the vectors $(1, 1, 1)^T$ and $(1, 2, 3)^T$.

Because U must be closed under multiplication by scalars, it must contain all multiples of $(1, 1, 1)^T$ and $(1, 2, 3)^T$, and since it must also be closed under addition of its vectors, it must contain the sums of all these multiples. In other words, U must contain all linear combinations $s(1, 1, 1)^T + t(1, 2, 3)^T$. This fact, however, is all we need: $U = \{\boldsymbol{u} \in \mathbb{R}^3 \mid \boldsymbol{u} = s(1, 1, 1)^T + t(1, 2, 3)^T;\ s, t \in \mathbb{R}\}$ is a subspace of $\mathbb{R}^3$, as we are going to show below. Furthermore, it is the smallest one that contains the vectors $(1, 1, 1)^T$ and $(1, 2, 3)^T$, since as we have just said, any subspace that contains these two vectors must contain all the linear combinations that this U consists of.

U is a subspace because, first, it is clearly nonempty. Then, second, let $\boldsymbol{u} = s_1(1, 1, 1)^T + t_1(1, 2, 3)^T$, $\boldsymbol{v} = s_2(1, 1, 1)^T + t_2(1, 2, 3)^T$ and let c be any scalar. Then both the sum $\boldsymbol{u} + \boldsymbol{v} = (s_1 + s_2)(1, 1, 1)^T + (t_1 + t_2)(1, 2, 3)^T$ and the product $c\boldsymbol{u} = cs_1(1, 1, 1)^T + ct_1(1, 2, 3)^T$ are linear combinations of $(1, 1, 1)^T$ and $(1, 2, 3)^T$; consequently U is closed under addition and under multiplication by scalars.

We say that $(1, 1, 1)^T$ and $(1, 2, 3)^T$ *span* U or that U is their span (see Definition 3.2.2 below). Geometrically, U is the plane through the origin containing the given vectors.

<<>>

EXAMPLE 3.2.4. Consider the set U in $\mathbb{R}^3$ of all solutions of the equation $x - 2y + z = 0$. If we solve this equation, we obtain the parametric form of U as the set of vectors $(x, y, z)^T = s(2, 1, 0)^T + t(-1, 0, 1)^T$. We can show this to be a subspace either by reasoning as in Example 3.2.3, or directly from the defining equation, without even solving it, as follows. Let $\boldsymbol{u} = (x_1, y_1, z_1)^T$ and $\boldsymbol{v} = (x_2, y_2, z_2)^T$ be two solutions of $x - 2y + z = 0$; that is, let $x_1 - 2y_1 + z_1 = 0$ and $x_2 - 2y_2 + z_2 = 0$ hold and let c be an arbitrary scalar. Then

$$(x_1 + x_2) - 2(y_1 + y_2) + (z_1 + z_2) =$$

$$(x_1 - 2y_1 + z_1) + (x_2 - 2y_2 + z_2) = 0 + 0 = 0, \tag{3.2.2}$$

and so $\boldsymbol{u} + \boldsymbol{v}$ is also a solution and obviously $c\boldsymbol{u}$ is as well. Thus U is closed under both operations, clearly nonempty, and therefore a subspace.

<<>>

The constructions illustrated in Examples 3.2.3 and 3.2.4 can be generalized, and constitute the two most important ways in which subspaces occur in applications. We state these as theorems:

THEOREM 3.2.1. Let n be any positive integer and $\boldsymbol{a}_1, \boldsymbol{a}_2, \ldots, \boldsymbol{a}_n$ be arbitrary vectors in a vector space X. Then the set $U = \{\boldsymbol{u} = \sum_{i=1}^{n} s_i\boldsymbol{a}_i \mid s_1, s_2, \ldots, s_n \in \mathbb{R}\}$ of all linear combinations of the given vectors is a subspace of X.

Proof Let

$$\boldsymbol{u} = \sum_{i=1}^{n} s_i\boldsymbol{a}_i \tag{3.2.3}$$

and

$$\boldsymbol{v} = \sum_{i=1}^{n} t_i\boldsymbol{a}_i \tag{3.2.4}$$

be arbitrary vectors in U and c any scalar. Then

$$\boldsymbol{u} + \boldsymbol{v} = \sum_{i=1}^{n} (s_i + t_i)\boldsymbol{a}_i \tag{3.2.5}$$

and

$$c\boldsymbol{u} = \sum_{i=1}^{n} cs_i\boldsymbol{a}_i \tag{3.2.6}$$

are evidently also linear combinations of the $\boldsymbol{u}_i$ vectors and, as such, members of U. Thus U is closed under both operations, is clearly nonempty, and is therefore a subspace of X.

<<>>

THEOREM 3.2.2. Let A be any $m \times n$ matrix. The set U of all solutions of the homogeneous equation $A\boldsymbol{x} = \boldsymbol{0}$ is a subspace of $\mathbb{R}^n$.

Proof Let $\boldsymbol{u}$ and $\boldsymbol{v}$ be arbitrary vectors in U, and c any scalar. Then $A\boldsymbol{u} = \boldsymbol{0}$ and $A\boldsymbol{v} = \boldsymbol{0}$ hold, and so $A(\boldsymbol{u} + \boldsymbol{v}) = A\boldsymbol{u} + A\boldsymbol{v} = \boldsymbol{0} + \boldsymbol{0} = \boldsymbol{0}$ and $A(c\boldsymbol{u}) = cA\boldsymbol{u} = c\boldsymbol{0} = \boldsymbol{0}$ as well. Thus U is closed under both operations, is nonempty since it always contains $\boldsymbol{0}$ even if nothing else, and is therefore a subspace of X.

<<>>

The subspaces occurring in these theorems have special names:

DEFINITION 3.2.2. Given a vector space X, the subspace of all linear combinations of any finite set of vectors $\boldsymbol{a}_i$ in X is called the *span* of the $\boldsymbol{a}_i$, or the subspace *spanned* or *generated* by the $\boldsymbol{a}_i$, and will be denoted by $\text{Span}\{\boldsymbol{a}_1, \boldsymbol{a}_2, \ldots, \boldsymbol{a}_n\}$.

DEFINITION 3.2.3. For any $m \times n$ matrix A, the set of all solutions of the homogeneous equation $A\boldsymbol{x} = \boldsymbol{0}$ is called the *solution space of the equation* or the *null space of A*, and will be denoted by $\text{Null}(A)$.

EXAMPLE 3.2.5. We can now easily describe all the subspaces of $\mathbb{R}^3$: If we consider all vectors as position vectors, then the subspaces are $\{\boldsymbol{0}\}$, the sets consisting of the position-vectors of the points of all the lines, and all the planes through the origin, and $\mathbb{R}^3$ itself. For, clearly, any single nonzero vector spans such a line, any two nonparallel nonzero vectors span such a plane, and any three noncoplanar vectors span $\mathbb{R}^3$.

<<>>

A frequently occurring problem involving these concepts is that of decomposing a given vector $\boldsymbol{b} \in \mathbb{R}^m$ into a linear combination of some other given vectors $\boldsymbol{a}_i \in \mathbb{R}^m$ if possible. This problem amounts to solving the linear vector equation

$$\boldsymbol{b} = \sum_{i=1}^{n} s_i \boldsymbol{a}_i \tag{3.2.7}$$

for the unknown coefficients s_i. If we consider the given vectors $\boldsymbol{a}_i$ as columns of a matrix A, as in Section 2.2, $\boldsymbol{b}$ as a column m-vector, and the

s_i as components of a column n-vector $\mathbf{s}$, then the equation above takes on the familiar form $A\mathbf{s} = \boldsymbol{b}$, because then,

$$\sum_{i=1}^{n} s_i \boldsymbol{a}_i = (\boldsymbol{a}_1\ \boldsymbol{a}_2\ \dots\ \boldsymbol{a}_n)\begin{bmatrix} s_1 \\ s_2 \\ \dots \\ s_n \end{bmatrix} = A\mathbf{s} = \boldsymbol{b}. \tag{3.2.8}$$

Let us look at some examples:

EXAMPLE 3.2.6. Write $\boldsymbol{b} = (7, 7, 9, 11)^T$ as a linear combination of $\boldsymbol{a}_1 = (2, 0, 3, 1)^T$, $\boldsymbol{a}_2 = (4, 1, 3, 2)^T$, and $\boldsymbol{a}_3 = (1, 3, -1, 3)^T$, if possible.

The system to be solved can be written

$$\begin{bmatrix} 2 & 4 & 1 \\ 0 & 1 & 3 \\ 3 & 3 & -1 \\ 1 & 2 & 3 \end{bmatrix}\begin{bmatrix} s_1 \\ s_2 \\ s_3 \end{bmatrix} = \begin{bmatrix} 7 \\ 7 \\ 9 \\ 11 \end{bmatrix}. \tag{3.2.9}$$

Gaussian elimination gives $s_1 = 6$, $s_2 = -2$, and $s_3 = 3$, and it is easy to check that $\boldsymbol{b} = 6\boldsymbol{a}_1 - 2\boldsymbol{a}_2 + 3\boldsymbol{a}_3$ is indeed true.

<<>>

EXAMPLE 3.2.7. Write $\boldsymbol{b} = (2, 8, 0)^T$ as a linear combination of $\boldsymbol{a}_1 = (1, 3, 1)^T$, $\boldsymbol{a}_2 = (2, 6, 2)^T$, and $\boldsymbol{a}_3 = (0, -1, 1)^T$, if possible.

The system to be solved can be written

$$\begin{bmatrix} 1 & 2 & 0 \\ 3 & 6 & -1 \\ 1 & 2 & 1 \end{bmatrix}\begin{bmatrix} s_1 \\ s_2 \\ s_3 \end{bmatrix} = \begin{bmatrix} 2 \\ 8 \\ 0 \end{bmatrix}. \tag{3.2.10}$$

Gaussian elimination gives $s_1 = 2 - 2t$, $s_2 = t$, and $s_3 = -2$, and it is easy to check that $\boldsymbol{b} = (2 - 2t)\boldsymbol{a}_1 + t\boldsymbol{a}_2 - 2\boldsymbol{a}_3$ is true for any value of the parameter t.

<<>>

As Example 3.2.7 shows, sometimes the decomposition of vectors into linear combinations is not unique. This uniqueness depends solely on the $\boldsymbol{a}_i$ vectors: given the $\boldsymbol{a}_i$, the decomposition is either unique for all vectors $\boldsymbol{b}$ for which it is possible, or it is not unique for any of them. This result follows from the fact that any solution of Equation (3.2.8) above is unique if in the row reduction there are only basic variables, and nonunique if there are also free variables, and this condition does not depend on $\boldsymbol{b}$. (For a direct proof, see Exercise 3.2.12.) The test for uniqueness of such decompositions is usually phrased in terms of the following definition, in which the zero vector is taken for $\boldsymbol{b}$.

DEFINITION 3.2.4. Let n be any positive integer and $\boldsymbol{a}_1, \boldsymbol{a}_2, \ldots, \boldsymbol{a}_n$ be arbitrary vectors in a vector space X. We call these vectors *linearly independent of each other* if the equation

$$\sum_{i=1}^{n} s_i \boldsymbol{a}_i = \boldsymbol{0} \qquad (3.2.11)$$

implies that the coefficients s_i are all zero. The corresponding linear combination is called the *trivial combination*. If the equations above have nontrivial solutions as well, then the $\boldsymbol{a}_i$ vectors are said to be *linearly dependent*.

Let us reemphasize that independence of the vectors $\boldsymbol{a}_1, \boldsymbol{a}_2, \ldots, \boldsymbol{a}_n$ is equivalent to the uniqueness of the decomposition, as in Equation (3.2.8), of any $\boldsymbol{b}$ for which such a decomposition is possible; and this is why the notion of independence is so important.

In the next two examples we reformulate the definition in case we have only two or three vectors to test for dependence and show what this relation means geometrically in $\mathbb{R}^2$ and $\mathbb{R}^3$.

EXAMPLE 3.2.8. Two vectors in any vector space are dependent if and only if one is a scalar multiple of the other.

According to Definition 3.2.4 any two vectors $\boldsymbol{a}_1$ and $\boldsymbol{a}_2$ are dependent if and only if there exist two scalars s_1 and s_2, not both zero, such that

$$s_1 \boldsymbol{a}_1 + s_2 \boldsymbol{a}_2 = \boldsymbol{0}. \qquad (3.2.12)$$

Say $s_1 \neq 0$. Then we can solve the equation above for $\boldsymbol{a}_1$, to obtain

$$\boldsymbol{a}_1 = -\frac{s_2}{s_1} \boldsymbol{a}_2, \qquad (3.2.13)$$

which exhibits $\boldsymbol{a}_1$ as a multiple of $\boldsymbol{a}_2$. If, on the other hand, $s_1 = 0$, then we must have $s_2 \neq 0$, and Equation (3.2.12) can be solved for $\boldsymbol{a}_2$ as a multiple of $\boldsymbol{a}_1$, namely as

$$\boldsymbol{a}_2 = 0 \boldsymbol{a}_1. \qquad (3.2.14)$$

Conversely, if one of the two vectors is a scalar multiple of the other, say

$$\boldsymbol{a}_2 = c \boldsymbol{a}_1, \qquad (3.2.15)$$

then we can rewrite this expression as

$$c \boldsymbol{a}_1 + (-1) \boldsymbol{a}_2 = \boldsymbol{0}. \qquad (3.2.16)$$

This statement shows that Equation (3.2.12) is solved by $s_1 = c$ and $s_2 = -1$, and so at least one of these is not zero. This is the condition in Definition 3.2.4 for the two vectors $\boldsymbol{a}_1$ and $\boldsymbol{a}_2$ to be dependent. The same conclusion follows by just exchanging the subscripts in the argument if $\boldsymbol{a}_1$ is a multiple of $\boldsymbol{a}_2$.

What does this result mean geometrically in $\mathbb{R}^2$ and $\mathbb{R}^3$? It means that two vectors are dependent if and only if they are parallel, or if we consider position vectors: collinear. (This statement includes the zero vector, which is considered to be both parallel and orthogonal to any vector.)

<<>>

EXAMPLE 3.2.9. Three vectors in any vector space are dependent if and only if one is a linear combination of the other two. We leave the proof of this statement as Exercise 3.2.11. Here we want to discuss only the geometric meaning in $\mathbb{R}^3$.

Say $\boldsymbol{a}_1$, $\boldsymbol{a}_2$, and $\boldsymbol{a}_3$ are nonzero linearly dependent position vectors in $\mathbb{R}^3$. Then, by Definition 3.2.4, there exist scalars s_1, s_2, and s_3, not all zero, such that $s_1\boldsymbol{a}_1 + s_2\boldsymbol{a}_2 + s_3\boldsymbol{a}_3 = \boldsymbol{0}$. Say, without loss of generality, that $s_3 \neq 0$. Then we can solve the above equation for $\boldsymbol{a}_3$ and obtain $\boldsymbol{a}_3$ as a linear combination of $\boldsymbol{a}_1$ and $\boldsymbol{a}_2$. Thus, by the geometric interpretation of multiplication of vectors by scalars and the parallelogram law for vector addition, we see that $\boldsymbol{a}_3$ is in the plane of $\boldsymbol{a}_1$ and $\boldsymbol{a}_2$. This result can also be stated symmetrically as saying that $\boldsymbol{a}_1$, $\boldsymbol{a}_2$, and $\boldsymbol{a}_3$ are coplanar if they are linearly dependent. The converse can also be seen easily, and both statements can be extended to include the zero vector. Thus the linear dependence of three position vectors in $\mathbb{R}^3$ means that they are coplanar.

<<>>

Notice that if $X = \mathbb{R}^m$, then, letting the $\boldsymbol{a}_i$ vectors be the columns of a matrix A, we can rewrite Equation (3.2.11) as

$$A\boldsymbol{s} = \boldsymbol{0} \tag{3.2.17}$$

and so the vectors $\boldsymbol{a}_1, \boldsymbol{a}_2, \ldots, \boldsymbol{a}_n$ of $X = \mathbb{R}^m$ are independent if and only if this equation has only the trivial solution.

EXAMPLE 3.2.10. Test the column vectors of the matrix

$$A = \begin{bmatrix} 2 & 3 & 5 \\ 1 & -2 & 3 \\ 3 & -4 & 2 \end{bmatrix} \tag{3.2.18}$$

for independence.

We need to solve the equation $A\boldsymbol{s} = \boldsymbol{0}$. Row reduction of A proceeds as follows:

$$\begin{bmatrix} 2 & 3 & 5 \\ 1 & -2 & 3 \\ 3 & -4 & 2 \end{bmatrix} \to \begin{bmatrix} 1 & -2 & 3 \\ 2 & 3 & 5 \\ 3 & -4 & 2 \end{bmatrix} \to \begin{bmatrix} 1 & -2 & 3 \\ 0 & 7 & -1 \\ 0 & 2 & -7 \end{bmatrix} \to \begin{bmatrix} 1 & -2 & 3 \\ 0 & 7 & -1 \\ 0 & 0 & -47/7 \end{bmatrix}. \quad (3.2.19)$$

Back substitution gives $(-47/7)s_3 = 0$ and so $s_3 = 0$, then $7s_2 = 0$, consequently $s_2 = 0$, and finally $s_1 = 0$. Thus the columns of A are independent.

<<>>

EXAMPLE 3.2.11. By the argument just above Definition 3.2.4, the vectors $\boldsymbol{a}_1$, $\boldsymbol{a}_2$, $\boldsymbol{a}_3$ of Example 3.2.6 are independent if the equation $A\boldsymbol{s} = \boldsymbol{b}$ has a unique solution for some $\boldsymbol{b}$. Since the solution is unique for the $\boldsymbol{b}$ in Example 3.2.6, the solution must be unique for any other decomposable $\boldsymbol{b}$, including $\boldsymbol{b} = \boldsymbol{0}$. Thus Equation (3.2.10) has only the trivial solution and so the $\boldsymbol{a}_i$ vectors are independent. We could, of course, have verified this independence directly by substituting the given vectors into Equation (3.2.17) and solving it.

<<>>

EXAMPLE 3.2.12. The vectors $\boldsymbol{a}_1$, $\boldsymbol{a}_2$, $\boldsymbol{a}_3$ in Example 3.2.7 are linearly dependent, since if the solution of $A\boldsymbol{s} = \boldsymbol{b}$ is not unique for the $\boldsymbol{b}$ in Example 3.2.7, then it is not unique for $\boldsymbol{b} = \boldsymbol{0}$, either, and so Equation (3.2.10) has nontrivial solutions. (Find some!)

<<>>

Notice how the back substitution in Example 3.2.10 results in zeroes for the unknown $\boldsymbol{s}_i$ values. The same procedure would prove all but the last of the following statements:

THEOREM 3.2.3. The columns of any upper triangular matrix with nonzero diagonal entries are independent. Similarly, the basic columns in any echelon matrix (that is, those containing the pivots) are independent and its nonzero rows are, too. Also, if A is a square matrix, then its columns are independent if and only if A is invertible, and the same holds for the rows.

Proof We prove only the last statement: If A is invertible, then multiplying $A\boldsymbol{s} = \boldsymbol{0}$ by A^{-1} from the left yields $\boldsymbol{s} = \boldsymbol{0}$, which means that the columns of A are independent. This result follows also from half of Corollary 2.3.1 on page 75, and the converse follows from the other half. Thus, the columns of A are independent if and only if A is invertible. Applying this condition to A^T in place of A, we find also that the rows of a square matrix A are independent if and only if A is invertible.

<<>>

In the next theorem we present a characterization of linear dependence for later use. We do this characterization here because it also provides good practice with the basic concepts.

THEOREM 3.2.4. Any two or more nonzero vectors $\boldsymbol{a}_1, \boldsymbol{a}_2, \ldots, \boldsymbol{a}_n$ in a vector space X are linearly dependent if and only if one of the vectors, say $\boldsymbol{a}_k$, for some $k \geq 2$, equals a linear combination of the previous vectors in the list.[4] Also, if $\boldsymbol{a}_k$ is equal to such a linear combination, then $\text{Span}(\mathscr{A} - \{\boldsymbol{a}_k\}) = \text{Span}(\mathscr{A})$, where $\mathscr{A} = \{\boldsymbol{a}_1, \boldsymbol{a}_2, \ldots, \boldsymbol{a}_n\}$.

Proof Suppose

$$\boldsymbol{a}_k = \sum_{i=1}^{k-1} s_i \boldsymbol{a}_i. \tag{3.2.20}$$

Then

$$\sum_{i=1}^{k-1} s_i \boldsymbol{a}_i + (-1)\boldsymbol{a}_k + \sum_{i=k+1}^{n} 0\boldsymbol{a}_i = \boldsymbol{0} \tag{3.2.21}$$

provides a nontrivial decomposition of $\boldsymbol{0}$, which proves the dependence of the vectors $\boldsymbol{a}_1, \boldsymbol{a}_2, \ldots, \boldsymbol{a}_n$.

Conversely, if the $\boldsymbol{a}_1, \boldsymbol{a}_2, \ldots, \boldsymbol{a}_n$ are linearly dependent, then

$$\sum_{i=1}^{n} s_i \boldsymbol{a}_i = \boldsymbol{0} \tag{3.2.22}$$

for some coefficients s_i, not all zero. Let s_k be the last nonzero coefficient. Then we must have $k \geq 2$, since otherwise s_1 would be nonzero and $s_1\boldsymbol{a}_1 = \boldsymbol{0}$ would lead to the contradictory fact $s_1 = 0$, because $\boldsymbol{a}_1$ was assumed to be nonzero. Multiplying both sides of Equation (3.2.22) by $1/s_k$, we get

$$\sum_{i=1}^{k-1} \frac{s_i}{s_k} \boldsymbol{a}_i + \boldsymbol{a}_k + \sum_{i=k+1}^{n} 0\boldsymbol{a}_i = \boldsymbol{0}, \tag{3.2.23}$$

and so

$$\boldsymbol{a}_k = \sum_{i=1}^{k-1} \left(-\frac{s_i}{s_k}\right) \boldsymbol{a}_i, \tag{3.2.24}$$

[4]We call a finite sequence or ordered n-tuple briefly a list, in contrast to a set, which is unordered.

which expresses $\boldsymbol{a}_k$ as a linear combination of the preceding vectors in the list.

To prove the last statement of the theorem, let $\boldsymbol{a}_k$ be a linear combination, as in Equation (3.2.20). If $\boldsymbol{b}$ is any vector in Span($\mathscr{A}$), then

$$\boldsymbol{b} = \sum_{i=1}^{n} t_i \boldsymbol{a}_i. \qquad [3.2.25]$$

Eliminating the kth term here by using Equation (3.2.20) gives $\boldsymbol{b}$ as a linear combination of the vectors of $\mathscr{A} - \{\boldsymbol{a}_k\}$. Conversely, if $\boldsymbol{b}$ is any linear combination of the vectors of $\mathscr{A} - \{\boldsymbol{a}_k\}$, then it is in Span($\mathscr{A}$) as well. Thus Span($\mathscr{A} - \{\boldsymbol{a}_k\}$) = Span($\mathscr{A}$).

<<>>

Exercises 3.2

In the next eight exercises, determine whether the given set is a subspace of the indicated vector space or not, and prove your statement.

• **3.2.1.** $U = \{\mathbf{x} \mid x_1 = x_2 = x_3\}$ in $\mathbb{R}^3$.

• **3.2.2.** $U = \{\mathbf{x} \mid x_1 = x_2^2\}$ in $\mathbb{R}^3$.

• **3.2.3.** $U = \{\mathbf{x} : |x_1| = |x_2| = |x_3|\}$ in $\mathbb{R}^3$.

• **3.2.4.** $U = \{\mathbf{x} \mid x_1 + x_2 + x_3 = 0\}$ in $\mathbb{R}^4$.

• **3.2.5.** $U = \{\mathbf{x} \mid x_1 = x_2 \text{ or } x_3 = 0\}$ in $\mathbb{R}^3$.

• **3.2.6.** $U = \{\mathbf{x} \mid x_1 = x_2 \text{ and } x_3 = 0\}$ in $\mathbb{R}^3$.

• **3.2.7.** $U = \{\mathbf{x} \mid x_1 \geq 0\}$ in $\mathbb{R}^n$.

• **3.2.8.** $U = \{\mathbf{x} : |\mathbf{x}| = |x_1| + |x_2|\}$ in $\mathbb{R}^3$.

3.2.9. Let U and V be subspaces of a vector space X. The set of all vectors belonging to both U and V is called the intersection of U and V and is denoted by $U \cap V$. Prove that $U \cap V$ is a subspace of X.

3.2.10. Let $\boldsymbol{a}$ be an arbitrarily given vector in $\mathbb{R}^n$. Show that the set of all vectors orthogonal to $\boldsymbol{a}$ is a subspace of $\mathbb{R}^n$.

3.2.11. Prove that three vectors in any vector space are dependent if and only if one is a linear combination of the other two. (Hint: Imitate the proof in Example 3.2.8.)

***3.2.12.** Show directly that, in any vector space X, if the decomposition of a vector $\boldsymbol{b}$ into a linear combination of given vectors $\boldsymbol{a}_i$ is not unique, then the decomposition of any other decomposable vector $\boldsymbol{c}$ is also not unique. (*Hint:* Write $\boldsymbol{c} = \boldsymbol{c} + \boldsymbol{b} - \boldsymbol{b}$ with different decompositions for the two $\boldsymbol{b}$ vectors.)

In the next four exercises, write the vectors $\boldsymbol{b}$ as linear combinations of the vectors $\boldsymbol{a}_i$ if possible.

3.2.13. $\boldsymbol{b} = (7, 32, 16, -3)^T$, $\boldsymbol{a}_1 = (4, 7, 2, 1)^T$, $\boldsymbol{a}_2 = (4, 0, -3, 2)^T$, $\boldsymbol{a}_3 = (1, 6, 3, -1)^T$.

• **3.2.14.** $\boldsymbol{b} = (7, 16, -3)^T$, $\boldsymbol{a}_1 = (4, 2, 1)^T$, $\boldsymbol{a}_2 = (4, -3, 2)^T$, $\boldsymbol{a}_3 = (1, 3, -1)^T$.

3.2.15. $\boldsymbol{b} = (7, 16, -3)^T$, $\boldsymbol{a}_1 = (4, 2, 1)^T$, $\boldsymbol{a}_2 = (4, -3 ,2)^T$, $\boldsymbol{a}_3 = (0, 5, -1)^T$.

3.2.16. $\boldsymbol{b} = (4, 7, 0)^T$, $\boldsymbol{a}_1 = (4, 2, 1)^T$, $\boldsymbol{a}_2 = (4, -3, 2)^T$, $\boldsymbol{a}_3 = (0, 5, -1)^T$.

In the next six exercises, determine whether the given vectors are independent or not, and prove your statement.

3.2.17. The four vectors in Exercise 3.2.14.

3.2.18. The three vectors $\boldsymbol{a}_1 = (4, 2, 1)^T$, $\boldsymbol{a}_2 = (4, -3, 2)^T$, $\boldsymbol{a}_3 = (1, 3, -1)^T$ from Exercise 3.2.14.

3.2.19. The three vectors $\boldsymbol{a}_1 = (4, 2, 1)^T$, $\boldsymbol{a}_2 = (4, -3, 2)^T$, $\boldsymbol{a}_3 = (0, 5, -1)^T$ from Exercise 3.2.15.

3.2.20. The two vectors $\boldsymbol{a}_1 = (4, 2, 1)^T$, $\boldsymbol{a}_2 = (4, -3, 2)^T$.

3.2.21. $\boldsymbol{a}_1 = (1, 0, 0, 1)^T$, $\boldsymbol{a}_2 = (0, 0, 1, 1)^T$, $\boldsymbol{a}_3 = (1, 1, 0, 0)^T$, $\boldsymbol{a}_4 = (1, 0, 1, 1)^T$.

3.2.22. $\boldsymbol{a}_1 = (2, 1)^T$, $\boldsymbol{a}_2 = (-3, 2)^T$, $\boldsymbol{a}_3 = (1, -1)^T$.

3.2.23. State three ways of characterizing the linear independence of n vectors.

3.2.24. Prove that in any vector space a finite set of vectors that contains the zero vector is a dependent set.

3.2.25. Prove that any set of more than three vectors in $\mathbb{R}^3$ is a dependent set. (*Hint:* Consider the $3 \times n$ matrix A with the given vectors as columns and apply Theorem 2.1.1 on page 45 to the equation $A\boldsymbol{s} = \boldsymbol{0}$.)

3.2.26. Prove that any set of three independent vectors in $\mathbb{R}^3$ spans $\mathbb{R}^3$. (*Hint:* Consider the matrix A with the given-vectors as columns and apply Theorem 2.3.5 on page 75.)

3.2.27. Prove that any three vectors in $\mathbb{R}^3$ that span $\mathbb{R}^3$ are independent. (*Hint:* Consider the matrix A with the given vectors as columns and apply Theorem 2.3.5 on page 75.)

3.2.28. Prove that a square matrix is singular if and only if its columns are dependent. (*Hint:* Apply Theorem 2.3.5 on page 75.)

***3.2.29.** Prove that in $\mathbb{R}^n$ a set of m vectors $\boldsymbol{a}_i$ is independent if and only if $0 < m \leq n$ and the matrix A with the given vectors as columns has rank m.

***3.2.30.** Prove that in $\mathbb{R}^n$ a set of m vectors $\boldsymbol{a}_i$ spans $\mathbb{R}^n$ if and only if $0 < n \le m$ and the matrix A with the given vectors as columns has rank n.

MATLAB Exercises 3.2

ML3.2.1. Solve Exercise 3.2.15 using MATLAB. (Hint: Reduce the associated system $\mathbf{b} = \sum_{i=1}^{n} s_i \boldsymbol{a}_i$ using **rref**.)

ML3.2.2. Solve Exercise 3.2.16 using MATLAB.

ML3.2.3. Use MATLAB to determine whether the four vectors of Exercise 3.2.13 span $\mathbb{R}^4$ or not.

a. Use the fact that the given vectors span $\mathbb{R}^4$ if and only if for their matrix A the equations $A\mathbf{x} = \boldsymbol{e}_i$ have a solution for each i.

b. For an alternative solution use the MATLAB command **rank**(A). Explain.

ML3.2.4. Solve Exercise 3.2.21 using MATLAB. Explain.

ML3.2.5. Use MATLAB to find a spanning set for the solution space of $A\mathbf{x} = \boldsymbol{0}$, where

$$A = \begin{bmatrix} 1 & -2 & 1 & 0 \\ 3 & 4 & 0 & 1 \\ -1 & 0 & 2 & 0 \\ 0 & -1 & 0 & 0 \end{bmatrix}.$$

(*Hint:* Find the general solution of $A\mathbf{x} = \boldsymbol{0}$ as in Example 2.1.8 on page 48.)

3.3. BASES

In any vector space or subspace we are frequently interested in finding a minimal set of vectors whose linear combinations make up the space as described below:

DEFINITION 3.3.1.

A finite subset $\mathcal{B}$ of a vector space X is called a *basis for X* if
1. $\mathcal{B}$ spans X, and
2. $\mathcal{B}$ is a set of independent vectors.[5,6]

[5]By a slight but useful abuse of language it is customary to say that the *set* $\mathcal{B}$ is independent and spans X rather than just that its vectors are and do so.

[6]There are many interesting, so-called infinite-dimensional vector spaces that do not possess finite spanning sets; that is, they have no basis in the sense above. The notion of a basis can, however, be generalized in various ways to cover such spaces as well but, except for a few elementary examples, those are beyond the scope of this book.

EXAMPLE 3.3.1. The standard vectors $\mathbf{e}_1, \mathbf{e}_2, \ldots, \mathbf{e}_n$ form a basis, the so-called standard basis for $\mathbb{R}^n$, since (1) every vector in $\mathbb{R}^n$ can be written as a linear combination of these in the usual way as $a = \sum_{i=1}^{n} a_i e_i$, and (2) the $\mathbf{e}_i$ are independent since their matrix is the unit matrix and $I\mathbf{s} = \mathbf{0}$ implies $\mathbf{s} = \mathbf{0}$.

<<>>

EXAMPLE 3.3.2. The columns of the matrix A of Example 3.2.10 on page 93 form a basis for $\mathbb{R}^3$, since (1) the same row operations as in Example 3.2.10 solve the equation $A\mathbf{s} = \mathbf{b}$ for any $\mathbf{b}$, and (2) the columns are independent as proved in Example 3.2.10.

<<>>

As in Example 3.3.2 and some earlier ones, we are often interested in the subspace generated by the columns of a matrix and give it a name:

DEFINITION 3.3.2.

> The subspace of $\mathbb{R}^m$ spanned by the columns of an $m \times n$ matrix A is called the *column space* of A and is usually denoted by Col(A).

EXAMPLE 3.3.3. Consider the columns of the matrix

$$A = \begin{bmatrix} 1 & 3 & 1 \\ 2 & 6 & 2 \\ 0 & -1 & 1 \end{bmatrix}. \tag{3.3.1}$$

This matrix can be reduced to the echelon matrix

$$E = \begin{bmatrix} 1 & 3 & 1 \\ 0 & 1 & -1 \\ 0 & 0 & 0 \end{bmatrix}. \tag{3.3.2}$$

By Theorem 3.2.3 on page 94, the first two columns of E are independent and may therefore serve as a basis for Col(E). Indeed,

$$s_1\begin{bmatrix} 1 \\ 0 \\ 0 \end{bmatrix} + s_2\begin{bmatrix} 3 \\ 1 \\ 0 \end{bmatrix} = \begin{bmatrix} 1 \\ -1 \\ 0 \end{bmatrix} \tag{3.3.3}$$

is solved by $s_2 = -1$ and $s_1 = 4$. Hence

$$4\begin{bmatrix}1\\0\\0\end{bmatrix} - \begin{bmatrix}3\\1\\0\end{bmatrix} = \begin{bmatrix}1\\-1\\0\end{bmatrix}, \tag{3.3.4}$$

and this equation exhibits the last column of E as a linear combination of the first two. Consequently, in any linear combination of all three columns the last one can be eliminated by substituting its expression from Equation (3.3.4).

We were, however, interested in the column space of A rather than that of E, and now we can easily find a basis for that too, once we have found a basis for $\mathrm{Col}(E)$. We have only to undo in Equation (3.3.4) the row operations that have led from A to E. Thus: interchange the last two rows of (3.3.4), multiply the new last row by -1, and add two times the first row to the second row. This procedure results in

$$4\begin{bmatrix}1\\2\\0\end{bmatrix} - \begin{bmatrix}3\\6\\-1\end{bmatrix} = \begin{bmatrix}1\\2\\1\end{bmatrix}. \tag{3.3.5}$$

Hence the last column of A is the same linear combination of the first two columns of A as was the case for the columns of E. The same argument with the right-hand side of Equation (3.3.3) replaced by the zero vector shows that the first two columns of A are also independent. Thus, those columns form a basis for $\mathrm{Col}(A)$.

Let us remark that had we carried the reduction of A further, the coefficients $s_1 = 4$ and $s_2 = -1$ would have shown up in the reduced echelon matrix

$$R = \begin{bmatrix}1 & 0 & 4\\0 & 1 & -1\\0 & 0 & 0\end{bmatrix}, \tag{3.3.6}$$

obtainable from (3.3.2) by subtracting three times the second row of E from its first row. In this matrix it is obvious that the last column equals four times the first column minus the second column.

<<>>

The procedure in the example above can be generalized for any matrix and leads to this theorem:

THEOREM 3.3.1. We can find a basis for the column space of any $m \times n$ matrix A by reducing A to an echelon matrix E and taking as basis vectors the columns of A that correspond to the basic columns of E.

Proof If E is obtained from A by row reduction, then the equation $E\boldsymbol{s} = \boldsymbol{0}$ has exactly the same set of solutions as $A\boldsymbol{s} = \boldsymbol{0}$. Therefore the columns of A are related by the same linear combinations as are the columns of E. For instance, if we set $s_i = 0$ for all free variables in $E\boldsymbol{s} = \boldsymbol{0}$, then all the basic variables must also be zero. This fact shows that the basic columns of E are independent, and then so too are the corresponding columns of A. Call these the *basic columns* of A.

Similarly, any linear combination showing the dependence of the nonbasic columns of E on the basic ones has its counterpart, *with the same coefficients*, for the corresponding columns of A. Furthermore, if $A\boldsymbol{s} = \boldsymbol{b}$ is any element of the column space of A and $E\boldsymbol{s} = \boldsymbol{c}$ is the corresponding reduced form, then we can eliminate the nonbasic columns of E and write $\boldsymbol{c}$ as a linear combination of the basic columns of E; and then $\boldsymbol{b}$ will be a linear combination, with the same coefficients, of the basic columns of A. Thus, the basic columns of A are independent and span Col(A).

<<>>

The column space of a matrix has an important application in the theory of linear systems:

THEOREM 3.3.2. The equation $A\boldsymbol{x} = \boldsymbol{b}$ is consistent if and only if the vector $\boldsymbol{b}$ lies in the column space of A.

Proof Since the expression $A\boldsymbol{x}$ equals the linear combination $\sum_{i=1}^{n} x_i \boldsymbol{a}_i$ of the columns of the matrix A, it is in the column space of A for any x; and so it can equal a vector b if and only if b is in that column space too.

<<>>

We can deal with the rows of a matrix much as with the columns:

DEFINITION 3.3.3. The subspace of $\mathbb{R}^n$ spanned by the transposed[7] rows of an $m \times n$ matrix A is called the *row space* of A and is usually denoted by Row(A).

EXAMPLE 3.3.4. Let us find a basis for the row space of the matrix A in Example 3.3.3:

[7]This transposition of the rows is an insignificant technicality that makes some formulas simpler by adhering to the convention of using only column vectors.

$$A = \begin{bmatrix} 1 & 3 & 1 \\ 2 & 6 & 2 \\ 0 & -1 & 1 \end{bmatrix}. \qquad (3.3.7)$$

The solution is very easy. Again, we consider the echelon matrix E corresponding to A:

$$E = \begin{bmatrix} 1 & 3 & 1 \\ 0 & 1 & -1 \\ 0 & 0 & 0 \end{bmatrix}. \qquad (3.3.8)$$

By Theorem 3.2.3 on page 94, the first two rows of E are independent and, when transposed, may serve as a basis for Row(E). The rows of A, however, are linear combinations of the rows of E, since they can be recovered from the latter by elementary row operations, and so the first two transposed rows of E may serve as a basis for Row(A) as well. In fact, Row(A) is the same as Row(E). Thus the vectors $(1, 3, 1)^T$ and $(0, 1, -1)^T$ form a basis for Row(A). It is easy to check that these two vectors are independent and the transposed rows of A are linear combinations of them.

<<>>

The procedure in the example above can again be generalized for any matrix and leads to the following theorem:

THEOREM 3.3.3. We can find a basis for the row space of any $m \times n$ matrix A by reducing A to an echelon matrix E and taking as basis vectors the transposed nonzero rows of E.[8]

For later use, let us exhibit the vectors of Row(A) as linear combinations of the rows transposed: Let $\boldsymbol{s} = (s_1, \ldots, s_m)^T$ be any vector in $\mathbb{R}^m$, and write $\boldsymbol{a}^i$ for the ith row of A. Then any vector $\boldsymbol{x} \in$ Row(A) can be written as

$$\boldsymbol{x} = \left(\sum_{i=1}^{m} s_i \boldsymbol{a}^i\right)^T = \sum_{i=1}^{m} s_i (\boldsymbol{a}^i)^T = \sum_{i=1}^{m} s_i (A^T)_i = A^T\boldsymbol{s}, \qquad (3.3.9)$$

where $(A^T)_i$ stands for the ith column of A^T. Since the expression on the right represents an arbitrary vector in the column space of A^T, we find that

$$\text{Row}(A) = \text{Col}(A^T), \qquad (3.3.10)$$

and similarly that

$$\text{Col}(A) = \text{Row}(A^T). \qquad (3.3.11)$$

[8]Notice that in this case the transposed rows of E themselves form a basis of the row space rather than the corresponding transposed rows of A, in contrast to the case for columns. (See Exercise 3.3.5.) The reason for the asymmetry lies in our use of *row* operations for the reduction to echelon form.

Applying Theorem 3.3.3 to A^T in place of A and using Equation (3.3.11), we obtain a new way of computing a basis for Col(A) :

COROLLARY 3.3.1. We can find a basis for the column space of any $m \times n$ matrix A by reducing A^T to an echelon matrix E and taking as basis vectors the transposed nonzero rows of E.

We have encountered still another subspace, associated with a matrix: its null space. How do we find a basis for it? The procedure is straightforward: We solve $A\mathbf{x} = \mathbf{0}$ in the usual manner by Gaussian elimination. The solution is always obtained as a linear combination of some vectors of $\mathbb{R}^n$, and these vectors are easily shown to be independent. They form a basis for Null(A), therefore, which is thus a subspace of $\mathbb{R}^n$. Let us look at an example:

EXAMPLE 3.3.5. Let

$$A = \begin{bmatrix} 0 & 1 & 2 & 3 \\ 4 & 2 & 6 & 6 \\ 8 & 2 & 8 & 6 \end{bmatrix}. \tag{3.3.12}$$

This expression can be reduced to the echelon form

$$E = \begin{bmatrix} 2 & 1 & 3 & 3 \\ 0 & 1 & 2 & 3 \\ 0 & 0 & 0 & 0 \end{bmatrix}. \tag{3.3.13}$$

The solution of $A\mathbf{x} = \mathbf{0}$ can be obtained by back substitution from Equation (3.3.13) as $x_3 = s$, $x_4 = t$, $x_2 = -2s - 3t$, and $x_1 = -s/2$. In vector form we may write

$$\mathbf{x} = s\begin{bmatrix} -1/2 \\ -2 \\ 1 \\ 0 \end{bmatrix} + t\begin{bmatrix} 0 \\ -3 \\ 0 \\ 1 \end{bmatrix}. \tag{3.3.14}$$

Thus Null(A) is spanned by the two vectors on the right of Equation (3.3.14). Furthermore, those are also independent, because if we set $\mathbf{x} = \mathbf{0}$, then the last two rows of (3.3.14) will correspond to the equations $1s + 0t = 0$ and $0s + 1t = 0$, from which $s = t = 0$ will follow. We have thus found a basis for Null(A). Writing

$$B = \begin{bmatrix} -1/2 & 0 \\ -2 & -3 \\ 1 & 0 \\ 0 & 1 \end{bmatrix}, \tag{3.3.15}$$

we can rephrase this result as saying that we have found a matrix B with independent columns such that $\mathrm{Col}(B) = \mathrm{Null}(A)$.

<<>>

Theorem 3.3.1 is often interpreted as giving a procedure for *reducing* a spanning set of a subspace U of $\mathbb{R}^m$ to a basis of U. The same construction can also be used to solve the related problem of *extending* an independent set $\{\boldsymbol{b}_1, \boldsymbol{b}_2, \ldots, \boldsymbol{b}_k\}$ in a subspace U to a basis. First, however, we want to prove a theorem showing that such an extension is always possible. This so-called Exchange Theorem will also be used to conclude that all bases in a vector space have the same number of vectors.

THEOREM 3.3.4.

In a vector space X, let $A = (\boldsymbol{a}_1, \boldsymbol{a}_2, \ldots, \boldsymbol{a}_n)$ be a list of nonzero vectors that span X, and $B = (\boldsymbol{b}_1, \boldsymbol{b}_2, \ldots, \boldsymbol{b}_k)$ be a list of independent vectors in X. Then $k \leq n$ holds, and k of the spanning vectors $\boldsymbol{a}_i$ can be exchanged for the vectors of B. That is, X is spanned by the k vectors of B together with some $n - k$ vectors of A.

Proof Consider the vectors $\boldsymbol{b}_1, \boldsymbol{a}_1, \boldsymbol{a}_2, \ldots, \boldsymbol{a}_n$. They span X because the $\boldsymbol{a}_i$ in themselves do, and adjoining $\boldsymbol{b}_1$ to the list does not change that. Furthermore, these vectors are linearly dependent since $\boldsymbol{b}_1$ can certainly be expressed as a linear combination of the spanning vectors $\boldsymbol{a}_i$, say as $\boldsymbol{b}_1 = \sum_{i=1}^{n} s_i \boldsymbol{a}_i$. Then $\boldsymbol{b}_1 - \sum_{i=1}^{n} s_i \boldsymbol{a}_i = \boldsymbol{0}$ has a nontrivial linear combination on the left, which shows that the vectors $\boldsymbol{b}_1, \boldsymbol{a}_1, \boldsymbol{a}_2, \ldots, \boldsymbol{a}_n$ are linearly dependent. Thus we may apply Theorem 3.2.4 on page 95 to this list and omit one of the $\boldsymbol{a}_i$ vectors so that the remaining n vectors will still span X. Call the remaining $\boldsymbol{a}_i$ vectors $\boldsymbol{a}'_1, \boldsymbol{a}'_2, \ldots, \boldsymbol{a}'_{n-1}$.

Next, consider the list $\boldsymbol{b}_1, \boldsymbol{b}_2, \boldsymbol{a}'_1, \boldsymbol{a}'_2, \ldots, \boldsymbol{a}'_{n-1}$ of vectors, which are linearly dependent just like those above, and apply to it Theorem 3.2.4 on page 95. Accordingly, one of the vectors $\boldsymbol{a}'_i$ can be omitted so that the remaining n vectors will still span X. (The omitted vector cannot be $\boldsymbol{b}_2$ because $\boldsymbol{b}_1$ and $\boldsymbol{b}_2$ were assumed to be independent and the omitted vector must depend on the previous ones in the list.)

We can proceed similarly with the rest of the vectors of B, exchanging an $\boldsymbol{a}_i$ for a $\boldsymbol{b}_j$ in each step, until we exhaust A or B. If A were exhausted first; that is, if we had $k > n$, then at some point all vectors of A would be exchanged for the k vectors of B, and the vectors $\boldsymbol{b}_1, \boldsymbol{b}_2, \ldots, \boldsymbol{b}_n$ would span X. But then $\boldsymbol{b}_{n+1}$ (like any other vector) would be a linear combination of $\boldsymbol{b}_1, \boldsymbol{b}_2, \ldots, \boldsymbol{b}_n$, and this fact contradicts the assumed independence of the $\boldsymbol{b}_j$ vectors. Thus A cannot be exhausted first; that is, we must have $k \leq n$, and in that case all the vectors of B can be brought into the spanning set.

<<>>

Although the theorem above ensures that an independent set can be extended to a basis, it does not give a practical method for doing so. But, in $\mathbb{R}^m$, the construction described in Theorem 3.3.1 does so, as follows. Suppose U is given as the span of $\{\boldsymbol{a}_1, \boldsymbol{a}_2, \ldots, \boldsymbol{a}_n\}$, and $\{\boldsymbol{b}_1, \boldsymbol{b}_2, \ldots, \boldsymbol{b}_k\}$ is an independent set in U with $k < n$. Then we construct according to Theorem 3.3.1 a basis for the column space of the matrix $C = (\boldsymbol{b}_1, \boldsymbol{b}_2, \ldots, \boldsymbol{b}_k, \boldsymbol{a}_1, \boldsymbol{a}_2, \ldots, \boldsymbol{a}_n)$. Because $U = \mathrm{Col}(C)$, and the row reduction proceeds from left to right, the independent vectors $\boldsymbol{b}_j$ will be in the basis of U so found. Let us see an example:

EXAMPLE 3.3.6. Let

$$U = \mathrm{Span}\left\{\begin{bmatrix}1\\0\\1\\0\end{bmatrix}, \begin{bmatrix}1\\1\\0\\1\end{bmatrix}, \begin{bmatrix}0\\1\\1\\0\end{bmatrix}\right\}, \tag{3.3.16}$$

where the columns on the right are the vectors $\boldsymbol{a}_1$, $\boldsymbol{a}_2$, $\boldsymbol{a}_3$, and let

$$\boldsymbol{b}_1 = \begin{bmatrix}3\\0\\1\\1\end{bmatrix}, \quad \boldsymbol{b}_2 = \begin{bmatrix}0\\3\\1\\1\end{bmatrix}. \tag{3.3.17}$$

It is easy to see that $\boldsymbol{b}_1 = 2\boldsymbol{a}_1 + \boldsymbol{a}_2 - \boldsymbol{a}_3$ and $\boldsymbol{b}_2 = -\boldsymbol{a}_1 + \boldsymbol{a}_2 + 2\boldsymbol{a}_3$, and so both $\boldsymbol{b}_1$ and $\boldsymbol{b}_2$ are in U. We want to extend the set $\{\boldsymbol{b}_1, \boldsymbol{b}_2\}$ to a basis for U.

We form the matrix

$$C = \begin{bmatrix}3 & 0 & 1 & 1 & 0\\0 & 3 & 0 & 1 & 1\\1 & 1 & 1 & 0 & 1\\1 & 1 & 0 & 1 & 0\end{bmatrix} \tag{3.3.18}$$

and reduce it to echelon form:

$$\begin{bmatrix}3 & 0 & 1 & 1 & 0\\0 & 3 & 0 & 1 & 1\\1 & 1 & 1 & 0 & 1\\1 & 1 & 0 & 1 & 0\end{bmatrix} \to \begin{bmatrix}1 & 1 & 0 & 1 & 0\\1 & 1 & 1 & 0 & 1\\3 & 0 & 1 & 1 & 0\\0 & 3 & 0 & 1 & 1\end{bmatrix} \to$$

$$\begin{bmatrix}1 & 1 & 0 & 1 & 0\\0 & 0 & 1 & -1 & 1\\0 & -3 & 1 & -2 & 0\\0 & 3 & 0 & 1 & 1\end{bmatrix} \to \begin{bmatrix}1 & 1 & 0 & 1 & 0\\0 & 0 & 1 & -1 & 1\\0 & -3 & 1 & -2 & 0\\0 & 3 & 0 & 1 & 1\end{bmatrix} \to$$

$$\begin{bmatrix}1 & 1 & 0 & 1 & 0\\0 & -3 & 1 & -2 & 0\\0 & 0 & 1 & -1 & 1\\0 & 0 & 1 & -1 & 1\end{bmatrix} \to \begin{bmatrix}1 & 1 & 0 & 1 & 0\\0 & -3 & 1 & -2 & 0\\0 & 0 & 1 & -1 & 1\\0 & 0 & 0 & 0 & 0\end{bmatrix}. \tag{3.3.19}$$

In the final form the pivots are in the first three columns, and so the corresponding columns of C—that is, $\boldsymbol{b}_1$, $\boldsymbol{b}_2$, $\boldsymbol{a}_1$—form a basis for $U = \mathrm{Col}(C)$.

<<>>

Exercises 3.3

In the first four exercises, find bases for Row(A), Col(A), Null(A).

• **3.3.1.** $A = \begin{bmatrix} 1 & 3 & 1 \\ 3 & 2 & 4 \\ 2 & -1 & 3 \end{bmatrix}$.

3.3.2. $A = \begin{bmatrix} 1 & 2 & 1 \\ 0 & 2 & 4 \\ 1 & 3 & 3 \end{bmatrix}$.

3.3.3. $A = \begin{bmatrix} 1 & 3 & 1 \\ 3 & 2 & 4 \\ 2 & -1 & 3 \\ 0 & 1 & 1 \end{bmatrix}$.

3.3.4. $A = \begin{bmatrix} 1 & 1 & 1 & 2 & -1 \\ 3 & 3 & 0 & 4 & 4 \end{bmatrix}$.

• **3.3.5.** Find a 3×3 matrix A whose first two rows transposed do not form a basis for Row(A), but only the first two transposed rows of any corresponding echelon matrix E do so.

3.3.6. Show that for the matrix in Exercise 3.3.1 the first two columns of any corresponding echelon matrix E do not form a basis for Col(A), but only the first two columns of A itself do so.

3.3.7. Determine whether each of the following vectors is in the column space of the matrix of Exercise 3.3.1, and if it is, then write it as a linear combination of the first two columns:

$$\boldsymbol{a} = (1, 4, 3)^T, \boldsymbol{b} = (-10, 1, 7)^T, \boldsymbol{c} = (9, -5, -10)^T, \boldsymbol{d} = (5, 9, 4)^T.$$

***3.3.8.** Show that for two matrices A and B we have $AB = O$ if and only if Col(B) is a subspace of Null(A).

***3.3.9.** Show that for any two matrices A and B such that AB exists and A is invertible we have Null(B) = Null(AB).

• **3.3.10.** Prove that in $\mathbb{R}^n$ any set of n independent vectors forms a basis. (*Hint:* Either consider the matrix A with the given vectors as columns and apply Theorem 2.3.5 on page 75, or use the Exchange Theorem.)

***3.3.11.** Prove that in $\mathbb{R}^n$ no set of fewer than n vectors spans $\mathbb{R}^n$. (*Hint:* Use the Exchange Theorem.)

***3.3.12.** Prove that in $\mathbb{R}^n$ any set of n vectors that span $\mathbb{R}^n$ forms a basis. (*Hint:* Either consider the matrix A with the given vectors as columns and apply Theorem 2.3.5 on page 75 or use the result of Exercise 3.3.11.)

***3.3.13.** Prove that in $\mathbb{R}^n$ any set of more than n vectors is a dependent set. (*Hint:* Use either Gaussian elimination or the Exchange Theorem.)

3.3.14. Let

$$U = \text{Span}\left\{\begin{bmatrix}1\\0\\0\\0\end{bmatrix}, \begin{bmatrix}1\\0\\0\\1\end{bmatrix}, \begin{bmatrix}0\\1\\1\\0\end{bmatrix}\right\}.$$

Check that the vectors

$$\boldsymbol{b}_1 = \begin{bmatrix}0\\1\\1\\1\end{bmatrix}, \quad \boldsymbol{b}_2 = \begin{bmatrix}1\\1\\1\\0\end{bmatrix}$$

are in U and extend the set $\{\boldsymbol{b}_1, \boldsymbol{b}_2\}$ to a basis for U.

MATLAB Exercises 3.3

ML3.3.1. Let A = **magic**(4).

a. Use **rref** on A and A^T to find a basis for the row space and the column space, respectively. Extract and transpose the appropriate submatrix in each case to obtain two matrices B and C whose columns form the bases. (Such a matrix is called a *basis-matrix*.)

b. The command D = **orth**(A) returns a basis-matrix for Col(A) (usually different from the one obtained by **rref**). Show that the columns of D and of the matrix C computed in Part (a) span the same space.

c. Let E = **orth**(A'). Show that the columns of E and of the matrix B computed in Part (a) span the same space.

d. The command N = **null**(A) returns a basis-matrix for Null(A). Compute B^TN and explain your result.

e. Compute M = **null**($A^T * A$) and explain your result.

ML3.3.2. Repeat Exercise ML3.3.1 for the matrix

$$A = \begin{bmatrix}1 & 3 & 5 & 7\\2 & 4 & 6 & 8\\0 & 2 & 4 & 6\\3 & 5 & 7 & 9\end{bmatrix}.$$

ML3.3.3. Let $A = \mathbf{round}(10 * \mathbf{rand}(3,4) - 5)$.

a. Compute the rank of A, $[A, A]$, $[A, A, A]$, $[A; A]$, $[A; A; A]$ and $[A, A; A, A]$ each.

b. Repeat Part (a) above for six instances of A.

c. Do you see any patterns? Make a conjecture and prove it.

3.4. DIMENSION, ORTHOGONAL COMPLEMENTS

We wish to define the dimension of a vector space as the number of vectors in a basis. To ensure the consistency of such a definition, we start with a theorem which says that all bases of a vector space must have the same number of vectors.

THEOREM 3.4.1. If a vector space X has two bases $\mathscr{A} = \{\boldsymbol{a}_1, \boldsymbol{a}_2, \ldots, \boldsymbol{a}_m\}$ and $\mathscr{B} = \{\boldsymbol{b}_1, \boldsymbol{b}_2, \ldots, \boldsymbol{b}_n\}$ with m and n positive integers, then $m = n$ must hold.

Proof By the definition of a basis, the set $\mathscr{A}$ is an independent set and the set $\mathscr{B}$ spans X. Hence, by Theorem 3.3.4 on page 104 (the Exchange Theorem), we must have $m \le n$. Reversing the roles of $\mathscr{A}$ and $\mathscr{B}$, we find that $n \le m$ too must hold. Thus $m = n$ follows.

<<>>

The theorem above enables us to make the following definition:

DEFINITION 3.4.1. If a vector space X has a basis of n vectors, where n is a positive integer, then n is called the *dimension* of X and we write $n = \dim(X)$. The dimension of the vector space $\{\boldsymbol{0}\}$ is defined to be zero and we say it has the empty set for a basis. If X has no finite basis, then it is said to be infinite-dimensional.

EXAMPLE 3.4.1. Not unexpectedly, for any positive integer n, the dimension of $\mathbb{R}^n$ is n.

<<>>

EXAMPLE 3.4.2. The subspace of $\mathbb{R}^3$ given by $U = \{\boldsymbol{u} \in \mathbb{R}^3 \mid \boldsymbol{u} = s(1, 1, 1)^T + t(1, 2, 3)^T\}$ has dimension two because the two vectors $(1, 1, 1)^T$ and $(1, 2, 3)^T$ form a basis for U.

<<>>

EXAMPLE 3.4.3. The space $\mathscr{P}$ of all polynomials in a variable x has the infinite set $\{1, x, x^2, \ldots\}$ for a basis, in the sense that any polynomial is a finite

linear combination of these vectors, and their independence (which too is defined using only finite linear combinations) follows because if a polynomial equals zero for every x, then all its coefficients must be zero. Thus $\mathcal{P}$ is an infinite-dimensional vector space.

<<>>

THEOREM 3.4.2. Let A be any $m \times n$ matrix with r denoting its rank. Then the dimension of its column space[9] and the dimension of its row space both equal r and the dimension of its null space equals $n - r$.

Proof Recall that the rank of a matrix was defined on page 43 as the number of nonzero rows in the corresponding echelon form, which is the same as the number of pivots in the latter. Since those rows transposed form a basis for the row space, we have $\dim(\text{Row}(A)) = r$. Also, since the column space of A has for a basis the columns of A corresponding to those of E with pivots, $\dim(\text{Col}(A)) = r$ holds, too. Last, the construction of a basis for the null space as shown in Example 3.3.5 on page 103 has a parameter for each of the $n - r$ free variables, and there is a basis vector corresponding to each of those, and vice versa.

<<>>

Notice how remarkable this theorem is. Looking at any matrix of any size, with rows and columns having very little to do with each other, who would have guessed that the row and column spaces have the same dimension?

The dimension of the null space of a matrix is sometimes called its *nullity* and part of Theorem 3.4.2 can be stated as follows:

COROLLARY 3.4.1. Let A be any $m \times n$ matrix. Then rank + nullity = n.

There is more to this expression than meets the eye: The row space and the null space of A are both subspaces of $\mathbb{R}^n$ and since their dimensions add up to n, we may, in a sense, expect *them* to add up to $\mathbb{R}^n$. They do indeed:

THEOREM 3.4.3. Let A be any $m \times n$ matrix and $\mathbf{x}$ any vector in $\mathbb{R}^n$. Then $\mathbf{x}$ can be decomposed uniquely into a sum of a vector $\mathbf{x}_0$ from Null(A) and a vector $\mathbf{x}_R$ from Row(A); that is, $\mathbf{x} = \mathbf{x}_0 + \mathbf{x}_R$, with $\mathbf{x}_0$ and $\mathbf{x}_R$ uniquely determined by $\mathbf{x}$. Furthermore, every vector of Null(A) is orthogonal to every vector of Row(A), and so, in particular, $\mathbf{x}_0$ and $\mathbf{x}_R$ are orthogonal to each other.

[9] The dimension of the column space is sometimes called the *column-rank of A*, and the first statement of the theorem is phrased as "column-rank = row-rank."

Proof Let us start with the orthogonality of the vectors of Row(A) and Null(A): Assume that $\boldsymbol{u}$ is in Null(A). Then we have $A\boldsymbol{u} = \boldsymbol{0}$; and if we write $\boldsymbol{a}^i$ for the ith row of A, this equation implies $\boldsymbol{a}^i\boldsymbol{u} = 0$ for each i; from which, for any linear combination of the rows, we get $\sum_{i=1}^{m} s_i\boldsymbol{a}^i\boldsymbol{u} = 0$. This equation can be written as $\boldsymbol{s}^T A\boldsymbol{u} = 0$, and furthermore, by the rule for the transpose of a product (which requires reversal of the factors), also as $\boldsymbol{u}^T A^T \boldsymbol{s} = 0$. Equation (3.3.9) shows that $\boldsymbol{v} = A^T\boldsymbol{s}$ is an arbitrary vector of Row(A), and so we have

$$\boldsymbol{u}^T\boldsymbol{v} = 0. \tag{3.4.1}$$

The matrix product of the row vector $\boldsymbol{u}^T$ and the column vector $\boldsymbol{v}$ corresponds to the dot product of the two column vectors $\boldsymbol{u}$ and $\boldsymbol{v}$, and so Equation 3.4.1 expresses the orthogonality of any $\boldsymbol{u} \in$ Null(A) to any $\boldsymbol{v} \in$ Row(A).

Let $\mathcal{B} = \{\boldsymbol{b}_1, \boldsymbol{b}_2, \ldots, \boldsymbol{b}_r\}$ be a basis for Row(A) and $\mathcal{C} = \{\boldsymbol{c}_1, \boldsymbol{c}_2, \ldots, \boldsymbol{c}_{n-r}\}$ a basis for Null(A), and $B = (\boldsymbol{b}_1\ \boldsymbol{b}_2 \ldots \boldsymbol{b}_r)$ and $C = (\boldsymbol{c}_1\ \boldsymbol{c}_2 \ldots \boldsymbol{c}_{n-r})$ the corresponding matrices with the given vectors as columns. Then, by Equation (3.4.1) applied to these basis vectors,

$$(\boldsymbol{c}_i)^T\boldsymbol{b}_j = 0 \tag{3.4.2}$$

holds for each i and j, which may be written in matrix form as:

$$C^T B = O. \tag{3.4.3}$$

We want to show that the n vectors of $\mathcal{B} \cup \mathcal{C}$ form a basis for $\mathbb{R}^n$. To this end, let us test the columns of the joint matrix

$$(B, C) = (\boldsymbol{b}_1\ \boldsymbol{b}_2 \ldots \boldsymbol{b}_r\ \boldsymbol{c}_1\ \boldsymbol{c}_2 \ldots \boldsymbol{c}_{n-r})$$

for independence: Assume

$$B\boldsymbol{s} + C\boldsymbol{t} = \boldsymbol{0} \tag{3.4.4}$$

for $\boldsymbol{s} \in \mathbb{R}^r$ and $\boldsymbol{t} \in \mathbb{R}^{n-r}$. Left-multiply this equation by $(C\boldsymbol{t})^T$ to obtain

$$(C\boldsymbol{t})^T B\boldsymbol{s} + (C\boldsymbol{t})^T C\boldsymbol{t} = 0, \tag{3.4.5}$$

and this equation can also be written as

$$\boldsymbol{t}^T C^T B\boldsymbol{s} + (C\boldsymbol{t})^T(C\boldsymbol{t}) = 0. \tag{3.4.6}$$

By Equation (3.4.3) the first term is zero, and the second term equals $|C\boldsymbol{t}|^2$. Because $\boldsymbol{0}$ is the only vector whose absolute value is zero, Equation (3.4.6) implies

$$C\boldsymbol{t} = \boldsymbol{0}. \tag{3.4.7}$$

We have assumed, however, that the columns of C form a basis for Null(A), and so Equation (3.4.7) has only the trivial solution $\boldsymbol{t} = \boldsymbol{0}$. Similarly, left-multiplying Equation (3.4.4) by $(B\boldsymbol{s})^T$ we would have obtained $\boldsymbol{s} = \boldsymbol{0}$. Thus, Equation (3.4.4) has only the trivial solution, and the columns of (B, C) are independent. Then they also form a basis for $\mathbb{R}^n$, for any set of n independent vectors in $\mathbb{R}^n$ does so.

Now, let us write any $\boldsymbol{x} \in \mathbb{R}^n$ in terms of the above basis as

$$\boldsymbol{x} = B\boldsymbol{s} + C\boldsymbol{t}. \tag{3.4.8}$$

Then

$$\boldsymbol{x}_R = B\boldsymbol{s} \text{ and } \boldsymbol{x}_0 = C\boldsymbol{t} \tag{3.4.9}$$

provide the claimed decomposition. Its uniqueness follows from the uniqueness of decompositions in any basis, ensured by the independence of the basis vectors.

<<>>

EXAMPLE 3.4.4. Let us again consider the matrix of Example 3.3.3:

$$A = \begin{bmatrix} 1 & 3 & 1 \\ 2 & 6 & 2 \\ 0 & -1 & 1 \end{bmatrix}, \tag{3.4.10}$$

and find the decomposition described above of the vector $\boldsymbol{x} = (1, 2, 3)^T$.

The echelon form

$$E = \begin{bmatrix} 1 & 3 & 1 \\ 0 & 1 & -1 \\ 0 & 0 & 0 \end{bmatrix} \tag{3.4.11}$$

provides us with the basis vectors $\boldsymbol{b}_1 = (1, 3, 1)^T$ and $\boldsymbol{b}_2 = (0, 1, -1)^T$ for the row space and with the single basis vector $\boldsymbol{c}_1 = (-4, 1, 1)^T$ for the null space. (Why?) It is easy to check that $\boldsymbol{c}_1$ is orthogonal to $\boldsymbol{b}_1$ and $\boldsymbol{b}_2$, as it should be. To decompose any $\boldsymbol{x} \in \mathbb{R}^3$ into row space and null space components, we should solve

$$s_1\boldsymbol{b}_1 + s_2\boldsymbol{b}_2 + t_1\boldsymbol{c}_1 = \mathbf{x} \tag{3.4.12}$$

for the unknown coefficients s_1, s_2, t_1. In our case this procedure amounts to solving the system

$$\begin{bmatrix} 1 & 0 & -4 \\ 3 & 1 & 1 \\ 1 & -1 & 1 \end{bmatrix} \begin{bmatrix} s_1 \\ s_2 \\ t_1 \end{bmatrix} = \begin{bmatrix} 1 \\ 2 \\ 3 \end{bmatrix}. \tag{3.4.13}$$

We can do this calculation in the usual way by Gaussian elimination:

$$\left[\begin{array}{ccc|c} 1 & 0 & -4 & 1 \\ 3 & 1 & 1 & 2 \\ 1 & -1 & 1 & 3 \end{array}\right] \to \left[\begin{array}{ccc|c} 1 & 0 & -4 & 1 \\ 0 & 1 & 13 & -1 \\ 0 & -1 & 5 & 2 \end{array}\right] \to \left[\begin{array}{ccc|c} 1 & 0 & -4 & 1 \\ 0 & 1 & 13 & -1 \\ 0 & 0 & 18 & 1 \end{array}\right] \tag{3.4.14}$$

and then back substitution gives $t_1 = 1/18$, $s_2 = -31/18$ and $s_1 = 22/18$. Thus $\mathbf{x} = (1, 2, 3)^T$ is decomposed into the row space and null space components

$$\mathbf{x}_R = \frac{22}{18}\begin{bmatrix} 1 \\ 3 \\ 1 \end{bmatrix} - \frac{31}{18}\begin{bmatrix} 0 \\ 1 \\ -1 \end{bmatrix} = \frac{1}{18}\begin{bmatrix} 22 \\ 35 \\ 53 \end{bmatrix} \text{ and } \mathbf{x}_0 = \frac{1}{18}\begin{bmatrix} -4 \\ 1 \\ 1 \end{bmatrix}. \tag{3.4.15}$$

Note that we could have obtained $\mathbf{x}_0$ by utilizing the orthogonality of $\boldsymbol{c}_1$ to $\boldsymbol{b}_1$ and $\boldsymbol{b}_2$, and taking the dot product of both sides of Equation (3.4.12) with $\boldsymbol{c}_1$ to obtain

$$t_1\boldsymbol{c}_1^T\boldsymbol{c}_1 = \boldsymbol{c}_1^T\mathbf{x}. \tag{3.4.16}$$

This equation evaluates to $18t_1 = 1$, and gives the same value $t_1 = 1/18$ as the previous computation but much more quickly. Then, once we have found $\mathbf{x}_0$, we can compute $\mathbf{x}_R$ as $\mathbf{x} - \mathbf{x}_0$. However, this shortcut works only if one of the subspaces Null(A) or Row(A) is one-dimensional, as in this example.

<<>>

Theorem 3.4.3 establishes two relations between Null(A) and Row(A) that can be generalized: first, that the sums of their vectors make up all of $\mathbb{R}^n$ and, second, that those vectors are orthogonal to each other. Since these concepts occur in other contexts as well, we make corresponding definitions for arbitrary subspaces:

DEFINITION 3.4.2.

Let U and V be subspaces of a vector space X. Then the set

$$\{\boldsymbol{u} + \boldsymbol{v} \mid \boldsymbol{u} \in U, \boldsymbol{v} \in V\}$$

is called the *sum* of U and V, and is denoted by $U + V$.

It is easy to see that $U + V$ is a subspace of X (Exercise 3.4.8).

DEFINITION 3.4.3.

Let U and V be subspaces of a vector space X with an inner product. They are said to be orthogonal to each other if every $\boldsymbol{u} \in U$ is orthogonal to every $\boldsymbol{v} \in V$. Furthermore, the set of all vectors $\boldsymbol{v}$ in X orthogonal to all vectors $\boldsymbol{u}$ in U is called the *orthogonal complement of U*, and is denoted by $U^{\perp}$ (read "U-perp"). In other words, $U^{\perp} = \{\boldsymbol{v} \in X \mid \boldsymbol{u} \cdot \boldsymbol{v} = 0 \text{ for all } \boldsymbol{u} \in U\}$.

Again, it is straightforward to verify that $U^{\perp}$ is a subspace of X. This demonstration is left as Exercise 3.4.13. Furthermore, with this notation we can write part of Theorem 3.4.3 as follows:

COROLLARY 3.4.2.

Let A be any $m \times n$ matrix. Then we have

$$\mathrm{Row}(A) + \mathrm{Null}(A) = \mathbb{R}^n \tag{3.4.17}$$

$$\mathrm{Row}(A) = \mathrm{Null}(A)^{\perp} \tag{3.4.18}$$

and

$$\mathrm{Null}(A) = \mathrm{Row}(A)^{\perp} \tag{3.4.19}$$

Let U be a subspace of $\mathbb{R}^n$. Then, by choosing a matrix A whose row space is U, the formulas above yield the following result:

COROLLARY 3.4.3.

Let U be a subspace of $\mathbb{R}^n$. Then we have

$$U + U^{\perp} = \mathbb{R}^n \tag{3.4.20}$$

It may seem strange that the column space of a matrix has been slighted in the discussion so far. The reason is that we usually write linear systems as $A\mathbf{x} = \boldsymbol{b}$ rather than as $\boldsymbol{y}^T A = \boldsymbol{b}^T$. Nevertheless, the column space is very important and occasionally we need yet another subspace associated with A. This addition will also remove the temporary asymmetry of the theory:

DEFINITION 3.4.4.

Let A be any $m \times n$ matrix. The set of all vectors $\boldsymbol{y} \in \mathbb{R}^m$ such that $\boldsymbol{y}^T A = \boldsymbol{0}^T$ is called the *left nullspace of A*, and we denote it by Left-null(A).[10]

For any matrix A, the four spaces Row(A), Col(A), Null(A), and Left-null(A) are sometimes referred to as the four *fundamental subspaces of A*.

[10]This notation is our own; there is no standard one.

As expected, we have relations between the left nullspace and the column space, analogous to those between the row space and the nullspace. Taking the transpose of both sides of the defining relation above, we get

$$A^T\boldsymbol{y} = \boldsymbol{0} \tag{3.4.21}$$

and so

$$\text{Left-null}(A) = \text{Null}(A^T). \tag{3.4.22}$$

Also, as we have seen in Equation (3.3.11) on page 102, we have

$$\text{Col}(A) = \text{Row}(A^T). \tag{3.4.23}$$

Consequently, the relations between the left nullspace and the column space can be obtained from the previous results simply by applying them to A^T. The main features are as follows:

THEOREM 3.4.4. Let A be any $m \times n$ matrix. Then

$$\dim(\text{Left-null}(A)) + \dim(\text{Col}(A)) = m, \tag{3.4.34}$$

$$\text{Col}(A) + \text{Left-null}(A) = \mathbb{R}^m, \tag{3.4.25}$$

$$\text{Left-null}(A)^{\perp} = \text{Col}(A), \tag{3.4.26}$$

$$\text{Col}(A)^{\perp} = \text{Left-null}(A), \tag{3.4.27}$$

and any vector $\boldsymbol{y} \in \mathbb{R}^m$ can be uniquely decomposed into the sum of a vector $\boldsymbol{y}_C$ in Col(A) and a vector $\boldsymbol{y}_L$ in Left-null(A).

The left nullspace arises naturally in the following type of problem:

Corresponding to nonparametric and parametric representations of lines and planes, the two fundamental ways of representing a subspace V of $\mathbb{R}^n$ are:

1. as the solution space of a homogeneous system $A\boldsymbol{x} = \boldsymbol{0}$; that is, as Null($A$) for some $m \times n$ matrix A, and
2. as the set of all linear combinations $\boldsymbol{x} = \sum_{i=1}^{p} s_i\boldsymbol{b}_i = B\boldsymbol{s}$ of some vectors $\boldsymbol{b}_i \in \mathbb{R}^n$, that is, as Col($B$) for some $n \times p$ matrix B.

The question is how to pass from one representation to the other. In the direction $1 \to 2$ we have solved this in Section 3.3 (see Example 3.3.5 on page 103). In the other direction, given the matrix B, we need to find a matrix A such that Null(A) = Col(B). Taking the orthogonal complement of each side and making use of Equations (3.4.18) and (3.4.27), we get

Row(A) = Left-null(B). Thus we can find a suitable A by finding a basis for Left-null(B) and using the transposes of these basis vectors as the rows of A.

EXAMPLE 3.4.5. Let V be the column space of the matrix

$$B = \begin{bmatrix} 0 & 2 & -1 \\ 1 & 0 & 0 \\ 0 & 0 & -2 \\ 0 & 1 & 0 \\ 0 & 0 & 1 \end{bmatrix} \qquad (3.4.28)$$

and write V as the nullspace of a matrix A.

As stated above, to solve this problem we need to find a basis for Left-null(B). This space is the same as Null(B^T), and so we need to find all solutions of $B^T\mathbf{x} = \mathbf{0}$. We can find them in the usual way by reducing B^T as follows:

$$\begin{bmatrix} 0 & 1 & 0 & 0 & 0 \\ 2 & 0 & 0 & 1 & 0 \\ -1 & 0 & -2 & 0 & 1 \end{bmatrix} \to \begin{bmatrix} 1 & 0 & 2 & 0 & -1 \\ 0 & 1 & 0 & 0 & 0 \\ 0 & 0 & -4 & 1 & 2 \end{bmatrix}. \qquad (3.4.29)$$

The free variables are x_4 and x_5, and we set them equal to parameters: $x_4 = s_1$ and $x_5 = s_2$. Solving for the other components, we get $4x_3 = s_1 + 2s_2$, $x_2 = 0$ and $x_1 = -s_1/2$. Hence we can write the solution vectors as

$$\mathbf{x} = \frac{1}{4}\begin{bmatrix} -2 & 0 \\ 0 & 0 \\ 1 & 2 \\ 4 & 0 \\ 0 & 4 \end{bmatrix}\mathbf{s}. \qquad (3.4.30)$$

The columns of the matrix on the right form a basis for Left-null(B), and so its transpose is a solution to the problem:

$$A = \begin{bmatrix} -2 & 0 & 1 & 4 & 0 \\ 0 & 0 & 2 & 0 & 4 \end{bmatrix}. \qquad [3.4.31]$$

We leave it to the reader to check that indeed Null(A) = Col(B).

<<>>

Still another numerical relation between subspaces of a matrix calls for deeper examination, namely that the row space and the column space both have the same dimension. If we consider any vector $\mathbf{x}_R$ in Row(A), then $A\mathbf{x}_R$ is in Col(A). Also, every vector in Col(A) can be obtained as $A\mathbf{x}_R$, since they are all of the form $A\mathbf{x}$ and, using the decomposition $\mathbf{x} = \mathbf{x}_0 + \mathbf{x}_R$ from Theorem 3.4.3, we find that $A\mathbf{x}_R = A\mathbf{x}$. Furthermore, $\mathbf{x}_R$ is uniquely determined by $A\mathbf{x}$, since if $A\mathbf{x}_R = A\mathbf{x}'_R$, then $A(\mathbf{x}_R - \mathbf{x}'_R) = \mathbf{0}$ and so $\mathbf{x}_R - \mathbf{x}'_R$ is in the nullspace as well as in the row space, and must be the zero vector therefore. This mapping of Row(A) to Col(A), given by $\mathbf{x}_R \to A\mathbf{x}_R$, can be represented, relative to any choice of bases in Row(A) and Col(A), by an $r \times r$ nonsingular matrix. We shall compute this matrix in Section 4.2. Also, the inverse of this mapping is discussed in Exercise 5.1.14 on page 176 for the special case of A with independent rows, without reference to any bases.

Exercises 3.4

For each of the matrices in the first three exercises,

a. Find the dimensions of the four subspaces associated with A.

b. Decompose $(1, 1, 1, 1, 1)^T$ into the sum of an $\mathbf{x}_0 \in$ Null(A) and an $\mathbf{x}_R \in$ Row(A).

• **3.4.1.** Let

$$A = \begin{bmatrix} 1 & 1 & 1 & 2 & -1 \\ 3 & 3 & 0 & 4 & 4 \end{bmatrix},$$

as in Exercise 3.3.4 on page 106.

3.4.2. Let

$$A = [1 \quad 1 \quad 1 \quad 2 \quad -1].$$

3.4.3. Let

$$A = [3 \quad 3 \quad 0 \quad 4 \quad 4].$$

For each of the matrices in the next three exercises,

a. Find the dimensions of the four subspaces associated with A.

b. Decompose $(1, 2, 3, 4, 5)^T$ into the sum of an $\mathbf{x}_0 \in$ Null(A) and an $\mathbf{x}_R \in$ Row(A).

3.4.4. Let

$$A = \begin{bmatrix} 1 & 1 & 1 & 2 & -1 \\ 0 & 2 & 0 & 0 & 4 \\ 0 & 0 & 0 & 2 & 2 \end{bmatrix}.$$

3.4.5. Let

$$A = \begin{bmatrix} 0 & 2 & 0 & 0 & 4 \\ 0 & 0 & 0 & 2 & 2 \end{bmatrix}.$$

3.4.6. Let

$$A = \begin{bmatrix} 1 & 1 & 1 & 2 & -1 \\ 0 & 2 & 0 & 0 & 4 \end{bmatrix}.$$

3.4.7. Show, without referring to Theorem 3.4.3, that $\text{Row}(A) \cap \text{Null}(A) = \{\boldsymbol{0}\}$ for any matrix A.

3.4.8. Let U and V be subspaces of a vector space X. Prove that $U + V$ is a subspace of X.

3.4.9. Let U and V be subspaces of a vector space X. Prove that $U \cap V$ is a subspace of $U + V$.

3.4.10. Show that if A and B have the same number of rows, then $\text{Col}(A) + \text{Col}(B) = \text{Col}[A \;\; B]$. (This fact can be used to find a basis for the sum of subspaces.)

3.4.11. Show that if A and B have the same number of columns, then

$$\text{Null}(A) \cap \text{Null}(B) = \text{Null}\begin{bmatrix} A \\ B \end{bmatrix}.$$

(This fact can be used to find a basis for the intersection of subspaces.)

• **3.4.12.** Let

$$A = \begin{bmatrix} 1 & 1 & 1 \\ 0 & 2 & 0 \\ 0 & 0 & 0 \end{bmatrix} \text{ and } B = \begin{bmatrix} 0 & 1 & -1 \\ 0 & 0 & 0 \\ 0 & 2 & 0 \end{bmatrix}.$$

a. Find a basis for each of $\text{Col}(A)$, $\text{Col}(B)$, and $\text{Col}(A + B)$.

b. Find a basis for each of $\text{Col}(A) + \text{Col}(B)$ and $\text{Col}(A) \cap \text{Col}(B)$.

c. Is $\text{Col}(A + B) = \text{Col}(A) + \text{Col}(B)$?

d. Verify the formula

$$\dim(U + V) = \dim U + \dim V - \dim(U \cap V)$$

in Exercise 3.4.23 for $U = \text{Col}(A)$ and $V = \text{Col}(B)$.

3.4.13. Let U be a subspace of an inner product space X. Prove that $U^{\perp}$ is a subspace of X.

3.4.14. Show that if A and B are matrices such that AB is defined and A is a nonsingular square matrix, then B and AB have the same rank. (*Hint:* Use the result of Exercise 3.3.9 on page 106.)

3.4.15. Use the result of Exercise 3.4.14 to show that row-equivalent matrices have the same rank. (Hint: Use also the results of Exercises 2.3.12, 2.3.13 and 2.3.14 on page 78, suitably generalized.)

3.4.16. Show that if A and B are matrices such that AB is defined, then $\text{Col}(AB)$ is a subspace of $\text{Col}(A)$, and $\text{Row}(AB)$ is a subspace of $\text{Row}(B)$. What does this result imply about the ranks and the nullities of these matrices?

3.4.17. Show that if A and B are matrices such that AB is defined and the columns of B are linearly dependent, then the columns of AB are linearly dependent as well. Is the converse true?

• **3.4.18.** Let A be the matrix of Exercise 3.4.4 above and let U be the subspace of $\mathbb{R}^3$ spanned by the first two columns of A, and V the subspace spanned by the last two columns. Find a basis for each of U, V, $U \cap V$, $U + V$, $U^\perp$, $V^\perp$.

3.4.19. Let

$$A = \begin{bmatrix} 1 & 0 & 1 & 1 & 1 \\ 0 & 2 & 0 & 0 & 1 \\ 0 & 0 & 1 & 2 & 2 \\ 0 & 0 & 0 & 1 & 1 \end{bmatrix}$$

and let U be the subspace of $\mathbb{R}^4$ spanned by the first three columns of A, and V the subspace spanned by the last three columns. Find a basis for each of U, V, $U \cap V$, $U + V$, $U^\perp$, $V^\perp$.

***3.4.20.** Let U and V be finite-dimensional subspaces of a vector space X. Prove that $U + V$ is the subspace generated by $U \cup V$. (*Hint:* Consider a basis for each subspace.)

***3.4.21.** For any subspaces U and V of a vector space X, we call $U + V$ a *direct sum* if $U \cap V = \{\mathbf{0}\}$, and denote it in that case by $U \oplus V$. Show that for any finite-dimensional subspaces with $U \cap V = \{\mathbf{0}\}$ we have $\dim(U \oplus V) = \dim U + \dim V$. (*Hint:* Consider a basis for each subspace.)

***3.4.22.** Show that for any finite-dimensional subspaces U and V of a vector space X the sum $U + V$ is direct if and only if every $\mathbf{x} \in U + V$ can be decomposed uniquely into a sum of a vector $\mathbf{u}$ from U and a vector $\mathbf{v}$ from V; that is, $\mathbf{x} = \mathbf{u} + \mathbf{v}$, with $\mathbf{u}$ and $\mathbf{v}$ uniquely determined by $\mathbf{x}$.

***3.4.23.** Let U and V be subspaces of $\mathbb{R}^n$. Show that

$$\dim(U + V) = \dim U + \dim V - \dim(U \cap V).$$

***3.4.24.** Generalize the definition of sum to more than two subspaces as summands.

***3.4.25.** Find and prove a formula analogous to the one in Exercise 3.4.23 for the sum of three subspaces of $\mathbb{R}^n$.

3.4.26. Let U and V be subspaces of $\mathbb{R}^n$. Show that if they are orthogonal to each other, then $U \cap V = \{\mathbf{0}\}$. Is the converse true? (Explain!)

***3.4.27.** Let U and V be subspaces of $\mathbb{R}^n$. Show that $(U \cap V)^\perp = U^\perp + V^\perp$. (*Hint:* Extend a basis for $U \cap V$ to bases for U, V, and $\mathbb{R}^n$.)

***3.4.28.** Let U and V be subspaces of $\mathbb{R}^n$. Show that $(U + V)^{\perp} = U^{\perp} \cap V^{\perp}$.

***3.4.29.** Let U and V be subspaces of $\mathbb{R}^n$. Show that $U \subset V$ implies $V^{\perp} \subset U^{\perp}$.

***3.4.30.** Let U be a subspace of $\mathbb{R}^n$. Show that $(U^{\perp})^{\perp} = U$.

3.4.31. Find a matrix A such that $\mathrm{Null}(A) = \mathrm{Col}(B)$, where

$$B = \begin{bmatrix} 1 & 1 & 1 \\ 0 & 2 & 0 \\ 0 & 0 & 0 \\ 2 & 1 & 0 \end{bmatrix}.$$

3.4.32. Find a matrix A such that $\mathrm{Null}(A) = \mathrm{Row}(B)$, where

$$B = \begin{bmatrix} 1 & 1 & 1 & 2 & -1 \\ 0 & 2 & 0 & 0 & 4 \\ 0 & 0 & 0 & 2 & 2 \end{bmatrix}.$$

• **3.4.33.** Prove the following alternative algorithm for obtaining a basis for the left nullspace of a matrix:

Let A be any real $m \times n$ matrix of rank r. Consider the block matrix $[A \ I]$, where I is the unit matrix of order m. Reduce it by elementary row operations to a form $\begin{bmatrix} U & L \\ O & M \end{bmatrix}$, in which U is an $r \times n$ echelon matrix and O the $(m - r) \times n$ zero matrix. Then the transposed rows of the $(m - r) \times m$ matrix M form a basis for the left nullspace of A. (*Hint:* For any $\boldsymbol{b} \in \mathrm{Col}(A)$ reduce $A\mathbf{x} = I\boldsymbol{b}$ by elementary row operations, until A is in an echelon form $\begin{bmatrix} U \\ O \end{bmatrix}$, with U having no zero rows. On the right-hand side denote the result of this reduction of the matrix I by $\begin{bmatrix} L \\ M \end{bmatrix}$, and consider the condition for consistency.)

3.4.34. Use the algorithm of Exercise 3.4.33 to solve Exercise 3.4.31.

3.4.35. Modify the algorithm of Exercise 3.4.33 to obtain a new algorithm for finding a basis for $\mathrm{Null}(A)$.

3.4.36. Show that the result of Exercise 3.4.33 can be used to solve $A\mathbf{x} = \boldsymbol{b}$ by the following new algorithm:

Let A be any real $m \times n$ matrix and $\boldsymbol{b} \in \mathrm{Col}(A)$. Reduce the block matrix $\begin{bmatrix} A^T & I \\ -\boldsymbol{b}^T & \boldsymbol{0} \end{bmatrix}$ to the form $\begin{bmatrix} U & L \\ \boldsymbol{0} & \mathbf{x}^T \end{bmatrix}$ by elementary row operations without ever exchanging the last row or multiplying it by a scalar, where U is an $n \times m$ echelon matrix and $\mathbf{x}^T$ a row n-vector. Then the latter is the transpose of a particular solution $\mathbf{x}$. The general solution is given by the sum of this

$\mathbf{x}$ and any linear combination of the transposed rows of L that correspond to the zero rows of U.

3.4.37. Use the algorithm of Exercise 3.4.36 to solve $A\mathbf{x} = \boldsymbol{b}$ with

$$A = \begin{bmatrix} 1 & -2 & 3 & 2 \\ 0 & 3 & -3 & 0 \\ -1 & -1 & 0 & -2 \end{bmatrix} \text{ and } \boldsymbol{b} = \begin{bmatrix} -5 \\ 6 \\ -1 \end{bmatrix}.$$

***3.4.38.** Prove the following generalization of the algorithm of Exercise 3.4.36 to invert a matrix:

Let A be any real, nonsingular $n \times n$ matrix. Consider the block matrix $\begin{bmatrix} A & I \\ -I & O \end{bmatrix}$, where I is the unit matrix of order n and O is the $n \times n$ zero matrix. Row-reduce this block matrix, without exchanging or multiplying any of the last n rows, to a form $\begin{bmatrix} U & L \\ O & M \end{bmatrix}$, in which U is upper triangular. Then $M = A^{-1}$.

3.4.39. Use the algorithm of Exercise 3.4.38 to invert

$$A = \begin{bmatrix} 1 & -2 \\ 3 & 4 \end{bmatrix}.$$

***3.4.40.** Prove that in any n-dimensional vector space X any set of n independent vectors forms a basis. (*Hint:* Use the Exchange Theorem.)

***3.4.41.** Prove that in any n-dimensional vector space X no set of fewer than n vectors spans X. (*Hint:* Use the Exchange Theorem.)

***3.4.42.** Prove that in any n-dimensional vector space X any set of n vectors that spans X forms a basis. (*Hint:* Use the results of Exercise 3.4.41 and Theorem 3.4.1.)

***3.4.43.** Prove that in any n-dimensional vector space X any set of more than n vectors is a dependent set. (*Hint:* Use the Exchange Theorem.)

MATLAB Exercises 3.4

In MATLAB, the method of Exercises 3.4.33 and 3.4.35 can be used to obtain bases for Left-null(A) and Null(A) by computing

$$\textbf{rref}([A, \textbf{eye}(\textbf{size}(A, 1))]),$$

and the same for A' in place of A. The command $N = \textbf{null}(A)$ also returns a (usually different) basis-matrix for Null(A).

ML3.4.1. Let $A = \textbf{magic}(4)$.

a. Use **rref** as explained above to find a basis for Left-null(A) and for Null(A). Extract and transpose the appropriate submatrix in each case to obtain two matrices B and C whose columns form the bases.

b. Use N = **null**(A) and L = **null**(A') to obtain two different bases, and show that they span the same subspaces as B and C, respectively.

c. Use **rref** as in Exercise ML3.3.1 on page 107 to compute bases for Row(A) and Col(A).

d. Decompose $\mathbf{x} = (1, 2, 3, 4)^T$ into a sum of an $\mathbf{x}_0 \in$ Null(A) and an $\mathbf{x}_R \in$ Row(A).

e. Decompose $\mathbf{x}$ into a sum of an $\mathbf{x}_L \in$ Left-null(A) and an $\mathbf{x}_C \in$ Col(A).

ML3.4.2. Repeat Exercise ML3.4.1 for the matrix A = **magic**(8) and

$$\mathbf{x} = (1, 2, 3, 4, 1, 2, 3, 4)^T.$$

ML3.4.3. Let A = **round**(10 * **rand**(3, 4) − 5).

a. Compute the nullity of A, A^T, A^TA, and AA^T each.

b. Repeat Part (a) for six instances of A.

c. Do you see any patterns? Make a conjecture and prove it.

3.5. CHANGE OF BASIS

The notion of a basis is a generalization of that of a coordinate system, and just as we sometimes need to change coordinate systems, so too do we sometimes need to change bases to make equations simpler. For example, changing bases will be indispensable in the so-called eigenvalue problems in Chapter 7, which arise in the evolution of many physical systems.

Let X be an n-dimensional vector space and $\mathscr{A} = \{\boldsymbol{a}_1, \boldsymbol{a}_2, \ldots, \boldsymbol{a}_n\}$ a basis for X. Then any vector $\mathbf{x}$ in X can be written uniquely as $\mathbf{x} = \sum_{i=1}^{n} x_{Ai}\boldsymbol{a}_i$. We have here an ordering of $\mathscr{A}$ implicit in the subscripts, and the ordered n-tuple $A = (\boldsymbol{a}_1, \boldsymbol{a}_2, \ldots, \boldsymbol{a}_n)$ is sometimes called an *ordered basis* for X. The transposed ordered n-tuple $\mathbf{x}_A = (x_{A1}, x_{A2}, \ldots, x_{An})^T$ is called the *coordinate vector of* $\mathbf{x}$ *relative to* A, and is a vector of $\mathbb{R}^n$.

We begin our discussion of changing bases in the special case of $X = \mathbb{R}^n$, and consider general vector spaces afterward. Working with $\mathbb{R}^n$ enables us to consider the ordered basis $A = (\boldsymbol{a}_1, \boldsymbol{a}_2, \ldots, \boldsymbol{a}_n)$ of $\mathbb{R}^n$ to be a matrix, called a *basis-matrix*. It is, of course, of size $n \times n$. Similarly, let $B = (\boldsymbol{b}_1, \boldsymbol{b}_2, \ldots, \boldsymbol{b}_n)$ be another basis-matrix of $\mathbb{R}^n$. Then any vector $\mathbf{x}$ of $\mathbb{R}^n$ can be decomposed as $\mathbf{x} = \sum_{i=1}^{n} x_{Ai}\boldsymbol{a}_i$ and also as $\mathbf{x} = \sum_{i=1}^{n} x_{Bi}\boldsymbol{b}_i$.

Using the basis-matrices we may write the equations above as

$$\mathbf{x} = A\mathbf{x}_A \quad \text{and} \quad \mathbf{x} = B\mathbf{x}_B. \tag{3.5.1}$$

For given A, B, and any $\mathbf{x}$, the Equations (3.5.1) can be solved for $\mathbf{x}_A$ and $\mathbf{x}_B$, and therefore, by Theorem 2.3.3 on page 74, the matrices A and B must be invertible. Thus we can solve $A\mathbf{x}_A = B\mathbf{x}_B$ as

$$\mathbf{x}_A = A^{-1}B\mathbf{x}_B. \tag{3.5.2}$$

The matrix

$$S = A^{-1}B \tag{3.5.3}$$

is usually referred to as the *transition matrix* or *change of basis matrix* from the basis A to the basis B, because Equation (3.5.3) can be solved to give this change as

$$B = AS. \tag{3.5.4}$$

The matrix S, being the product of two invertible matrices, must also be invertible, and we have $S^{-1} = B^{-1}A$.

Using the definition of S we can write Equation (3.5.2) as

$$\mathbf{x}_A = S\mathbf{x}_B \tag{3.5.5}$$

and multiplying both sides by S^{-1} we can also write

$$\mathbf{x}_B = S^{-1}\mathbf{x}_A. \tag{3.5.6}$$

Notice the formal difference between Equations (3.5.4) and (3.5.6): whereas the basis-matrix A transforms with S on the right, the corresponding coordinate vector $\mathbf{x}_A$ transforms with S^{-1} on the left.

Let us summarize the discussion above in a theorem:

THEOREM 3.5.1. If A and B are $n \times n$ matrices whose columns form two bases of $\mathbb{R}^n$, then A and B are invertible and there exists an invertible matrix S such that $B = AS$, and so $S = A^{-1}B$. Furthermore, the coordinate vectors $\mathbf{x}_A$ and $\mathbf{x}_B$ of any $\mathbf{x} \in \mathbb{R}^n$ are related by Equations (3.5.5) and (3.5.6).

COROLLARY 3.5.1. In the important particular case of a transition from the standard basis I to a basis B, replacing A in Equation (3.5.4) by I, we get $\mathbf{x}_A = \mathbf{x}_I = \mathbf{x}$, $B = IS = S$ and Equations (3.5.5) and (3.5.6) become $\mathbf{x} = B\mathbf{x}_B$ and $\mathbf{x}_B = B^{-1}\mathbf{x}$. Let us emphasize that where the standard vectors transform with B, the coordinate vectors transform with B^{-1}; that is,

$$\boldsymbol{b}_i = B\boldsymbol{e}_i \quad \text{and} \quad \mathbf{x}_B = B^{-1}\mathbf{x}. \tag{3.5.7}$$

EXAMPLE 3.5.1. In $\mathbb{R}^2$ let us change from the standard basis $\{\boldsymbol{i}, \boldsymbol{j}\}$ to the basis $\{b_1, b_2\}$, where $b_1 = \frac{\sqrt{2}}{2}\begin{bmatrix}1\\1\end{bmatrix}$ and $b_2 = \frac{\sqrt{2}}{2}\begin{bmatrix}-1\\1\end{bmatrix}$.

This new basis is obtained from the old one by a 45° rotation. (See Figure 3.1.)

FIGURE 3.1

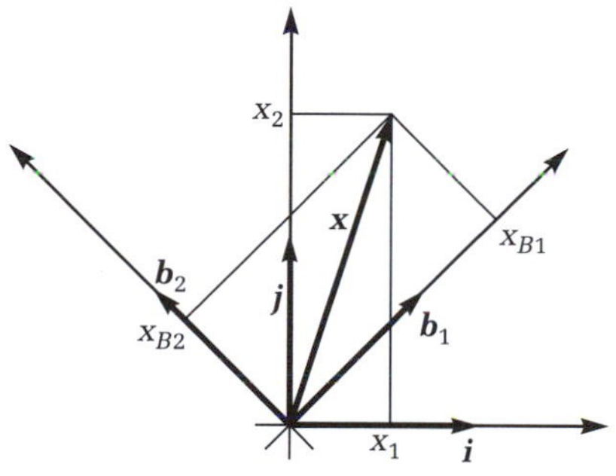

Then the transition matrix is

$$B = \frac{\sqrt{2}}{2}\begin{bmatrix}1 & -1\\1 & 1\end{bmatrix} \tag{3.5.8}$$

and

$$B^{-1} = \frac{\sqrt{2}}{2}\begin{bmatrix}1 & 1\\-1 & 1\end{bmatrix}. \tag{3.5.9}$$

Hence for any vector $\mathbf{x} = (x_1, x_2)^T$ we have

$$\mathbf{x}_B = B^{-1}\mathbf{x} = \frac{\sqrt{2}}{2}\begin{bmatrix}1 & 1\\-1 & 1\end{bmatrix}\begin{bmatrix}x_1\\x_2\end{bmatrix} = \frac{\sqrt{2}}{2}\begin{bmatrix}x_1 + x_2\\x_2 - x_1\end{bmatrix}. \tag{3.5.10}$$

<<>>

EXAMPLE 3.5.2. (a) Find the transition matrix S for the change of basis in $\mathbb{R}^3$ from $(\boldsymbol{a}_1, \boldsymbol{a}_2, \boldsymbol{a}_3)$ to $(\boldsymbol{b}_1, \boldsymbol{b}_2, \boldsymbol{b}_3)$, where $\boldsymbol{a}_1 = (1, 2, 0)^T$, $\boldsymbol{a}_2 = (0, 1, 3)^T$, $\boldsymbol{a}_3 = (0, 0, 1)^T$, $\boldsymbol{b}_1 = (1, 0, 1)^T$, $\boldsymbol{b}_2 = (1, 1, 0)^T$, $\boldsymbol{b}_3 = (0, 1, 1)^T$.
(b) Use S to write $\mathbf{x} = -\boldsymbol{b}_1 + 2\boldsymbol{b}_2 + 3\boldsymbol{b}_3$ as $x_{A1}\boldsymbol{a}_1 + x_{A2}\boldsymbol{a}_2 + x_{A3}\boldsymbol{a}_3$.

Solution: The basis-matrix A is given by

$$A = \begin{bmatrix} 1 & 0 & 0 \\ 2 & 1 & 0 \\ 0 & 3 & 1 \end{bmatrix} \tag{3.5.11}$$

and the basis-matrix B by

$$B = \begin{bmatrix} 1 & 1 & 0 \\ 0 & 1 & 1 \\ 1 & 0 & 1 \end{bmatrix}. \tag{3.5.12}$$

Thus

$$A^{-1} = \begin{bmatrix} 1 & 0 & 0 \\ -2 & 1 & 0 \\ 6 & -3 & 1 \end{bmatrix} \tag{3.5.13}$$

and so the change of basis-matrix is

$$S = A^{-1}B = \begin{bmatrix} 1 & 1 & 0 \\ -2 & -1 & 1 \\ 7 & 3 & -2 \end{bmatrix}. \tag{3.5.14}$$

Consequently,

$$\mathbf{x}_A = S\mathbf{x}_B = \begin{bmatrix} 1 & 1 & 0 \\ -2 & -1 & 1 \\ 7 & 3 & -2 \end{bmatrix} \begin{bmatrix} -1 \\ 2 \\ 3 \end{bmatrix} = \begin{bmatrix} 1 \\ 3 \\ -7 \end{bmatrix} \tag{3.5.15}$$

and

$$\mathbf{x} = \boldsymbol{a}_1 + 3\boldsymbol{a}_2 - 7\boldsymbol{a}_3. \tag{3.5.16}$$

<<>>

All the formulas in the previous discussion except those involving A^{-1} and B^{-1} can also be given a meaning in an *arbitrary finite dimensional vector space* V, not just in $\mathbb{R}^n$, by considering A and B as ordered bases of V and writing $A = (\boldsymbol{a}_1, \boldsymbol{a}_2, \ldots, \boldsymbol{a}_n)$ and $B = (\boldsymbol{b}_1, \boldsymbol{b}_2, \ldots, \boldsymbol{b}_n)$.

We now start with the formula $B = AS$, which may be regarded as shorthand for

$$\mathbf{b}_j = \sum_{i=1}^{n} \boldsymbol{a}_i s_{ij} \text{ for } j = 1, \dots, n. \tag{3.5.17}$$

This equation expresses each vector $\mathbf{b}_j$ as a linear combination of the $\boldsymbol{a}_i$ vectors, and shows that s_{ij} is the ith component of the coordinate vector $\mathbf{b}_{jA}$ of $\mathbf{b}_j$ relative to the basis A. Thus for the jth column of S we have

$$\mathbf{s}_j = \mathbf{b}_{jA}. \tag{3.5.18}$$

The expressions $\mathbf{x} = A\mathbf{x}_A$ and $\mathbf{x} = B\mathbf{x}_B$ of any vector relative to the bases A and B can now be regarded as shorthand for

$$\mathbf{x} = \sum_{i=1}^{n} x_{Ai}\boldsymbol{a}_i \tag{3.5.19}$$

and

$$\mathbf{x} = \sum_{j=1}^{n} x_{Bj}\mathbf{b}_j. \tag{3.5.20}$$

Substituting the expression of $\mathbf{b}_j$ from Equation (3.5.17) into this, we get

$$\mathbf{x} = \sum_{i=1}^{n} \sum_{j=1}^{n} s_{ij} x_{Bj} \boldsymbol{a}_i. \tag{3.5.21}$$

Because the $\boldsymbol{a}_i$ vectors form a basis, the coefficients in the two expressions (3.5.19) and (3.5.21) of $\mathbf{x}$ relative to the basis A must be equal:

$$x_{Ai} = \sum_{j=1}^{n} s_{ij} x_{Bj}, \tag{3.5.22}$$

or equivalently

$$\mathbf{x}_A = S\mathbf{x}_B. \tag{3.5.23}$$

We can also show that the matrix S must again be invertible and S^{-1} gives the transformation in the reverse direction. To this end, write the $\boldsymbol{a}_k$ vectors as linear combinations of the $\mathbf{b}_j$ vectors:

$$\boldsymbol{a}_k = \sum_{j=1}^{n} \mathbf{b}_j t_{jk} \text{ for } k = 1, \dots, n. \tag{3.5.24}$$

If we substitute the expression (3.5.17) of the $\mathbf{b}_j$ vectors into this equation and equate corresponding coefficients of the $\boldsymbol{a}_i$ vectors on the two sides, then we get

$$\sum_{j=1}^{n} s_{ij}t_{jk} = \delta_{ik} \text{ for } i, k = 1, \ldots, n. \tag{3.5.25}$$

In matrix notation this relation becomes $ST = I$, and so S must be invertible and $T = S^{-1}$. Multiplying both sides of Equation (3.5.23) by S^{-1}, we get

$$\mathbf{x}_B = S^{-1}\mathbf{x}_A. \tag{3.5.26}$$

Also, using $T = S^{-1}$, we can abbreviate Equation (3.5.24) as $A = BS^{-1}$.

Thus we have proved the following generalization of Theorem 3.5.1:

THEOREM 3.5.2. If $A = (\boldsymbol{a}_1, \boldsymbol{a}_2, \ldots, \boldsymbol{a}_n)$ and $B = (\boldsymbol{b}_1, \boldsymbol{b}_2, \ldots, \boldsymbol{b}_n)$ are two ordered bases of a vector space V, then the matrix S whose columns are given by $\boldsymbol{s}_j = \boldsymbol{b}_{jA}$ is invertible and relates the coordinate vectors $\mathbf{x}_A$ and $\mathbf{x}_B$ of any $\boldsymbol{x} \in V$ by $\mathbf{x}_A = S\mathbf{x}_B$ and $\mathbf{x}_B = S^{-1}\mathbf{x}_A$.

EXAMPLE 3.5.3. Consider the two-dimensional subspace V of $\mathbb{R}^3$ spanned by the columns of

$$A = \begin{bmatrix} 1 & 0 \\ 2 & 1 \\ 2 & -1 \end{bmatrix}. \tag{3.5.27}$$

Another basis for V is given by the columns of

$$B = \begin{bmatrix} 1 & 1 \\ 0 & 1 \\ 4 & 3 \end{bmatrix}. \tag{3.5.28}$$

Find the transition matrix S from the basis A to the basis B.

In this case we have no problem with considering A and B to be matrices. Furthermore, by Equation (3.5.17), we must have $AS = B$ with S being 2×2. We can solve this equation for the unknown matrix S by Gauss-Jordan elimination (see Exercise 3.5.8), pretty much as we obtained the inverse of a matrix:

$$\left[\begin{array}{cc|cc} 1 & 0 & 1 & 1 \\ 2 & 1 & 0 & 1 \\ 2 & -1 & 4 & 3 \end{array}\right] \to \left[\begin{array}{cc|cc} 1 & 0 & 1 & 1 \\ 0 & 1 & -2 & -1 \\ 0 & -1 & 2 & 1 \end{array}\right] \to \left[\begin{array}{cc|cc} 1 & 0 & 1 & 1 \\ 0 & 1 & -2 & -1 \\ 0 & 0 & 0 & 0 \end{array}\right]. \tag{3.5.29}$$

Thus,

$$S = \begin{bmatrix} 1 & 1 \\ -2 & -1 \end{bmatrix}. \tag{3.5.30}$$

<<>>

EXAMPLE 3.5.4. In quantum mechanics the simplest solutions of the differential equation of the harmonic oscillator involve the so-called Hermite polynomials. For the purpose of finding more general solutions it is important to be able to rewrite a given polynomial as a linear combination of Hermite polynomials. Although this problem is handled generally by introducing an appropriate inner product in $\mathcal{P}$ and exploiting the orthogonality of the Hermite polynomials, in simple cases like the following one we can also use the present method.

The first four Hermite polynomials are given by $H_0(x) = 1$, $H_1(x) = 2x$, $H_2(x) = 4x^2 - 2$ and $H_3(x) = 8x^3 - 12x$. We want to find a formula that gives an arbitrary third-degree polynomial as a linear combination of these polynomials. In the space

$$\mathcal{P}_3 = \{\boldsymbol{p} = P : P(x) = p_0 + p_1 x + p_2 x^2 + p_3 x^3;\ p_0, p_1, p_2, p_3 \in \mathbb{R}\}$$

we choose the basis A to consist of the monomials; that is, $\boldsymbol{a}_i = x^i$ for $i = 0, \ldots, 3$, and the basis B to consist of the first four Hermite polynomials; that is, $\boldsymbol{b}_i = H_i$ for $i = 0, \ldots, 3$. (We could show directly that these are independent and so form a basis for $\mathcal{P}_3$, but this result follows, by Theorem 2.3.3 on page 74, also from the fact shown below that we can find the coordinates of any $\boldsymbol{p}$ relative to B.) Then, according to Theorem 3.5.2, the columns of the matrix S are given by the coordinates of the $\boldsymbol{b}_i$ vectors relative to A. These can be read off the definitions of the Hermite polynomials, to give (in ascending order of degrees)

$$S = \begin{bmatrix} 1 & 0 & -2 & 0 \\ 0 & 2 & 0 & -12 \\ 0 & 0 & 4 & 0 \\ 0 & 0 & 0 & 8 \end{bmatrix}. \tag{3.5.31}$$

The coordinate vector of any $\boldsymbol{p}$ relative to A is $\boldsymbol{p}_A = (p_0, p_1, p_2, p_3)^T$ and its coordinate vector relative to B is given by $\boldsymbol{p}_B = S^{-1}\boldsymbol{p}_A$. Thus multiplication of $\boldsymbol{p}_A$ by

$$S^{-1} = \frac{1}{8}\begin{bmatrix} 8 & 0 & 4 & 0 \\ 0 & 4 & 0 & 6 \\ 0 & 0 & 2 & 0 \\ 0 & 0 & 0 & 1 \end{bmatrix} \tag{3.5.32}$$

will give the coordinates of any $\boldsymbol{p}$ relative to B. For instance, for the polynomial $P(x) = 1 + 2x + 16x^3$, we have

$$\frac{1}{8}\begin{bmatrix} 8 & 0 & 4 & 0 \\ 0 & 4 & 0 & 6 \\ 0 & 0 & 2 & 0 \\ 0 & 0 & 0 & 1 \end{bmatrix}\begin{bmatrix} 1 \\ 2 \\ 0 \\ 16 \end{bmatrix} = \begin{bmatrix} 1 \\ 13 \\ 0 \\ 2 \end{bmatrix}. \tag{3.5.33}$$

Thus

$$P(x) = H_0(x) + 13H_1(x) + 2H_3(x);$$

that is,

$$1 + 2x + 16x^3 = 1 + 13(2x) + 2(8x^3 - 12x),$$

which is obviously true.

<<>>

We have just seen how vectors can be described in terms of their components relative to a basis, and how a change of basis affects those components. We can similarly define the components of a matrix relative to a basis and examine how they change with a change of basis. We consider square matrices only, since it is mainly for those that this procedure becomes necessary in applications. (Foe nonsquare matrices see Exercise 3.5.13.)

Let M be an $n \times n$ matrix and consider the associated mapping

$$\boldsymbol{y} = M\mathbf{x} \tag{3.5.34}$$

of $\mathbb{R}^n$ to itself. Let us write $\mathbf{x}$ and $\boldsymbol{y}$ in terms of the ordered basis $A = (\boldsymbol{a}_1, \boldsymbol{a}_2, \ldots, \boldsymbol{a}_n)$ of $\mathbb{R}^n$ as $\mathbf{x} = A\mathbf{x}_A$ and $\boldsymbol{y} = A\boldsymbol{y}_A$. Substituting from these into $\boldsymbol{y} = M\mathbf{x}$, we get $A\boldsymbol{y}_A = MA\mathbf{x}_A$, and multiplying both sides by A^{-1},

$$\boldsymbol{y}_A = A^{-1}MA\mathbf{x}_A. \tag{3.5.35}$$

We call the matrix $M_A = A^{-1}MA$ the matrix *representing M with respect to the basis A*, because M_A represents the *same mapping* of $\mathbf{x}$ to $\boldsymbol{y}$ in terms of the basis A as M does in terms of the standard basis. Similarly, if we introduce a second ordered basis $B = (\boldsymbol{b}_1, \boldsymbol{b}_2, \ldots, \boldsymbol{b}_n)$ into $\mathbb{R}^n$, then $M_B = B^{-1}MB$ represents M with respect to the basis B. With these notations we have

$$\boldsymbol{y}_A = M_A\mathbf{x}_A \tag{3.5.36}$$

and

$$\boldsymbol{y}_B = M_B\mathbf{x}_B. \tag{3.5.37}$$

Substituting $\mathbf{x}_A = S\mathbf{x}_B$ and $\boldsymbol{y}_A = S\boldsymbol{y}_B$ into Equation (3.5.36) we obtain

$$S\boldsymbol{y}_B = M_AS\mathbf{x}_B \tag{3.5.38}$$

and then

$$\boldsymbol{y}_B = S^{-1}M_AS\mathbf{x}_B. \tag{3.5.39}$$

Comparing Equations (3.5.37) and (3.5.39) we find (why?) that

$$M_B = S^{-1}M_AS. \tag{3.5.40}$$

Again, we restate our findings as a theorem:

THEOREM 3.5.3. If A and B are $n \times n$ matrices whose columns form two bases of $\mathbb{R}^n$ and M is an $n \times n$ matrix representing the mapping $\boldsymbol{y} = M\mathbf{x}$ of $\mathbb{R}^n$ to itself, then $M_A = A^{-1}MA$ is the matrix representing M with respect to the basis A and so too is $M_B = B^{-1}MB$ with respect to the basis B. Furthermore, M_A and M_B are related by Equation (3.5.40), where $S = A^{-1}B$.

The transformation of M_A to M_B given by Equation (3.5.40) with any invertible matrix S is called a *similarity transformation,* and M_A and M_B are said to be similar to each other. Observe that the formula $M_B = B^{-1}MB$ describes a similarity transformation, too. It transforms the representation M of our mapping in the standard basis to its representation M_B relative to the basis B. In this case the transition matrix S happens to be B, as we have seen in Corollary 3.5.1.

EXAMPLE 3.5.5. In $\mathbb{R}^2$ let us change from the standard basis $\{\boldsymbol{i}, \boldsymbol{j}\}$ to the basis $\{b_1, b_2\}$, where $b_1 = \frac{\sqrt{2}}{2}\begin{bmatrix}1\\1\end{bmatrix}$ and $b_2 = \frac{\sqrt{2}}{2}\begin{bmatrix}-1\\1\end{bmatrix}$, as in Example 3.5.1.

Let us see how the matrix

$$M = \begin{bmatrix}0 & 1\\1 & 0\end{bmatrix} \tag{3.5.41}$$

changes with the change from the standard basis to the rotated one described above. This matrix transforms any vector $\mathbf{x} = (x_1, x_2)^T$ to $M\mathbf{x} = (x_1, x_2)^T$, and so it represents the reflection across the $x_2 = x_1$ line. Then

$$M_B = B^{-1}MB = \frac{1}{2}\begin{bmatrix}1 & 1\\-1 & 1\end{bmatrix}\begin{bmatrix}0 & 1\\1 & 0\end{bmatrix}\begin{bmatrix}1 & -1\\1 & 1\end{bmatrix} = \begin{bmatrix}1 & 0\\0 & -1\end{bmatrix}. \tag{3.5.42}$$

Applied to any vector $\mathbf{x}_B$ this gives

$$\boldsymbol{y}_B = M_B\mathbf{x}_B = \begin{bmatrix}1 & 0\\0 & -1\end{bmatrix}\begin{bmatrix}x_{B1}\\x_{B2}\end{bmatrix} = \begin{bmatrix}x_{B1}\\-x_{B2}\end{bmatrix}. \tag{3.5.43}$$

This result shows that M_B represents the reflection across the x_{B1}-axis, as it should, for that axis is the $x_2 = x_1$ line.

<<>>

EXAMPLE 3.5.6. In $\mathbb{R}^2$ let us change from the standard basis $\{\boldsymbol{i}, \boldsymbol{j}\}$ to the basis $\{\boldsymbol{a}_1, \boldsymbol{a}_2\}$, where $\boldsymbol{a}_1 = \boldsymbol{i} = (1, 0)^T$ and $\boldsymbol{a}_2 = (1, 1)^T$. Then the transition matrix is

$$A = \begin{bmatrix} 1 & 1 \\ 0 & 1 \end{bmatrix} \tag{3.5.44}$$

and

$$A^{-1} = \begin{bmatrix} 1 & -1 \\ 0 & 1 \end{bmatrix}. \tag{3.5.45}$$

Hence for any vector $\mathbf{x} = (x_1, x_2)^T$ we have

$$\mathbf{x}_A = A^{-1}\mathbf{x} = \begin{bmatrix} 1 & -1 \\ 0 & 1 \end{bmatrix} \begin{bmatrix} x_1 \\ x_2 \end{bmatrix} = \begin{bmatrix} x_1 - x_2 \\ x_2 \end{bmatrix} \tag{3.5.46}$$

and

$$\mathbf{x} = (x_1 - x_2)\boldsymbol{a}_1 + x_2\boldsymbol{a}_2. \tag{3.5.47}$$

We can check that indeed,

$$(x_1 - x_2)\begin{bmatrix} 1 \\ 0 \end{bmatrix} + x_2\begin{bmatrix} 1 \\ 1 \end{bmatrix} = \begin{bmatrix} x_1 - x_2 \\ 0 \end{bmatrix} + \begin{bmatrix} x_2 \\ x_2 \end{bmatrix} = \begin{bmatrix} x_1 \\ x_2 \end{bmatrix} = \mathbf{x}. \tag{3.5.48}$$

Next, let us see how the rotation matrix

$$M = \begin{bmatrix} \cos\theta & -\sin\theta \\ \sin\theta & \cos\theta \end{bmatrix} \tag{3.5.49}$$

of Example 2.2.1 on page 54 is represented in the new basis. We find that

$$\begin{aligned} M_A = A^{-1}MA &= \begin{bmatrix} 1 & -1 \\ 0 & 1 \end{bmatrix} \begin{bmatrix} \cos\theta & -\sin\theta \\ \sin\theta & \cos\theta \end{bmatrix} \begin{bmatrix} 1 & 1 \\ 0 & 1 \end{bmatrix} \\ &= \begin{bmatrix} \cos\theta - \sin\theta & -2\sin\theta \\ -\sin\theta & -\sin\theta + \cos\theta \end{bmatrix}. \end{aligned} \tag{3.5.50}$$

Exercises 3.5

3.5.1. Let $\boldsymbol{a}_1 = (1, 2)^T$ and $\boldsymbol{a}_2 = (-2, 1)^T$ be the vectors of the ordered basis A for $\mathbb{R}^2$.

a. Find the transition matrix S for the change from the standard basis to A.

b. Use S to find the components of $\mathbf{x} = (3, 5)^T$ in the basis A.

• **3.5.2.** Let $\boldsymbol{a}_1 = (1, 1, 2)^T$, $\boldsymbol{a}_2 = (0, -2, 1)^T$ and $\boldsymbol{a}_3 = (1, 1, 0)^T$ be the vectors of the ordered basis A for $\mathbb{R}^3$.

a. Find the transition matrix S for the change from the standard basis to A.

b. What is the transition matrix for the change from the basis A to the standard basis?

c. Use S to find the components of $\mathbf{x} = (3, 4, 5)^T$ in the basis A.

3.5.3. **a.** Find the transition matrix S corresponding to the change of basis in $\mathbb{R}^2$ from $(\boldsymbol{a}_1, \boldsymbol{a}_2)$ to $(\boldsymbol{b}_1, \boldsymbol{b}_2)$, where $\boldsymbol{a}_1 = (1, 2)^T$, $\boldsymbol{a}_2 = (-2, 1)^T$, $\boldsymbol{b}_1 = (3, 2)^T$, $\boldsymbol{b}_2 = (1, 1)^T$.

b. Use S to write $\mathbf{x} = 3\boldsymbol{b}_1 - 2\boldsymbol{b}_2$ as $x_{A1}\boldsymbol{a}_1 + x_{A2}\boldsymbol{a}_2$.

c. Use S to write $\mathbf{x} = 2\boldsymbol{a}_1 + 4\boldsymbol{a}_2$ as $x_{B1}\boldsymbol{b}_1 + x_{B2}\boldsymbol{b}_2$.

3.5.4. **a.** Find the transition matrix S corresponding to the change of basis in $\mathbb{R}^3$ from $(\boldsymbol{a}_1, \boldsymbol{a}_2, \boldsymbol{a}_3)$ to $(\boldsymbol{b}_1, \boldsymbol{b}_2, \boldsymbol{b}_3)$, where $\boldsymbol{a}_1 = (1, 2, 0)^T$, $\boldsymbol{a}_2 = (-2, 1, 0)^T$, $\boldsymbol{a}_3 = (0, 0, 1)^T$, $\boldsymbol{b}_1 = (3, 2, 0)^T$, $\boldsymbol{b}_2 = (1, 1, 0)^T$, $\boldsymbol{b}_3 = (0, 1, 1)^T$.

b. Use S to write $\mathbf{x} = 2\boldsymbol{a}_1 + 4\boldsymbol{a}_2 + 3\boldsymbol{a}_3$ as $x_{B1}\boldsymbol{b}_1 + x_{B2}\boldsymbol{b}_2 + x_{B3}\boldsymbol{b}_3$.

• **3.5.5.** In $\mathbb{R}^3$ let $x_{A1} = x_2 - x_3$, $x_{A2} = x_3 - x_1$, $x_{A3} = x_1 + x_2$ give the transformation of the coordinates of a vector $\mathbf{x}$ upon a change from the standard basis $\{\boldsymbol{i}, \boldsymbol{j}, \boldsymbol{k}\}$ to a new basis $\{\boldsymbol{a}_1, \boldsymbol{a}_2, \boldsymbol{a}_3\}$. Find the new basis vectors $\boldsymbol{a}_1, \boldsymbol{a}_2, \boldsymbol{a}_3$.

• **3.5.6.** In $\mathbb{R}^2$ the matrix

$$M = \begin{bmatrix} 1 & 0 \\ 0 & 0 \end{bmatrix} \tag{3.5.51}$$

represents the projection onto the x_1-axis. Find its representation M_A relative to the basis $(\boldsymbol{a}_1, \boldsymbol{a}_2)$ obtained from the standard basis by a rotation through an angle θ.

3.5.7. Let $\boldsymbol{a}_1 = \mathbf{e}_3$, $\boldsymbol{a}_2 = \mathbf{e}_1$, and $\boldsymbol{a}_3 = \mathbf{e}_2$ be the vectors of the ordered basis A for $\mathbb{R}^3$.

a. Find the transition matrix S for the change from the standard basis to A.

b. Use S to find the components of $\mathbf{x} = (3, 4, 5)^T$ in the basis A.

c. Find the representation M_A of the matrix

$$M = \begin{bmatrix} 1 & 2 & 3 \\ 1 & 2 & 0 \\ 1 & 0 & 0 \end{bmatrix} \tag{3.5.52}$$

relative to the basis A.

3.5.8. Explain why the reduction of the augmented matrix $[A|B]$ in Example 3.5.3 solves $AS = B$.

• **3.5.9.** Let V be the subspace of $\mathbb{R}^3$ spanned by the columns of

$$A = \begin{bmatrix} 1 & 0 \\ 2 & 4 \\ 3 & -1 \end{bmatrix} \text{ and } B = \begin{bmatrix} 2 & 3 \\ 0 & -2 \\ 7 & 11 \end{bmatrix}.$$

a. Find the transition matrix from the basis A to the basis B.

b. Find the transition matrix from B to A.

3.5.10. Find a basis C for the space V of Exercise 3.5.9 so that $\boldsymbol{c}_1 = \boldsymbol{a}_1$ and $\boldsymbol{c}_2 \perp \boldsymbol{a}_1$.

3.5.11. **a.** Find a formula that gives an arbitrary third-degree polynomial as a linear combination of the first four, so-called Legendre polynomials:

$$L_0(x) = 1,\ L_1(x) = \mathbf{x},\ L_2(x) = \frac{1}{2}(3x^2 - 1), \text{ and } L_3(x) = \frac{1}{2}(5x^3 - 3x).$$

b. Express the polynomial $P(x) = 1 - 2x + 3x^2 - 4x^3$ as a linear combination of these.

3.5.12. Let L and M be $n \times n$ matrices and A the matrix of an ordered basis of $\mathbb{R}^n$. Show that

a. $(L + M)_A = L_A + M_A$, and

b. $(LM)_A = L_A M_A$.

3.5.13. Suppose M is an $m \times n$ matrix, representing a mapping from $\mathbb{R}^n$ to $\mathbb{R}^m$. If we change from the standard basis I_n in $\mathbb{R}^n$ to an ordered basis A, and from I_m in $\mathbb{R}^m$ to some B, then find the representation $M_{A,B}$ of M relative to the new bases.

• **3.5.14.** Show that all matrices of the form

$$M(t) = \begin{bmatrix} 1 & t \\ 0 & 2 \end{bmatrix} \tag{3.5.53}$$

are similar to one another. (*Hint:* For any values t and t', try to find an invertible matrix S, depending on t and t', such that $SM(t) = M(t')S$ holds.)

3.5.15. Show that $\begin{bmatrix} 1 & 1 \\ 0 & 1 \end{bmatrix}$ is not similar to $\begin{bmatrix} 1 & 0 \\ 0 & 1 \end{bmatrix}$.

3.5.16. Is $\begin{bmatrix} 1 & 0 \\ 0 & 2 \end{bmatrix}$ similar to $\begin{bmatrix} 2 & 0 \\ 0 & 1 \end{bmatrix}$?

3.5.17. Show that if A and B are similar matrices, then

a. A^T and B^T are similar,

b. A^k and B^k are similar for any positive integer k, and

c. If additionally A is invertible, then so too is B, and A^{-1} and B^{-1} are similar as well.

***3.5.18.** Show that similarity of matrices is an equivalence relation. (A relation $\sim$ is called an equivalence relation on a set X if it is:

1. Reflexive, that is, $A \sim A$ holds for all A in X,

2. Symmetric, that is, $A \sim B$ implies $B \sim A$ for all A, B in X, and

3. Transitive, that is, $A \sim B$ and $B \sim C$ imply $A \sim C$ for all A, B, C in X.)

3.5.19. Show that if A and B are similar matrices, then $\operatorname{rank}(A) = \operatorname{rank}(B)$. (*Hint:* Show first that if A and B are similar, then their nullspaces have the same dimension.)

3.5.20. The sum of the diagonal elements of a square matrix is called its *trace*: for an $n \times n$ matrix

$$\operatorname{Tr}(A) = \sum_{i=1}^{n} a_{ii}.$$

a. Show that if A and B are $n \times n$ matrices, then $\operatorname{Tr}(AB) = \operatorname{Tr}(BA)$.

b. Apply the result of Part (a) to S and BS^{-1} to show that if A and B are similar so that $A = SBS^{-1}$, then $\operatorname{Tr}(A) = \operatorname{Tr}(B)$.

c. Is $\begin{bmatrix} 1 & 2 \\ 4 & 3 \end{bmatrix}$ similar to $\begin{bmatrix} 1 & 2 \\ 3 & 4 \end{bmatrix}$?

MATLAB Exercises 3.5

ML3.5.1. Let

$$A = \begin{bmatrix} 1 & 0 & 1 & 0 \\ 3 & 4 & 0 & 1 \\ 0 & 0 & 2 & 0 \\ 0 & -1 & 0 & 0 \end{bmatrix}$$

be a basis matrix for $\mathbb{R}^4$. Find the coordinate vector $\mathbf{x}_A$ relative to this basis for each of the $\mathbf{x}$ vectors $\mathbf{e}_1$, $\mathbf{e}_4$, $(1, 1, 1, 1)^T$, and $(1, 2, 3, 4)^T$.

ML3.5.2. **a.** Find the transition matrix S from the basis A in Exercise ML3.5.1 to the standard basis.

b. Use this S to compute $\mathbf{x}$ if $\mathbf{x}_A$ is $\mathbf{e}_1$, $\mathbf{e}_4$, $(1, 1, 1, 1)^T$, or $(1, 2, 3, 4)^T$.

ML3.5.3. Let

$$B = \begin{bmatrix} 1 & 0 & 1 & 4 \\ 3 & 4 & 0 & 3 \\ 2 & 4 & 4 & 6 \\ 0 & -1 & 0 & 1 \end{bmatrix}$$

be a basis-matrix for $\mathbb{R}^4$.

a. Find the transition matrix S from the basis A in Exercise ML3.5.1 to this basis.

b. Find the transition matrix S' from the basis B to the basis A.

c. Use S or S' to compute $\mathbf{x}_B$ if $\mathbf{x}_A$ is $\mathbf{e}_1$, $\mathbf{e}_4$, $(1, 1, 1, 1)^T$, or $(1, 2, 3, 4)^T$.

d. Find the representative matrix A_B of the matrix A relative to the basis B.

e. Find the representative matrix B_A of the matrix B relative to the basis A.

ML3.5.4. Let

$$A = \begin{bmatrix} 1 & 0 & 1 & 4 & -1 & 0 \\ 3 & 4 & 0 & 3 & 2 & 1 \\ 1 & 2 & 2 & 3 & 0 & 3 \\ 0 & 0 & 2 & 2 & 0 & 0 \\ 2 & 4 & 4 & 6 & 0 & 6 \\ 0 & -1 & 0 & 1 & 1 & 0 \end{bmatrix}.$$

Use the MATLAB command **rref** to find a basis-matrix for Col(A). The command C = **orth**(A) creates another basis matrix C for Col(A). Find the transition matrix S from the basis B to the basis C. (Hint: By Theorem 3.5.2 you need to solve $BS = C$.)

<<4>> LINEAR TRANSFORMATIONS

4.1. REPRESENTATION OF LINEAR TRANSFORMATIONS BY MATRICES

Beginning with our first discussion of matrix operations, we have seen matrices being used to represent mappings or transformations.[1] For instance, the rotation matrix on page 54 was introduced for that purpose, and the notion of similarity of matrices in Section 3.5 resulted from the fact that similar matrices represent the same transformation in different bases. In this section we explore more systematically the connections between matrices and mappings.

The first question that arises quite naturally is: What kinds of mappings can be represented by matrices? The answer is fairly simple:

If A is an $m \times n$ matrix, then the equation $\boldsymbol{y} = A\boldsymbol{x}$ describes a mapping of $\mathbb{R}^n$ to $\mathbb{R}^m$. This mapping has two fundamental properties: If $\mathbf{x}_1$ and $\mathbf{x}_2 \in \mathbb{R}^n$ are mapped to $\boldsymbol{y}_1$ and $\boldsymbol{y}_2 \in \mathbb{R}^m$ respectively, then first $\mathbf{x}_1 + \mathbf{x}_2$ is mapped to $\boldsymbol{y}_1 + \boldsymbol{y}_2$ and second, $c\mathbf{x}_1$ is mapped to $c\boldsymbol{y}_1$, for any scalar c. Thus any mapping representable by a matrix A via the equation $\boldsymbol{y} = A\boldsymbol{x}$ must have these two properties. Luckily these properties are also sufficient; that is, any mapping with these two properties is representable by a matrix in this manner, as we shall see shortly. Such mappings have a special name:

DEFINITION 4.1.1. A mapping $\boldsymbol{T}$ of a vector space U to a vector space V is called *linear* if it preserves vector addition and multiplication by scalars; that is, if for any $\mathbf{x}_1$ and $\mathbf{x}_2 \in U$ and any scalar c

$$\boldsymbol{T}(\mathbf{x}_1 + \mathbf{x}_2) = \boldsymbol{T}(\mathbf{x}_1) + \boldsymbol{T}(\mathbf{x}_2) \tag{4.1.1}$$

and

$$\boldsymbol{T}(c\mathbf{x}_1) = c\boldsymbol{T}(\mathbf{x}_1) \tag{4.1.2}$$

hold. A mapping $\boldsymbol{T}$ satisfying the first condition is said to be *additive,* and one satisfying the second condition, *homogeneous.*[2]

[1]The two terms are used interchangeably.

[2]Notice that the use of the word *linear* above is more restrictive than is customary in calculus, where a linear function is defined as one of the form $A\boldsymbol{x} + \boldsymbol{b}$ rather than just $A\boldsymbol{x}$.

We can combine these two requirements into a single, more useful form:

LEMMA 4.1.1. A mapping $\boldsymbol{T}$ of a vector space U to a vector space V is linear if and only if it preserves all linear combinations; that is, if for any vectors $\mathbf{x}_1, \mathbf{x}_2, \ldots, \mathbf{x}_n \in U$ and any scalars $c_1, c_2, \ldots, c_n$ the equation

$$\boldsymbol{T}\left(\sum_{i=1}^{n} c_i \mathbf{x}_i\right) = \sum_{i=1}^{n} c_i \boldsymbol{T}(\mathbf{x}_i) \tag{4.1.3}$$

holds.

We leave the proof of this result to the reader as Exercise 4.1.1.

Lemma 4.1.1 can be combined with the definition of a basis to obtain the following observation:

LEMMA 4.1.2. A linear mapping $\boldsymbol{T}$ of a finite-dimensional vector space U to a vector space V is completely determined by its action on the vectors of any basis of U. In other words, if $\boldsymbol{a}_1, \boldsymbol{a}_2, \ldots, \boldsymbol{a}_n$ form a basis for U and $\boldsymbol{T}(\boldsymbol{a}_1), \boldsymbol{T}(\boldsymbol{a}_2), \ldots, \boldsymbol{T}(\boldsymbol{a}_n)$ are arbitrarily prescribed vectors of V, then $\boldsymbol{T}(\mathbf{x})$ is uniquely determined for every $\mathbf{x} \in U$.

Proof By the definition of a basis, if $\boldsymbol{a}_1, \boldsymbol{a}_2, \ldots, \boldsymbol{a}_n$ form a basis for U, then we can write any $\mathbf{x} \in U$ uniquely as

$$\mathbf{x} = \sum_{i=1}^{n} x_{Ai} \boldsymbol{a}_i \tag{4.1.4}$$

and then, from this equation and Lemma 4.1.1, $\boldsymbol{T}(\mathbf{x})$ must satisfy

$$\boldsymbol{T}(\mathbf{x}) = \sum_{i=1}^{n} x_{Ai} \boldsymbol{T}(\boldsymbol{a}_i). \tag{4.1.5}$$

We leave it as Exercise 4.1.2 to show that this definition of $\boldsymbol{T}(\mathbf{x})$ makes $\boldsymbol{T}$ linear.

<<>>

Let us specialize now to mappings from $\mathbb{R}^n$ to $\mathbb{R}^m$ and to the standard basis for $\mathbb{R}^n$. Denote the transform $\boldsymbol{T}(\mathbf{e}_i)$ of the standard vector $\mathbf{e}_i$ by $\boldsymbol{t}_i$, for $i = 1, 2, \ldots, n$, and the $m \times n$ matrix with these as columns by $[T]$. Then the equations above yield

$$\mathbf{x} = \sum_{i=1}^{n} x_i \mathbf{e}_i \tag{4.1.6}$$

and

$$T(\mathbf{x}) = \sum_{i=1}^{n} x_i T(\mathbf{e}_i) = \sum_{i=1}^{n} x_i \mathbf{t}_i = (\mathbf{t}_1\ \mathbf{t}_2\ \dots\ \mathbf{t}_n)\begin{bmatrix} x_1 \\ x_2 \\ \dots \\ x_n \end{bmatrix} = [T]\mathbf{x}. \quad (4.1.7)$$

Let us summarize our result as a theorem:

THEOREM 4.1.1. If T is a linear mapping from $\mathbb{R}^n$ to $\mathbb{R}^m$ and $[T]$ is the $m \times n$ matrix whose columns are the vectors $\mathbf{t}_i = T(\mathbf{e}_i)$ of $\mathbb{R}^m$ for each standard vector $\mathbf{e}_i$ of $\mathbb{R}^n$, then the mapping T corresponds to multiplication by the matrix $[T]$, so that

$$T(\mathbf{x}) = [T]\mathbf{x} \quad (4.1.8)$$

holds for every $\mathbf{x} \in \mathbb{R}^n$. The matrix $[T]$ can also be obtained by factoring $T(\mathbf{x})$ as in Equation (4.1.7).

Let us now look at some examples of linear transformations and their matrix representations:

EXAMPLE 4.1.1. Let T denote the transformation from $\mathbb{R}^3$ to $\mathbb{R}^2$ given by

$$T(\mathbf{x}) = \begin{bmatrix} x_1 - x_2 \\ x_1 + x_3 \end{bmatrix} \quad (4.1.9)$$

and find the matrix $[T]$ that represents this transformation.

By Theorem 4.1.1, all we need to do is find the transforms of the standard vectors $\mathbf{e}_1 = (1, 0, 0)^T$, $\mathbf{e}_2 = (0, 1, 0)^T$, and $\mathbf{e}_3 = (0, 0, 1)^T$. Substituting these for $\mathbf{x}$, one after the other, into Equation (4.1.9) we obtain

$$\mathbf{t}_1 = T(\mathbf{e}_1) = \begin{bmatrix} 1 - 0 \\ 1 + 0 \end{bmatrix} = \begin{bmatrix} 1 \\ 1 \end{bmatrix}, \quad (4.1.10)$$

$$\mathbf{t}_2 = T(\mathbf{e}_2) = \begin{bmatrix} 0 - 1 \\ 0 + 0 \end{bmatrix} = \begin{bmatrix} -1 \\ 0 \end{bmatrix}, \quad (4.1.11)$$

and

$$\mathbf{t}_3 = T(\mathbf{e}_3) = \begin{bmatrix} 0 + 0 \\ 0 + 1 \end{bmatrix} = \begin{bmatrix} 0 \\ 1 \end{bmatrix}. \quad (4.1.12)$$

Thus

$$[T] = \begin{bmatrix} 1 & -1 & 0 \\ 1 & 0 & 1 \end{bmatrix} \quad (4.1.13)$$

is the matrix that represents the given transformation. It is easy to see that $[T]\mathbf{x}$ is indeed the same as $\boldsymbol{T}(\mathbf{x})$ for any $\mathbf{x}$:

$$[T]\mathbf{x} = \begin{bmatrix} 1 & -1 & 0 \\ 1 & 0 & 1 \end{bmatrix} \begin{bmatrix} x_1 \\ x_2 \\ x_3 \end{bmatrix} = \begin{bmatrix} x_1 - x_2 \\ x_1 + x_3 \end{bmatrix} = \boldsymbol{T}(\mathbf{x}). \tag{4.1.14}$$

As remarked at the end of Theorem 4.1.1, we may obtain $[T]$ alternatively by factoring $\boldsymbol{T}(\mathbf{x})$ as follows:

$$\boldsymbol{T}(\mathbf{x}) = \begin{bmatrix} x_1 - x_2 \\ x_1 + x_3 \end{bmatrix} = \begin{bmatrix} 1x_1 - 1x_2 + 0x_3 \\ 1x_1 + 0x_2 + 1x_3 \end{bmatrix} = \begin{bmatrix} 1 & -1 & 0 \\ 1 & 0 & 1 \end{bmatrix} \begin{bmatrix} x_1 \\ x_2 \\ x_3 \end{bmatrix} \tag{4.1.15}$$

and we can read off here the same matrix for $[T]$ that we have found before.

<<>>

EXAMPLE 4.1.2. Let us find the matrix that represents reflection across the $y = x$ line in $\mathbb{R}^2$.

Again we need only find the action of the transformation on the standard vectors. This is obviously a transformation from $\mathbb{R}^2$ to itself, and from the description we find that

$$\boldsymbol{t}_1 = \boldsymbol{T}(\boldsymbol{e}_1) = \boldsymbol{e}_2 = \begin{bmatrix} 0 \\ 1 \end{bmatrix} \tag{4.1.16}$$

and

$$\boldsymbol{t}_2 = \boldsymbol{T}(\boldsymbol{e}_2) = \boldsymbol{e}_1 = \begin{bmatrix} 1 \\ 0 \end{bmatrix}. \tag{4.1.17}$$

Thus

$$[T] = \begin{bmatrix} 0 & 1 \\ 1 & 0 \end{bmatrix}. \tag{4.1.18}$$

<<>>

Example 4.1.3. Let us find the matrix $[T]$ that represents reflection across the y-axis in $\mathbb{R}^2$ followed by a two-fold stretch in the x direction.

The matrix $[R]$ for the reflection can be obtained in a way similar to that used in Example 4.1.2, from

$$\boldsymbol{r}_1 = \boldsymbol{R}(\boldsymbol{e}_1) = -\boldsymbol{e}_1 = \begin{bmatrix} -1 \\ 0 \end{bmatrix} \tag{4.1.19}$$

and

$$\boldsymbol{r}_2 = \boldsymbol{R}(\boldsymbol{e}_2) = \boldsymbol{e}_2 = \begin{bmatrix} 0 \\ 1 \end{bmatrix} \tag{4.1.20}$$

as

$$[R] = \begin{bmatrix} -1 & 0 \\ 0 & 1 \end{bmatrix}. \tag{4.1.21}$$

Similarly, the matrix of the stretch S is

$$[S] = \begin{bmatrix} 2 & 0 \\ 0 & 1 \end{bmatrix}, \tag{4.1.22}$$

and the matrix $[T]$ of the composite transformation is obtained, by the definition of the matrix product, from

$$\boldsymbol{S}(\boldsymbol{R}(\mathbf{x})) = \boldsymbol{S}([R]\mathbf{x}) = [S]([R]\mathbf{x}) = [S][R](\mathbf{x}) \tag{4.1.23}$$

as

$$[T] = [S][R] = \begin{bmatrix} 2 & 0 \\ 0 & 1 \end{bmatrix}\begin{bmatrix} -1 & 0 \\ 0 & 1 \end{bmatrix} = \begin{bmatrix} -2 & 0 \\ 0 & 1 \end{bmatrix}. \tag{4.1.24}$$

Notice that the action of $[S]$ follows that of $[R]$ even though $[S]$ is written first in going from left to right, because $[S][R]$ is defined so that $([S][R])\mathbf{x} = [S]([R]\mathbf{x})$ holds for any vector $\mathbf{x}$.

<<>>

Linear transformations can be represented by matrices in terms of any bases, not just the standard bases, and such representations are possible in any finite-dimensional vector spaces not just in $\mathbb{R}^n$ and $\mathbb{R}^m$. For the latter, however, we have the following simple theorem:

THEOREM 4.1.2.

Let $\boldsymbol{T}$ be a linear transformation from a finite-dimensional vector space U to a finite-dimensional vector space V and $(\boldsymbol{a}_1, \boldsymbol{a}_2, \ldots, \boldsymbol{a}_n)$ an ordered basis[3] for U, and $(\boldsymbol{b}_1, \boldsymbol{b}_2, \ldots, \boldsymbol{b}_m)$ an ordered basis for V. Write the vectors of U and V in terms of these bases as

[3]For terminology and notation see Section 3.5, page 121.

$$\boldsymbol{x} = \sum_{j=1}^{n} x_{Aj}\boldsymbol{a}_j \text{ and } \boldsymbol{y} = \sum_{i=1}^{n} y_{Bi}\boldsymbol{b}_i. \tag{4.1.25}$$

Then there exists a matrix $T_{A,B}$ that represents $\boldsymbol{T}$ relative to these ordered bases so that $\boldsymbol{y} = \boldsymbol{T}(\mathbf{x})$ becomes

$$y_B = T_{A,B}\mathbf{x}_A. \tag{4.1.26}$$

Proof We have

$$\boldsymbol{T}(\mathbf{x}) = \boldsymbol{T}\left(\sum_{j=1}^{n} x_{Aj}\boldsymbol{a}_j\right) = \sum_{j=1}^{n} x_{Aj}\boldsymbol{T}(\boldsymbol{a}_j), \tag{4.1.27}$$

and $\boldsymbol{T}(\boldsymbol{a}_j)$ being an element of V can be written with appropriate coefficients as

$$\boldsymbol{T}(\boldsymbol{a}_j) = \sum_{i=1}^{m} [T_{A,B}]_{ij}\boldsymbol{b}_i. \tag{4.1.28}$$

Then

$$\begin{aligned} \boldsymbol{y} = \boldsymbol{T}(\mathbf{x}) &= \sum_{j=1}^{n} x_{Aj} \sum_{i=1}^{m} [T_{A,B}]_{ij}\boldsymbol{b}_i \\ &= \sum_{i=1}^{m}\left(\sum_{j=1}^{n} [T_{A,B}]_{ij}x_{Aj}\right)\boldsymbol{b}_i = \sum_{i=1}^{m} [T_{A,B}\mathbf{x}_A]_i\boldsymbol{b}_i. \end{aligned} \tag{4.1.29}$$

Because, on the other hand,

$$\boldsymbol{y} = \sum_{i=1}^{m} y_{Bi}\boldsymbol{b}_i,$$

Equation (4.1.26) must hold.

<<>>

COROLLARY 4.1.1. If $U = \mathbb{R}^n$ and $V = \mathbb{R}^m$, then the ordered bases can be considered as matrices $A = (\boldsymbol{a}_1, \boldsymbol{a}_2, \ldots, \boldsymbol{a}_n)$ and $B = (\boldsymbol{b}_1, \boldsymbol{b}_2, \ldots, \boldsymbol{b}_m)$, and we have

$$T_{A,B} = B^{-1}[T]A. \tag{4.1.30}$$

Proof In this case,

$$B\boldsymbol{y}_B = \boldsymbol{y} = \boldsymbol{T}(\mathbf{x}) = \boldsymbol{T}(A\mathbf{x}_A) = [T]A\mathbf{x}_A, \tag{4.1.31}$$

and multiplying through by B^{-1} we get the statement.

<<>>

EXAMPLE 4.1.4. Let $\boldsymbol{T}$ denote the transformation from $\mathbb{R}^3$ to $\mathbb{R}^2$ given by

$$\boldsymbol{T}(\mathbf{x}) = \begin{bmatrix} x_1 - x_2 \\ x_1 + x_3 \end{bmatrix}, \tag{4.1.32}$$

as in Example 4.1.1, and consider the ordered bases given by the columns of

$$A = \begin{bmatrix} 1 & 0 & 1 \\ 1 & 1 & 0 \\ 0 & 1 & 1 \end{bmatrix} \tag{4.1.33}$$

and

$$B = \begin{bmatrix} 1 & 1 \\ -1 & 1 \end{bmatrix}. \tag{4.1.34}$$

Find the matrix $T_{A,B}$ that represents this transformation.

Taking $[T]$ from Example 4.1.1, computing the inverse of B, and substituting into Equation (4.1.30), we get

$$T_{A,B} = \frac{1}{2}\begin{bmatrix} 1 & -1 \\ 1 & 1 \end{bmatrix}\begin{bmatrix} 1 & -1 & 0 \\ 1 & 0 & 1 \end{bmatrix}\begin{bmatrix} 1 & 0 & 1 \\ 1 & 1 & 0 \\ 0 & 1 & 1 \end{bmatrix} = \frac{1}{2}\begin{bmatrix} -1 & -2 & -1 \\ 1 & 0 & 3 \end{bmatrix}. \tag{4.1.35}$$

<<>>

The formalism we discussed earlier of representing a transformation $\boldsymbol{T}$ from $\mathbb{R}^n$ to $\mathbb{R}^m$ by a matrix $[T]$ is a particular case of the present formalism using the standard ordered bases with matrix $A = I_n$ and $B = I_m$ respectively. Thus in the present notation $[T] = T_{I_n,I_m}$, and $[T]$ is the matrix representing $\boldsymbol{T}$ relative to the standard ordered bases.

Sometimes the matrix $T_{A,B}$ can be obtained only directly from its definition by Equation (4.1.28), as in the following example:

EXAMPLE 4.1.5. Let

$$U = \{\boldsymbol{p} : \boldsymbol{p} = p_0 + p_1x + \cdots + p_nx^n\}$$

be the space $\mathcal{P}_n$ of polynomials of degree n or less and

$$V = \{\boldsymbol{q} : \boldsymbol{q} = q_0 + q_1x + \cdots + q_{n-1}x^{n-1}\}$$

the space $\mathcal{P}_{n-1}$ of polynomials of degree $n-1$ or less. Define the *differentiation map* $\boldsymbol{D}$ from U to V by

$$\boldsymbol{D}(\boldsymbol{p}) = \boldsymbol{D}(p_0 + p_1x + \cdots + p_nx^n) =$$
$$p_1 + 2p_2x + \cdots + np_nx^{n-1}. \tag{4.1.36}$$

It is easy to see that $\boldsymbol{D}$ is linear.

Consider the ordered bases $A = (1, x, \ldots, x^n)$ and $B = (1, x, \ldots, x^{n-1})$. Then with the previous notation we have $\boldsymbol{a}_j = x^{j-1}$ and $\boldsymbol{b}_i = x^{i-1}$. Furthermore,

$$\boldsymbol{D}(\boldsymbol{a}_j) = (j-1)x^{j-2} \quad \text{for } j = 1, 2, \ldots, n+1.$$

Thus

$$\boldsymbol{D}(\boldsymbol{a}_1) = 0\boldsymbol{b}_1 + 0\boldsymbol{b}_2 + \cdots + 0\boldsymbol{b}_n,$$

$$\boldsymbol{D}(\boldsymbol{a}_2) = 1\boldsymbol{b}_1 + 0\boldsymbol{b}_2 + \cdots + 0\boldsymbol{b}_n,$$

$$\boldsymbol{D}(\boldsymbol{a}_3) = 0\boldsymbol{b}_1 + 2\boldsymbol{b}_2 + \cdots + 0\boldsymbol{b}_n,$$

etc. According to Equation (4.1.28) the coefficients of the $\boldsymbol{b}_i$ here form the columns of the $n \times (n+1)$ matrix that represents $\boldsymbol{D}$ relative to these bases, which we now denote by $D_{A,B}$. Thus

$$D_{A,B} = \begin{bmatrix} 0 & 1 & 0 & \cdots & 0 \\ 0 & 0 & 2 & \cdots & 0 \\ \vdots & \vdots & \vdots & & \vdots \\ 0 & 0 & 0 & \cdots & n \end{bmatrix}. \tag{4.1.37}$$

The coordinate vector $\boldsymbol{p}_A = (p_0, p_1, \ldots, p_n)^T$ of $\boldsymbol{p} = p_0 + p_1x + \cdots + p_nx^n$ is transformed according to Equation (4.1.26) into

$$\boldsymbol{q}_B = D_{A,B}\boldsymbol{p}_A = (p_1, 2p_2, \ldots, np_n)^T. \tag{4.1.38}$$

Note that a standard basis is defined only for $\mathbb{R}^n$, and for $\mathcal{P}_n$ and $\mathcal{P}_{n-1}$ the bases A and B come closest to the notion of a standard basis. Also note that we could have defined $\boldsymbol{D}$ as a mapping from $\mathcal{P}_n$ to itself, and in that case $D_{A,B}$ would have had to be amended by an extra all-zero row.

<<>>

Exercises 4.1

• **4.1.1.** **a.** Prove Lemma 4.1.1.

b. Prove that $\boldsymbol{T}(\boldsymbol{0}) = \boldsymbol{0}$ for every linear transformation $\boldsymbol{T}$.

4.1.2. Show that if $\boldsymbol{T}(\mathbf{x})$ is defined by Equation (4.1.5), then $\boldsymbol{T}$ is a linear transformation, that is, it satisfies Equations (4.1.1) and (4.1.2).

***4.1.3.** Is it true that a mapping from $\mathbb{R}^n$ to $\mathbb{R}^m$ is linear if and only if it preserves straight lines, that is, if and only if, given any $\mathbf{x}_0$ and $\boldsymbol{a} \in \mathbb{R}^n$, the vectors $\mathbf{x} = \mathbf{x}_0 + t\boldsymbol{a}$, for every scalar t, are mapped into vectors $\boldsymbol{y} = \boldsymbol{y}_0 + t\boldsymbol{b}$ in $\mathbb{R}^m$? If this statement is true, prove it, and if not, then prove an appropriate modification.

4.1.4. Determine whether or not each of the following transformations is linear or not and explain:

a. $\boldsymbol{T} : \mathbb{R}^n \to \mathbb{R}$ such that $\boldsymbol{T}(\mathbf{x}) = \boldsymbol{a} \cdot \mathbf{x}$, with $\boldsymbol{a}$ a fixed vector of $\mathbb{R}^n$.

b. $\boldsymbol{T} : \mathbb{R}^n \to \mathbb{R}^n$ such that $\boldsymbol{T}(\mathbf{x}) = (A - \lambda I)\mathbf{x}$, with A a fixed $n \times n$ matrix and λ any number. (This kind of operation will play an important role in Chapter 7.)

c. $\boldsymbol{T} : \mathbb{R}^n \to \mathbb{R}^m$ such that $\boldsymbol{T}(\mathbf{x}) = A\mathbf{x} + \boldsymbol{b}$, with A a fixed $m \times n$ matrix and $\boldsymbol{b}$ a fixed nonzero vector of $\mathbb{R}^m$.

d. $\boldsymbol{T} : \mathbb{R}^n \to \mathbb{R}$ such that $\boldsymbol{T}(\mathbf{x}) = |\mathbf{x}|$.

e. $\boldsymbol{T} : \mathbb{R}^n \to \mathbb{R}$ such that $\boldsymbol{T}(\mathbf{x}) = \mathbf{x}^T\boldsymbol{a}$, with $\boldsymbol{a}$ a fixed vector of $\mathbb{R}^n$.

f. $\boldsymbol{T} : \mathbb{R}^n \to \mathbb{R}^m$ such that $\boldsymbol{T}(\mathbf{x}) = (\boldsymbol{a}^T\mathbf{x})\boldsymbol{b}$, with $\boldsymbol{a}$ a fixed vector of $\mathbb{R}^n$ and $\boldsymbol{b}$ a fixed vector of $\mathbb{R}^m$.

• **4.1.5.** Find the matrix $[T]$ that represents the transformation from $\mathbb{R}^2$ to $\mathbb{R}^3$ given by

$$\boldsymbol{T}(\mathbf{x}) = \begin{bmatrix} x_1 - x_2 \\ 2x_1 + 3x_2 \\ 3x_1 + 2x_2 \end{bmatrix}.$$

4.1.6. Find the matrix $[T]$ that represents the transformation from $\mathbb{R}^3$ to $\mathbb{R}^3$ given by

$$\boldsymbol{T}(\mathbf{x}) = \begin{bmatrix} x_1 - x_2 \\ x_2 - x_3 \\ x_3 - x_1 \end{bmatrix}.$$

4.1.7. Find the matrix $[T]$ that represents the transformation of Exercise 4.1.4(f). (This matrix is called the *tensor product* of $\boldsymbol{a}$ and $\boldsymbol{b}$ and is usually denoted by $\boldsymbol{a} \otimes \boldsymbol{b}$.) It can also be expressed as an outer product. How?

• **4.1.8.** Find the linear transformation $\boldsymbol{T}$ from $\mathbb{R}^2$ to $\mathbb{R}^3$ (that is, $\boldsymbol{T}(\mathbf{x})$ for any $\mathbf{x}$) that maps the vector $(1, 1)^T$ to $(1, 1, 1)^T$ and $(1, -1)^T$ to $(1, -1, -1)^T$, and find the corresponding matrix $[T]$.

4.1.9. Find the linear transformation $\boldsymbol{T}$ from $\mathbb{R}^3$ to $\mathbb{R}^2$ that maps the vector $(1, 1, 1)^T$ to $(1, 1)^T$, the vector $(1, -1, -1)^T$ to $(1, -1)^T$, and $(1, 1, 0)^T$ to $(1, 0)^T$, and find the corresponding matrix $[T]$.

4.1.10. Let $A = (\boldsymbol{a}_1, \boldsymbol{a}_2, \ldots, \boldsymbol{a}_n)$ be a basis-matrix of $\mathbb{R}^n$ and $\boldsymbol{b}_1, \boldsymbol{b}_2, \ldots, \boldsymbol{b}_n$ arbitrary vectors of $\mathbb{R}^m$. Let $\boldsymbol{T}$ be the linear transformation from $\mathbb{R}^n$ to $\mathbb{R}^m$ for which $\boldsymbol{T}(\boldsymbol{a}_i) = \boldsymbol{b}_i$. Find the corresponding matrix $[T]$. (*Hint:* Use $(\boldsymbol{T}(\boldsymbol{a}_1), \boldsymbol{T}(\boldsymbol{a}_2), \ldots, \boldsymbol{T}(\boldsymbol{a}_n)) = [T]A$.)

4.1.11. Find the matrix $[T]$ that represents a twofold stretch of $\mathbb{R}^2$ in the $y = x$ direction. (*Hint:* Rotate by 45°, stretch, and rotate by −45°.)

4.1.12. Find the matrix $[T]$ that represents the reflection of $\mathbb{R}^2$ across the $ax + by = 0$ line.

4.1.13. Use the result of Exercise 4.1.12 to show that the composition of two reflections of $\mathbb{R}^2$ across any two lines through the origin equals a rotation. Describe this rotation geometrically.

4.1.14. Show that any linear transformation transforms parallel lines into parallel lines.

***4.1.15.** Find all linear transformations $\boldsymbol{T}$ from $\mathbb{R}^2$ to $\mathbb{R}^2$ that map perpendicular lines into perpendicular lines.

4.1.16. Verify Equation (4.1.26) for the result of Example 4.1.4.

4.1.17. Find the matrix $T_{A,B}$ that represents the transformation $\boldsymbol{T}$ of Exercise 4.1.6 relative to the ordered bases given by

$$A = B = \begin{bmatrix} 1 & 0 & 1 \\ 1 & 1 & 0 \\ 0 & 1 & 1 \end{bmatrix}.$$

4.1.18. Find on the space $\mathcal{P}_n$ the representative matrix of the operation $\boldsymbol{D}$ relative to the ordered basis

$$A = B = (1, 1 + x, 1 + x + x^2, \ldots, 1 + x + \cdots + x^n).$$

(Here $\boldsymbol{D}$ is the transformation of Example 4.1.5.)

4.1.19. Let $U = \{\boldsymbol{p} : \boldsymbol{p} = p_0 + p_1x + \cdots + p_nx^n\}$ be the space $\mathcal{P}_n$ of polynomials of degree n or less and $V = \{\boldsymbol{q} : \boldsymbol{q} = q_0 + q_1x + \cdots + q_{n+1}x^{n+1}\}$ the space $\mathcal{P}_{n+1}$ of polynomials of degree $n + 1$ or less. Define the *integration map* $\boldsymbol{T}$ from U to V by

$$\boldsymbol{T}(\boldsymbol{p}) = \boldsymbol{T}(p_0 + p_1x + \cdots + p_nx^n) =$$

$$p_0x + \frac{p_1x^2}{2} + \cdots + \frac{p_nx^{n+1}}{n+1}. \qquad (4.1.39)$$

Find the matrix $T_{A,B}$ that represents this transformation relative to the ordered bases $A = (1, x, \ldots, x^n)$ and $B = (1, x, \ldots, x^{n+1})$.

4.1.20. For the same spaces with the same bases as in Exercise 4.1.19 find the representative matrix of the transformation $\boldsymbol{X}$ corresponding to multiplication by x; that is, of $\boldsymbol{X}$ such that

$$\boldsymbol{X}(p_0 + p_1x + \cdots + p_nx^n) = p_0x + p_1x^2 + \cdots + p_nx^{n+1}.$$

MATLAB Exercises 4.1

ML4.1.1. Let

$$A = \begin{bmatrix} 1 & 0 & 1 & 0 \\ 3 & 4 & 0 & 1 \\ 0 & 0 & 2 & 0 \\ 0 & -1 & 0 & 0 \end{bmatrix}$$

be a basis-matrix for $\mathbb{R}^4$. Find the matrix $[T]$ of the linear transformation that transforms the columns of this matrix into the corresponding columns of

$$B = \begin{bmatrix} 1 & 0 & 1 & 0 \\ 2 & 2 & 0 & 2 \\ 0 & 0 & 0 & 0 \\ 0 & -1 & 0 & 0 \end{bmatrix}.$$

ML4.1.2. Let $A =$ **magic**(4).

a. Use **rref** on A and A^T to find a basis B for its row space and a basis C for its column space.

b. Find the matrix $[T]$ of the linear transformation from Row(A) to Col(A) that maps the basis B to the basis C, relative to the standard basis.

c. Find the matrix $T_{B,C}$ of the same linear transformation relative to the bases B and C.

ML4.1.3. In MATLAB, polynomials are stored as row vectors of their coefficients in order of descending powers. For example, the polynomial $f(x) = 3x^2 - 2x + 1$ can be entered as $\boldsymbol{f}$ = [3 −2 1]. The product of polynomials is computed by the function **conv**.

Thus, if $g(x) = 2x^2 + x$ and $\mathbf{g} = [2\ \ 1\ \ 0]$, then $\mathbf{conv}(\boldsymbol{f}, \mathbf{g})$ produces $[6\ \ {-1}\ \ 2\ \ 1\ \ 0]$, corresponding to $fg(x) = 6x^4 - x^3 + 2x^2 + x$.

a. Show by hand that if $\boldsymbol{f}$ is as above and $\boldsymbol{p} \in \mathcal{P}_4$, then the mapping $\boldsymbol{T}$ from $\mathcal{P}_4$ to $\mathcal{P}_6$ given by $\boldsymbol{p} \to \mathbf{conv}(\boldsymbol{f}, \boldsymbol{p})$ is linear.

b. Use MATLAB and judiciously selected values of $\boldsymbol{p}$ to find the matrix $[T]$ of this transformation, that is, a matrix $[T]$ such that $[T]\boldsymbol{p} = \mathbf{conv}(\boldsymbol{f}, \boldsymbol{p})$ for every $\boldsymbol{p} \in \mathcal{P}_4$.

4.2. PROPERTIES OF LINEAR TRANSFORMATIONS

Because linear transformations are particular types of functions, all the various concepts relevant to functions in general apply to linear transformations, plus they have some properties that are significant for linear functions only. We first list these one by one for an arbitrary linear transformation $\boldsymbol{T}$ from any U to any V and then discuss their relationships to corresponding properties of the representative matrices $T_{A,B}$:

DEFINITION 4.2.1.

Let $\boldsymbol{T}$ be a linear transformation from a vector space U to a vector space V.

1. The *domain of* $\boldsymbol{T}$ is the set of those $\mathbf{x}$ values for which $\boldsymbol{T}(\mathbf{x})$ is defined. It is almost always taken to be the whole space U, since the linearity of $\boldsymbol{T}$ implies that its domain must be a subspace of some vector space (see Exercise 4.2.1), and we may as well call that subspace U.
2. The *range of* $\boldsymbol{T}$ is the subset $\{\mathbf{y} : \mathbf{y} = \boldsymbol{T}(\mathbf{x}), \mathbf{x} \in U\}$ of V. In fact, because of $\boldsymbol{T}$'s linearity, Range($\boldsymbol{T}$) is a subspace of V. (See Exercise 4.2.6).
3. The linear transformation $\boldsymbol{T}$ from U to V is said to be a mapping *onto* V if Range($\boldsymbol{T}$) = V.
4. The linear transformation $\boldsymbol{T}$ from U to V is said to be *one-to-one* if $\boldsymbol{T}(\mathbf{x}) = \boldsymbol{T}(\mathbf{y})$ implies $\mathbf{x} = \mathbf{y}$ for all $\mathbf{x}, \mathbf{y}$. (Actually $\boldsymbol{T}(\mathbf{x}) = \boldsymbol{T}(\mathbf{y})$ is then equivalent to $\mathbf{x} = \mathbf{y}$, because the converse implication is always true by the definition of mappings.)
5. The linear transformation $\boldsymbol{T}$ from U to V is said to be an *isomorphism* if it is both one-to-one and onto. Furthermore, U and V are said to be *isomorphic* to each other if there exists an isomorphism from U to V. (In general, an isomorphism between algebraic structures means a mapping that is one-to-one, onto and preserves all algebraic operations. In case of vector spaces the linearity of $\boldsymbol{T}$ expresses the preservation of the algebraic operations.)
6. The *kernel of* $\boldsymbol{T}$, denoted by Ker($\boldsymbol{T}$), is the subset $\{\mathbf{x} : \boldsymbol{T}(\mathbf{x}) = \mathbf{0}, \mathbf{x} \in U\}$ of U. It is always a subspace of U. (See Exercise 4.2.7.)
7. The *rank of* $\boldsymbol{T}$ is the dimension of its range.

Example 4.2.1 The transformation $\boldsymbol{T}$ of Example 4.1.1 on page 137 from $\mathbb{R}^3$ to $\mathbb{R}^2$, given by

$$\boldsymbol{T}(\mathbf{x}) = \begin{bmatrix} x_1 - x_2 \\ x_1 + x_3 \end{bmatrix}, \tag{4.2.1}$$

has range $\mathbb{R}^2$ and is a mapping *onto* $\mathbb{R}^2$ because, for *any* $\boldsymbol{y} = (y_1, y_2)^T$, the solution of $\boldsymbol{T}(\mathbf{x}) = \boldsymbol{y}$ is

$$\mathbf{x} = \begin{bmatrix} y_2 - s \\ y_2 - y_1 - s \\ s \end{bmatrix} \tag{4.2.2}$$

and so the mapping $\boldsymbol{T}$ transforms any $\mathbf{x}$ of this form into the arbitrarily given $\boldsymbol{y}$. Furthermore, since all the $\mathbf{x}$ vectors of Equation (4.2.2), for fixed $\boldsymbol{y}$ and different values of the parameter s, are mapped to the same $\boldsymbol{y}$, $\boldsymbol{T}$ is not one-to-one. Equation (4.2.2) also shows, with $\boldsymbol{y} = \boldsymbol{0}$, that $\text{Ker}(\boldsymbol{T}) = \{\mathbf{x} : \mathbf{x} = (-s, -s, s)^T, s \in \mathbb{R}\}$.

<<>>

EXAMPLE 4.2.2. Consider an arbitrary n-dimensional vector space X and a basis $\mathscr{A} = \{\boldsymbol{a}_1, \boldsymbol{a}_2, \ldots, \boldsymbol{a}_n\}$ for X. The mapping $\boldsymbol{T}$ of $\mathbb{R}^n$ to X, given by associating with each coordinate vector $\mathbf{x}_A$ the corresponding $\mathbf{x} \in X$, is an isomorphism. Indeed, the defining equation

$$\boldsymbol{T}(\mathbf{x}_A) = \sum_{i=1}^{n} x_{Ai}\boldsymbol{a}_i$$

shows that $\boldsymbol{T}$ is linear, and $\mathscr{A}$ being a basis has the spanning property, which shows that $\boldsymbol{T}$ is onto, and the linear independence of the basis-vectors shows that $\boldsymbol{T}$ is one-to-one.

The corresponding mapping in the reverse direction; that is, from X to $\mathbb{R}^n$, given by $\boldsymbol{T}^{-1}(\mathbf{x}) = \mathbf{x}_A$ is called the *inverse* of $\boldsymbol{T}$ and is also an isomorphism. (See Exercise 4.2.10.)

That X and $\mathbb{R}^n$ are isomorphic shows that for each value of n there is essentially only one n-dimensional vector space: $\mathbb{R}^n$. For instance, a five-dimensional subspace of $\mathbb{R}^{10}$, the space of all polynomials of degree less than five together with the zero-polynomial, and any five-dimensional subspace of any other vector space, have exactly the same vector space structure as $\mathbb{R}^5$.

<<>>

EXAMPLE 4.2.3. Consider the transformation $\boldsymbol{T}$ from $\mathbb{R}^3$ to $\mathbb{R}^3$ given by

$$\boldsymbol{T}(\mathbf{x}) = \begin{bmatrix} x_1 - x_2 \\ x_1 + x_2 \\ 2x_1 - 3x_2 \end{bmatrix}. \tag{4.2.3}$$

We can also write this transformation as

$$\boldsymbol{T}(\mathbf{x}) = x_1 \begin{bmatrix} 1 \\ 1 \\ 2 \end{bmatrix} + x_2 \begin{bmatrix} -1 \\ 1 \\ -3 \end{bmatrix}. \tag{4.2.4}$$

Thus the range of $\boldsymbol{T}$ is the two-dimensional subspace of such vectors in $\mathbb{R}^3$, and so $\boldsymbol{T}$ is not onto. Also, our $\boldsymbol{T}$ is represented by the matrix

$$[T] = \begin{bmatrix} 1 & -1 & 0 \\ 1 & 1 & 0 \\ 2 & -3 & 0 \end{bmatrix}, \tag{4.2.5}$$

and we can see that Range($\boldsymbol{T}$) = Col($[T]$).

Furthermore, $\boldsymbol{T}$ is not one-to-one, because Ker($\boldsymbol{T}$) = $\{\mathbf{x} : \mathbf{x} = (0, 0, s)^T, s \in \mathbb{R}\}$ and all the infinitely many vectors in Ker($\boldsymbol{T}$) are mapped to the single vector $\boldsymbol{0}$.

<<>>

We can see from Example 4.2.3 and in general from Theorem 4.1.1 on page 137 that for a transformation from $\mathbb{R}^n$ to $\mathbb{R}^m$ the range of $\boldsymbol{T}$ is exactly the same as the column space of the representative matrix $[T]$ relative to the standard bases, and the kernel of $\boldsymbol{T}$ the same as the null space of $[T]$; and so the rank of $\boldsymbol{T}$ equals the rank of $[T]$. The same sort of relationships hold for general spaces and bases as well:

THEOREM 4.2.1. Let $\boldsymbol{T}$ be a linear transformation from a vector space U to a vector space V, $A = (\boldsymbol{a}_1, \boldsymbol{a}_2, \ldots, \boldsymbol{a}_n)$ an ordered basis for U and $B = (\boldsymbol{b}_1, \boldsymbol{b}_2, \ldots, \boldsymbol{b}_m)$ an ordered basis for V, and $T_{A,B}$ the $m \times n$ matrix that represents $\boldsymbol{T}$ relative to these bases. Then rank($T_{A,B}$) = rank($\boldsymbol{T}$) and Equation (4.1.29) on page 140 establishes an isomorphism from Col($T_{A,B}$) to Range($\boldsymbol{T}$).

Proof The matrix $T_{A,B}$ is $m \times n$ and, say, rank($T_{A,B}$) = r for some $r \le m, n$. Let $C = (\boldsymbol{c}_1\ \boldsymbol{c}_2\ \ldots\ \boldsymbol{c}_r)$ be an $m \times r$ matrix whose columns form a basis for Col($T_{A,B}$). Then, for any $\mathbf{x} \in U$, we have a vector $\boldsymbol{t}_C = (\boldsymbol{t}_{C1}, \boldsymbol{t}_{C2}, \ldots, \boldsymbol{t}_{Cr})^T \in \mathbb{R}^r$ such that

$$T_{A,B}\mathbf{x}_A = \sum_{k=1}^{r} t_{Ck}\boldsymbol{c}_k = C\boldsymbol{t}_C \tag{4.2.6}$$

holds. Substituting this result into Equation (4.1.29) on page 140, we get

$$\boldsymbol{T}(\mathbf{x}) = \sum_{i=1}^{m} (T_{A,B}\mathbf{x}_A)_i \boldsymbol{b}_i = \left(\sum_{i=1}^{m} c_{i1}\boldsymbol{b}_i, \ \ldots, \ \sum_{i=1}^{m} c_{ir}\boldsymbol{b}_i \right) \begin{bmatrix} t_{C1} \\ \vdots \\ t_{Cr} \end{bmatrix}. \tag{4.2.7}$$

Thus $\boldsymbol{T}(\mathbf{x})$ lies in the span of the r vectors $\sum_{i=1}^{m} c_{ik}\boldsymbol{b}_i$.

These vectors not only span the range of $\boldsymbol{T}$ but are also independent, because from Equation (4.2.7) we can write

$$\boldsymbol{T}(\mathbf{x}) = \sum_{i=1}^{m}\sum_{k=1}^{r} c_{ik}t_{Ck}\boldsymbol{b}_i = \sum_{i=1}^{m}(C\boldsymbol{t}_C)_i\boldsymbol{b}_i, \tag{4.2.8}$$

and if $\boldsymbol{T}(\mathbf{x}) = \boldsymbol{0}$ holds, then, by the independence of the $\boldsymbol{b}_j$, we must have $C\boldsymbol{t}_C = \boldsymbol{0}$. Hence, by the independence of the columns of C, we obtain $\boldsymbol{t}_C = \boldsymbol{0}$. Thus the $\boldsymbol{r}$ vectors

$$\sum_{i=1}^{m} c_{ik}\boldsymbol{b}_i \quad \text{for } k = 1, 2, \ldots, r$$

form a basis for the range of $\boldsymbol{T}$, and so the rank of $\boldsymbol{T}$ equals the rank r of $T_{A,B}$. The first half of Equation (4.2.7) maps every vector $T_{A,B}\mathbf{x}_A$ of the column space of $T_{A,B}$ to a vector $\boldsymbol{T}(\mathbf{x})$ in the range of $\boldsymbol{T}$. Furthermore, by the independence of the $\boldsymbol{b}_j$, this mapping of the column space of $T_{A,B}$ to the range of $\boldsymbol{T}$ is one-to-one, and then by the result of Exercise 4.2.9 it is also an isomorphism.

<<>>

We can define sum and scalar multiple of linear transformations in the obvious way:

DEFINITION 4.2.2.

Let $\boldsymbol{S}$ and $\boldsymbol{T}$ be linear transformations from a vector space U to a vector space V and c any scalar. Then $\boldsymbol{Q} = \boldsymbol{S} + \boldsymbol{T}$ and $\boldsymbol{R} = c\boldsymbol{T}$ are defined as the linear transformations from U to V that satisfy

$$\boldsymbol{Q}(\mathbf{x}) = \boldsymbol{S}(\mathbf{x}) + \boldsymbol{T}(\mathbf{x}) \tag{4.2.9}$$

and

$$\boldsymbol{R}(\mathbf{x}) = c\boldsymbol{T}(\mathbf{x}) \tag{4.2.10}$$

for every $\mathbf{x} \in U$.

EXAMPLE 4.2.4. Let $\boldsymbol{S}$ and $\boldsymbol{T}$ be the linear transformations from $\mathbb{R}^3$ to $\mathbb{R}^2$ given by

$$\boldsymbol{S}(\mathbf{x}) = \begin{bmatrix} x_2 + x_3 \\ x_3 \end{bmatrix} \tag{4.2.11}$$

and

$$T(\mathbf{x}) = \begin{bmatrix} x_1 - x_2 \\ x_1 + x_3 \end{bmatrix} \tag{4.2.12}$$

Then $Q = S + T$ and $R = cT$ are the linear transformations given by

$$Q(\mathbf{x}) = \begin{bmatrix} x_1 + x_3 \\ x_1 + 2x_3 \end{bmatrix} \tag{4.2.13}$$

and

$$R(\mathbf{x}) = \begin{bmatrix} c(x_1 - x_2) \\ c(x_1 + x_3) \end{bmatrix}. \tag{4.2.14}$$

<<>>

With the definitions above the foregoing proof also establishes the following corollary:

COROLLARY 4.2.1. Let $L(U, V)$ denote the set of all linear transformations from an n-dimensional vector space U to an m-dimensional vector space V. Then $L(U, V)$ is a vector space and the mapping $T \to T_{A,B}$ is an isomorphism from $L(U, V)$ to the vector space of all $m \times n$ matrices. Hence $L(U, V)$ is mn-dimensional.

Just as for functions of a single variable, we define the composition of linear transformations as follows:

DEFINITION 4.2.3. Let R be a linear transformation from a vector space U to a vector space V and S a linear transformation from V to a vector space W. Then $T = S \circ R$ is defined as the linear transformation from U to W that satisfies

$$T(\mathbf{x}) = S(R(\mathbf{x})) \tag{4.2.15}$$

for every $\mathbf{x} \in U$.

As we have seen in connection with Example 4.1.3 on page 138, composition of transformations induces the multiplication of the corresponding standard representative matrices; that is, we have

$$[T] = [S][R]. \tag{4.2.16}$$

To obtain such a relation was, of course, the motivation behind the definition of matrix multiplication. We have analogous relations for the representative matrices relative to arbitrary bases, but we do not go into this subject any further.

From the first half of Equation (4.2.7) we can also see, by the independence of the $\boldsymbol{b}_i$, that $T_{A,B}\mathbf{x}_A = \boldsymbol{0}$ is equivalent to $\boldsymbol{T}(\boldsymbol{x}) = \boldsymbol{0}$. Then, for any $\mathbf{x}_A$ in the null space of $T_{A,B}$, the linear mapping of $\mathbf{x}_A$ to $\boldsymbol{x}$ given by

$$\boldsymbol{x} = \sum_{j=1}^{n} x_{Aj}\boldsymbol{a}_j \tag{4.2.17}$$

is a mapping onto the kernel of $\boldsymbol{T}$. Thus, by the result of Exercise 4.2.8, we have the following theorem:

THEOREM 4.2.2. Let $\boldsymbol{T}$ be a linear transformation from a vector space U to a vector space V, $A = (\boldsymbol{a}_1, \boldsymbol{a}_2, \ldots, \boldsymbol{a}_n)$ an ordered basis for U and $B = (\boldsymbol{b}_1, \boldsymbol{b}_2, \ldots, \boldsymbol{b}_m)$ an ordered basis for V, and $T_{A,B}$ the matrix that represents $\boldsymbol{T}$ relative to these bases. Then the mapping given by Equation (4.2.17) restricted to the null space of $T_{A,B}$ is an isomorphism to the kernel of $\boldsymbol{T}$. Thus $\dim(\text{Null}(T_{A,B})) = \dim(\text{Ker}(\boldsymbol{T}))$.

From Theorems 4.2.1 and 4.2.2 above and Corollary 3.4.1 on page 109 we have the following:

COROLLARY 4.2.2. For any linear transformation $\boldsymbol{T}$ the relation

$$\text{rank}(\boldsymbol{T}) + \dim(\text{Ker}(\boldsymbol{T})) = \dim(\text{Domain}(\boldsymbol{T})) \tag{4.2.18}$$

holds.[4]

This corollary reflects the fact that the action of the transformation $\boldsymbol{T}$ splits the domain into two parts, analogously to the decomposition for matrices in Theorem 3.4.3 on page 109:

THEOREM 4.2.3. Let $\boldsymbol{T}$ be a linear transformation of rank r from a vector space U to a vector space V and let us denote its range and kernel by R and K respectively. If $\overline{K}$ is a subspace of U complementary to K; that is, one that satisfies both

$$U = K + \overline{K} \tag{4.2.19}$$

and

$$K \cap \overline{K} = \{\boldsymbol{0}\}, \tag{4.2.20}$$

then $\boldsymbol{T}$ maps K to $\{\boldsymbol{0}\}$, while mapping the r-dimensional $\overline{K}$ isomorphically onto the subspace R of V.

[4]The dimension of the kernel of $\boldsymbol{T}$ is sometimes also called the *nullity of* $\boldsymbol{T}$.

The subspace $\overline{K}$ is generally not unique (see Exercise 4.2.12), but if U has an inner product, then $\overline{K}$ may be taken as the unique orthogonal complement of K.

Proof Let $\dim(U) = n$, and let $\{\boldsymbol{a}_1, \boldsymbol{a}_2, \ldots, \boldsymbol{a}_{n-r}\}$ be a basis for K. Then we can extend this set to a basis $\mathscr{A} = \{\boldsymbol{a}_1, \boldsymbol{a}_2, \ldots, \boldsymbol{a}_n\}$ for U. Let $\overline{K} = \text{Span}\{\boldsymbol{a}_{n-r+1}, \ldots, \boldsymbol{a}_n\}$. Then any $\mathbf{x}$ in U can be uniquely decomposed as

$$\mathbf{x} = \sum_{i=1}^{n-r} x_{Ai}\boldsymbol{a}_i + \sum_{i=n-r+1}^{n} x_{Ai}\boldsymbol{a}_i. \tag{4.2.21}$$

Writing $\mathbf{x}_K$ for the first sum and $\mathbf{x}_{\overline{K}}$ for the second, we have

$$\boldsymbol{T}(\mathbf{x}) = \boldsymbol{T}(\mathbf{x}_K) + \boldsymbol{T}(\mathbf{x}_{\overline{K}}) = \boldsymbol{0} + \boldsymbol{T}(\mathbf{x}_{\overline{K}}) = \boldsymbol{T}(\mathbf{x}_{\overline{K}}) \tag{4.2.22}$$

Because, by the definition of range, any $\boldsymbol{y}$ of R can be written as $\boldsymbol{T}(\mathbf{x})$ for some $\mathbf{x}$, Equation (4.2.22) shows that it can also be written as $\boldsymbol{T}(\mathbf{x}_{\overline{K}})$. Thus $\boldsymbol{T}$ maps $\overline{K}$ onto R. Because both $\overline{K}$ and R have dimension r, the result of Exercise 4.2.8 shows that $\boldsymbol{T}$ is an isomorphism of $\overline{K}$ onto R.

$\overline{K}$ is generally not unique, because the extension of the basis $\{\boldsymbol{a}_1, \boldsymbol{a}_2, \ldots, \boldsymbol{a}_{n-r}\}$ is not unique, and different extensions result in different subspaces $\overline{K}$. Orthogonal complements, however, are unique as in Section 3.4.

<<>>

At the end of Section 3.4 we mentioned, in effect, that for any $\mathbf{x}_R$ in the row space of a matrix M the mapping given by $\mathbf{x}_R \to M\mathbf{x}_R$ is an isomorphism from Row(M) to Col(M). As promised there, let us now show how to represent this isomorphism by an $r \times r$ matrix.

THEOREM 4.2.4. Let M be an $m \times n$ matrix of rank r, and $A = (\boldsymbol{a}_1\ \boldsymbol{a}_2 \ldots \boldsymbol{a}_r)$ the matrix of an ordered basis for Row(M) and $B = (\mathbf{b}_1\ \mathbf{b}_2 \ldots \mathbf{b}_r)$ one for Col(M). Then

$$M_{A,B} = (B^TB)^{-1}B^TMA \tag{4.2.23}$$

is the $r \times r$ matrix that represents, relative to the bases A and B, the mapping given by $\mathbf{x}_R \to M\mathbf{x}_R$.

Proof Let M be an $m \times n$ matrix of rank r, and $A = (\boldsymbol{a}_1\ \boldsymbol{a}_2 \ldots \boldsymbol{a}_r)$ the matrix of an ordered basis for Row(M) and $B = (\mathbf{b}_1\ \mathbf{b}_2 \ldots \mathbf{b}_r)$ one for Col(M). Then any $\mathbf{x}$ in Row(M) can be written as

$$\mathbf{x} = \sum_{i=1}^{r} x_{Ai}\boldsymbol{a}_i = A\mathbf{x}_A \tag{4.2.24}$$

and any $\boldsymbol{y}$ in Col(M) similarly as

$$\boldsymbol{y} = \sum_{i=1}^{r} y_{Bi}\boldsymbol{b}_i = B\boldsymbol{y}_B. \tag{4.2.25}$$

Let $\boldsymbol{M}$: Row(M) $\rightarrow$ Col(M) be the isomorphism given by $\boldsymbol{y} = M\mathbf{x}$; that is, by

$$B\boldsymbol{y}_B = MA\mathbf{x}_A. \tag{4.2.26}$$

Because we know that $\boldsymbol{M}$ is an isomorphism, there must exist a unique solution $\boldsymbol{y}_B$ of this equation, and comparing that solution with Equation (4.1.26) on page 140 we can obtain $M_{A,B}$. This solution is best found by Gaussian Elimination, but we can also write a formula for it. However, in trying to solve Equation (4.2.26) for $\boldsymbol{y}_B$ explicitly, we encounter a problem, namely that B is an $m \times r$ matrix and generally not square. Thus it has no inverse unless $r = m$. We can, however, apply the following trick: We left-multiply Equation (4.2.26) by B^T and now B^TB is $r \times r$ and will be shown to have an inverse by Lemma 5.1.3 on page 170. Thus we obtain

$$\boldsymbol{y}_B = (B^TB)^{-1}B^TMA\mathbf{x}_A. \tag{4.2.27}$$

Comparing this result with Equation (4.1.26) on page 140, we find

$$M_{A,B} = (B^TB)^{-1}B^TMA \tag{4.2.28}$$

for the matrix that represents $\boldsymbol{M}$ relative to the bases A for Row(M) and B for Col(M).

<<>>

EXAMPLE 4.2.5. Let the given matrix be

$$M = \begin{bmatrix} 1 & 1 & 1 & 2 \\ 0 & 1 & 0 & 1 \\ 1 & 2 & 1 & 3 \end{bmatrix}. \tag{4.2.29}$$

The reduced echelon form of this matrix is

$$E = \begin{bmatrix} 1 & 0 & 1 & 0 \\ 0 & 1 & 0 & 1 \\ 0 & 0 & 0 & 0 \end{bmatrix}. \tag{4.2.30}$$

Thus the vectors $\boldsymbol{a}_1 = (1, 0, 1, 1)^T$ and $\boldsymbol{a}_2 = (0, 1, 0, 1)^T$ form a basis for the row space of M, and $\boldsymbol{b}_1 = (1, 0, 1)^T$ and $\boldsymbol{b}_2 = (1, 1, 2)^T$ a basis for its column space.

Equation (4.2.26) now becomes

$$\begin{bmatrix} 1 & 1 \\ 0 & 1 \\ 1 & 2 \end{bmatrix} \begin{bmatrix} y_{B1} \\ y_{B2} \end{bmatrix} = \begin{bmatrix} 1 & 1 & 1 & 2 \\ 0 & 1 & 0 & 1 \\ 1 & 2 & 1 & 3 \end{bmatrix} \begin{bmatrix} 1 & 0 \\ 0 & 1 \\ 1 & 0 \\ 1 & 1 \end{bmatrix} \begin{bmatrix} x_{A1} \\ x_{A2} \end{bmatrix} \tag{4.2.31}$$

and

$$\begin{bmatrix} 1 & 1 \\ 0 & 1 \\ 1 & 2 \end{bmatrix} \begin{bmatrix} y_{B1} \\ y_{B2} \end{bmatrix} = \begin{bmatrix} 4 & 3 \\ 1 & 2 \\ 5 & 5 \end{bmatrix} \begin{bmatrix} x_{A1} \\ x_{A2} \end{bmatrix}. \tag{4.2.32}$$

This equation can be reduced to

$$\begin{bmatrix} 1 & 0 \\ 0 & 1 \\ 0 & 0 \end{bmatrix} \begin{bmatrix} y_{B1} \\ y_{B2} \end{bmatrix} = \begin{bmatrix} 3 & 1 \\ 1 & 2 \\ 0 & 0 \end{bmatrix} \begin{bmatrix} x_{A1} \\ x_{A2} \end{bmatrix}. \tag{4.2.33}$$

Thus

$$M_{A,B} = \begin{bmatrix} 3 & 1 \\ 1 & 2 \end{bmatrix}. \tag{4.2.34}$$

To check this result, consider, for example, $\mathbf{x} = \boldsymbol{a}_1$. Then $\mathbf{x}_A = (1, 0)^T$ and, on the one hand, $\boldsymbol{M}$ maps $\boldsymbol{x}$ to

$$M\mathbf{x} = \begin{bmatrix} 1 & 1 & 1 & 2 \\ 0 & 1 & 0 & 1 \\ 1 & 2 & 1 & 3 \end{bmatrix} \begin{bmatrix} 1 \\ 0 \\ 1 \\ 1 \end{bmatrix} = \begin{bmatrix} 4 \\ 1 \\ 5 \end{bmatrix} \tag{4.2.35}$$

and, on the other hand, $M_{A,B}$ maps $\mathbf{x}_A$ to

$$\boldsymbol{y}_B = M_{A,B}\mathbf{x}_A = \begin{bmatrix} 3 & 1 \\ 1 & 2 \end{bmatrix} \begin{bmatrix} 1 \\ 0 \end{bmatrix} = \begin{bmatrix} 3 \\ 1 \end{bmatrix}. \tag{4.2.36}$$

This coordinate vector corresponds to the y in the range of $\boldsymbol{M}$ given by

$$\boldsymbol{y} = y_{B1}\boldsymbol{b}_1 + y_{B2}\boldsymbol{b}_2 = 3\begin{bmatrix} 1 \\ 0 \\ 1 \end{bmatrix} + 1\begin{bmatrix} 1 \\ 1 \\ 2 \end{bmatrix} = \begin{bmatrix} 4 \\ 1 \\ 5 \end{bmatrix}, \tag{4.2.37}$$

the same as $M\mathbf{x}$ before.

<<>>

Exercises 4.2

4.2.1. Show that if a transformation $\boldsymbol{T}$ defined on a nonempty subset A of a vector space U satisfies Equations (4.1.1) and (4.1.2) on page 135 for all $\mathbf{x}_1$ and $\mathbf{x}_2 \in A$, then its domain A must be a subspace of U.

• **4.2.2.** For each of the transformations in Exercise 4.1.4, determine the range, the kernel, and whether it is one-to-one or onto. (These concepts apply to nonlinear transformations as well.)

***4.2.3.** Prove that any linear transformation $\boldsymbol{T}$ is one-to-one if and only if it has the property that $\text{Ker}(\boldsymbol{T}) = \{\boldsymbol{0}\}$.

4.2.4. For each of the transformations in Exercises 4.1.5 through 4.1.9, determine the range, the kernel, and whether it is one-to-one or onto.

4.2.5. Let $N = M^T$, where M is the matrix of Equation (4.2.29). Find a basis A for Row(N) and B for Col(N), and find the representative matrix $N_{A,B}$ for the mapping $\boldsymbol{N}$ from Row(N) to Col(N) given by $\boldsymbol{y} = N\mathbf{x}$.

4.2.6. Prove that if $\boldsymbol{T}$ is a linear transformation from a vector space U to a vector space V, then Range($\boldsymbol{T}$) is a subspace of V.

4.2.7. Prove that if $\boldsymbol{T}$ is a linear transformation from a vector space U to a vector space V, then Ker($\boldsymbol{T}$) is a subspace of U.

***4.2.8.** Prove that if $\boldsymbol{T}$ is a linear transformation from a vector space U onto a vector space V, and $\dim(U) = \dim(V)$, then $\boldsymbol{T}$ is an isomorphism. (*Hint:* First, show that if $\mathcal{B} = \{\boldsymbol{b}_1, \boldsymbol{b}_2, \ldots, \boldsymbol{b}_m\}$ is a basis for V and $\boldsymbol{T}(\boldsymbol{a}_i) = \boldsymbol{b}_i$ for each i, then the $\boldsymbol{a}_i$ form a basis for U.)

***4.2.9.** Prove that if $\boldsymbol{T}$ is a one-to-one linear transformation from a vector space U to a vector space V, and $\dim(U) = \dim(V)$, then $\boldsymbol{T}$ is an isomorphism. (Hint: First, show that if $\mathcal{A} = \{\boldsymbol{a}_1, \boldsymbol{a}_2, \ldots, \boldsymbol{a}_n\}$ is a basis for U, then the vectors $\boldsymbol{T}(\boldsymbol{a}_i) = \boldsymbol{b}_i$ for each i form a basis for V.)

4.2.10. Prove that if $T_{A,B}$ represents an isomorphism, then it is nonsingular. What, then, does $(T_{A,B})^{-1}$ represent?

4.2.11. Find, on the space $\mathcal{P}_n$, the representative matrix of the operation $\boldsymbol{X} \circ \boldsymbol{D} - \boldsymbol{D} \circ \boldsymbol{X}$ relative to the ordered basis $A = B = (1, x, \ldots, x^n)$. (Here, $\boldsymbol{X}$ is the transformation of multiplication by x as in Exercise 4.1.20 on page 145 and $\boldsymbol{D}$ the differentiation map.) Determine the range and the kernel of this mapping and whether it is one-to-one or onto.

4.2.12. Let K be the x-axis in $\mathbb{R}^3$; that is, $K = \{\mathbf{x} : \mathbf{x} = x\mathbf{e}_1,\ x \in \mathbb{R}\}$. Show that the complementary subspace $\overline{K}$ of this K as defined in Theorem 4.2.3 is not unique, by exhibiting two different ones that both satisfy Equations (4.2.19) and (4.2.20).

MATLAB Exercises 4.2

ML4.2.1. Let

$$A = \begin{bmatrix} 0 & 1 & 0 \\ 4 & 0 & 1 \\ 0 & 2 & 0 \\ 1 & 0 & 0 \end{bmatrix}$$

be a basis-matrix for a subspace U of $\mathbb{R}^4$. Let

$$[T] = \begin{bmatrix} 0 & 3 & 1 & 0 \\ 4 & 2 & 0 & 1 \\ 0 & 0 & 2 & 0 \\ 1 & 3 & 0 & 4 \\ 2 & 2 & 1 & 0 \\ 1 & 1 & 0 & 0 \end{bmatrix}$$

be the matrix of a linear transformation $\boldsymbol{T}$ from U into $\mathbb{R}^6$. Find the range and kernel of $\boldsymbol{T}$. (Notice that $\boldsymbol{T}$ is considered only on U and not on $\mathbb{R}^4$.)

ML4.2.2. **a.** Let $\boldsymbol{T}$ denote the mapping of Exercise ML4.1.3 on page 145 from $\mathcal{P}_4$ to $\mathcal{P}_6$ given by $\boldsymbol{p} \to \mathbf{conv}(\boldsymbol{f}, \boldsymbol{p})$, with $\boldsymbol{f} = [3 \quad -2 \quad 1]$. Find the range and kernel of $\boldsymbol{T}$.

b. What are the range and kernel of the analogous mapping with $\boldsymbol{f} = [3 \quad -2 \quad 0]$?

4.3. APPLICATIONS OF LINEAR TRANSFORMATIONS IN COMPUTER GRAPHICS

One of the most important tasks in computer graphics is the programming of motion. For example, we may want to show a robot as it moves across the screen, as its arm rotates, or as its legs move. Whether we want to represent two-dimensional motion, or motion in three dimensions, the computation is usually done by applying the matrices that represent the desired transformations, to the position vectors of points of the picture. (Which points to transform is a technical matter, which we need not consider, for it varies from one system to another.) A computer can do these computations so fast that it can create the illusion of continuous motion by displaying 50 or 60 slightly altered versions of a picture per second.

There is, however, a problem with one of the needed transformations: translation is not linear. This fact can be seen in either $\mathbb{R}^2$ or $\mathbb{R}^3$ by writing the translation of any vector $\boldsymbol{p}$ by the fixed vector $\boldsymbol{t}$ as $\boldsymbol{p}' = \boldsymbol{p} + \boldsymbol{t}$, and then for any scalar c we have $(c\boldsymbol{p})' = c\boldsymbol{p} + \boldsymbol{t}$; which, if this were a linear transformation, should equal $c\boldsymbol{p}' = c(\boldsymbol{p} + \boldsymbol{t})$ for any c and not just for specific values.

It would be very convenient if translations could also be brought under the umbrella of linear transformations, and this can indeed be done. To this end, in the two-dimensional case, we regard $\mathbb{R}^2$ as the $x_3 = 1$ plane in $\mathbb{R}^3$; that is, associate with every vector $(x_1, x_2)^T \in \mathbb{R}^2$ the vector $(x_1, x_2, 1)^T \in \mathbb{R}^3$. The components x_1, x_2, 1 are so-called *homogeneous coordinates* of the point given by (x_1, x_2), which in general mean any triple of the form (x_1x_3, x_2x_3, x_3) with $x_3 \neq 0$. These coordinates were originally introduced in the middle of the 19th century to unify the treatment of parallel and nonparallel lines in projective geometry but have recently found a new application for the problem at hand. A similar construction can be given in three dimensions as well.

Let us then consider translation in $\mathbb{R}^2$. We can make it into a linear transformation in $\mathbb{R}^3$, as spelled out in the following theorem:

THEOREM 4.3.1. Let $\boldsymbol{t} = (t_1, t_2, 0)^T$ be a fixed vector in $\mathbb{R}^3$ and let $\boldsymbol{p}' = \boldsymbol{p} + \boldsymbol{t}$ represent translation by $\boldsymbol{t}$ in the $x_3 = 1$ plane; that is, for vectors of the form $\boldsymbol{p} = (p_1, p_2, 1)^T$. Then there is a unique linear transformation $\boldsymbol{T}: \mathbb{R}^3 \to \mathbb{R}^3$ that coincides in the $x_3 = 1$ plane with the given translation or, in other words, one for which $\boldsymbol{T}(\boldsymbol{p}) = \boldsymbol{p}' = \boldsymbol{p} + \boldsymbol{t}$, when $\boldsymbol{p}$ is of the form above. This $\boldsymbol{T}$ has the representative matrix

$$T(t_1, t_2) = \begin{bmatrix} 1 & 0 & t_1 \\ 0 & 1 & t_2 \\ 0 & 0 & 1 \end{bmatrix}. \tag{4.3.1}$$

Proof It is clear that the matrix above provides a linear transformation that has the desired property of coinciding in the $x_3 = 1$ plane with translation by $\boldsymbol{t}$ (just apply $T(t_1, t_2)$ to $\boldsymbol{p} = (p_1, p_2, 1)^T$); and so all we need to prove is that it is unique.

Thus let $\boldsymbol{L}: \mathbb{R}^3 \to \mathbb{R}^3$ be a linear transformation such that $\boldsymbol{L}(\boldsymbol{p}) = \boldsymbol{p}' = \boldsymbol{p} + \boldsymbol{t}$ for all $\boldsymbol{p} = (p_1, p_2, 1)^T$, with $\boldsymbol{t}$ being a fixed vector of the form $\boldsymbol{t} = (t_1, t_2, 0)^T$. Let $\boldsymbol{x} \in \mathbb{R}^3$ be any vector with $x_3 \neq 0$. Then

$$\boldsymbol{x} = x_1\boldsymbol{e}_1 + x_2\boldsymbol{e}_2 + x_3\boldsymbol{e}_3 = x_3\left[\frac{x_1}{x_3}\boldsymbol{e}_1 + \frac{x_2}{x_3}\boldsymbol{e}_2 + \boldsymbol{e}_3\right] \tag{4.3.2}$$

and, using the assumed linearity of $\boldsymbol{L}$ and the fact that the vector in the brackets on the right side of Equation (4.3.2) is in the $x_3 = 1$ plane, where $\boldsymbol{L}$ is translation by $\boldsymbol{t}$, we get

$$\boldsymbol{L}(\mathbf{x}) = x_3\boldsymbol{L}\left[\frac{x_1}{x_3}\boldsymbol{e}_1 + \frac{x_2}{x_3}\boldsymbol{e}_2 + \boldsymbol{e}_3\right] = x_3\left[\left(\frac{x_1}{x_3} + t_1\right)\boldsymbol{e}_1 + \left(\frac{x_2}{x_3} + t_2\right)\boldsymbol{e}_2 + \boldsymbol{e}_3\right] =$$

$$(x_1 + t_1x_3)\boldsymbol{e}_1 + (x_2 + t_2x_3)\boldsymbol{e}_2 + x_3\boldsymbol{e}_3 = \begin{bmatrix} 1 & 0 & t_1 \\ 0 & 1 & t_2 \\ 0 & 0 & 1 \end{bmatrix}\begin{bmatrix} x_1 \\ x_2 \\ x_3 \end{bmatrix} = T(t_1, t_2)\mathbf{x}. \quad (4.3.3)$$

We have thus shown $\boldsymbol{L}(\mathbf{x}) = T(t_1, t_2)\mathbf{x}$ for any $\mathbf{x} \in \mathbb{R}^3$ with $x_3 \neq 0$. If, on the other hand, $x_3 = 0$, then we can write

$$\mathbf{x} = x_1\boldsymbol{e}_1 + x_2\boldsymbol{e}_2 = (x_1\boldsymbol{e}_1 + x_2\boldsymbol{e}_2 + \boldsymbol{e}_3) - \boldsymbol{e}_3, \quad (4.3.4)$$

where both terms on the right satisfy $x_3 = 1$. Consequently, we can use the linearity of $\boldsymbol{L}$ and the fact that for such vectors it coincides with translation by $\boldsymbol{t}$, to write

$$\boldsymbol{L}(\mathbf{x}) = \boldsymbol{L}(x_1\boldsymbol{e}_1 + x_2\boldsymbol{e}_2 + \boldsymbol{e}_3) - \boldsymbol{L}(\boldsymbol{e}_3) = (x_1\boldsymbol{e}_1 + x_2\boldsymbol{e}_2 + \boldsymbol{e}_3 + \boldsymbol{t}) - (\boldsymbol{e}_3 + \boldsymbol{t})$$

$$= x_1\boldsymbol{e}_1 + x_2\boldsymbol{e}_2 = \mathbf{x} = \begin{bmatrix} 1 & 0 & t_1 \\ 0 & 1 & t_2 \\ 0 & 0 & 1 \end{bmatrix}\begin{bmatrix} x_1 \\ x_2 \\ 0 \end{bmatrix} = T(t_1, t_2)\mathbf{x}. \quad (4.3.5)$$

<<>>

The matrices of the other basic geometric transformations can easily be represented by corresponding matrices in homogeneous coordinates, as follows:

The matrix of rotation by an angle θ around the origin of $\mathbb{R}^2$ becomes the matrix

$$R(\theta) = \begin{bmatrix} \cos\theta & -\sin\theta & 0 \\ \sin\theta & \cos\theta & 0 \\ 0 & 0 & 1 \end{bmatrix} \quad (4.3.6)$$

representing rotation in $\mathbb{R}^3$ around the x_3-axis. Similarly, the matrix

$$S(a, b) = \begin{bmatrix} a & 0 & 0 \\ 0 & b & 0 \\ 0 & 0 & 1 \end{bmatrix} \quad (4.3.7)$$

represents scaling by the factor a in the x_1-direction and by the factor b in the x_2-direction.

EXAMPLE 4.3.1. Let us consider a greatly simplified robot consisting of a rectangular body and an arm that is just a line segment as shown in Figure 4.1. We want to find the matrix that represents lifting the arm by an angle θ from its normally horizontal position. Since the arm rotates about the point (2, 4), and the matrix $R(\theta)$ above represents rotation about the origin, we first apply $T(-2, -4)$ to move the point (2, 4) to the origin, then rotate, and then use $T(2, 4)$ to move the "shoulder" back to (2, 4).

FIGURE 4.1

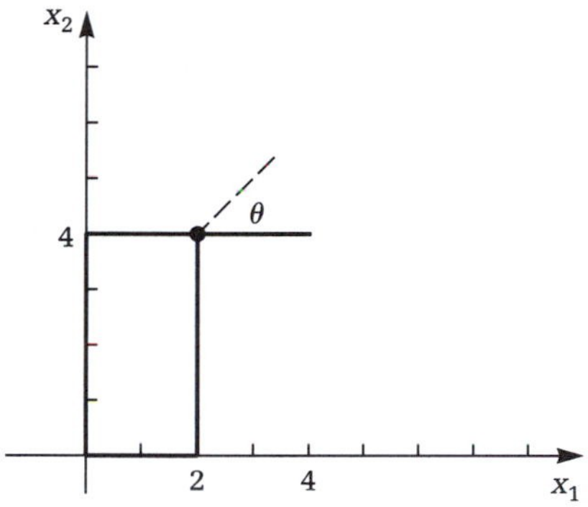

Thus the required matrix is given by

$$\overline{R}(\theta) = T(2, 4)R(\theta)T(-2, -4)$$

$$= \begin{bmatrix} 1 & 0 & 2 \\ 0 & 1 & 4 \\ 0 & 0 & 1 \end{bmatrix} \begin{bmatrix} \cos\theta & -\sin\theta & 0 \\ \sin\theta & \cos\theta & 0 \\ 0 & 0 & 1 \end{bmatrix} \begin{bmatrix} 1 & 0 & -2 \\ 0 & 1 & -4 \\ 0 & 0 & 1 \end{bmatrix}$$

$$= \begin{bmatrix} \cos\theta & -\sin\theta & -2\cos\theta + 4\sin\theta + 2 \\ \sin\theta & \cos\theta & -2\sin\theta - 4\cos\theta + 4 \\ 0 & 0 & 1 \end{bmatrix}. \tag{4.3.8}$$

This matrix is to be applied to the homogeneous position vectors of the points of the arm. Notice that it leaves the center of rotation fixed, as it should; that is,

$$\begin{bmatrix} \cos\theta & -\sin\theta & -2\cos\theta + 4\sin\theta + 2 \\ \sin\theta & \cos\theta & -2\sin\theta - 4\cos\theta + 4 \\ 0 & 0 & 1 \end{bmatrix} \begin{bmatrix} 2 \\ 4 \\ 1 \end{bmatrix} = \begin{bmatrix} 2 \\ 4 \\ 1 \end{bmatrix}. \tag{4.3.9}$$

It changes the homogeneous coordinates of the end point (4, 4) into

$$\begin{bmatrix} \cos\theta & -\sin\theta & -2\cos\theta + 4\sin\theta + 2 \\ \sin\theta & \cos\theta & -2\sin\theta - 4\cos\theta + 4 \\ 0 & 0 & 1 \end{bmatrix} \begin{bmatrix} 4 \\ 4 \\ 1 \end{bmatrix} = \begin{bmatrix} 2\cos\theta + 2 \\ 2\sin\theta + 4 \\ 1 \end{bmatrix}. \quad (4.3.10)$$

This result could of course have been obtained much more easily by elementary means, but those would have required custom tailoring for different problems. In contrast, the generality of the procedure shown above makes it more suitable for computer calculations.

<<>>

In $\mathbb{R}^3$ we could develop the handling of translations in a similar manner, but we leave that for the exercises and deduce the matrix of a particular rotation instead.

EXAMPLE 4.3.2. In $\mathbb{R}^3$ let us find the matrix of the rotation by an angle θ about the vector $\boldsymbol{p} = (1, 1, 1)^T$. (The pictures of rotating objects on computer screens are obtained by applying this kind of a matrix, with small changes in θ, to the position vectors of points of the object about 50 times a second.)

The plan we want to follow is the following: First, we apply a matrix R_1 that rotates the whole space by $\pi/4$ around the z-axis. This step rotates the vector $\boldsymbol{p}$ into the yz-plane, so that $R_1\boldsymbol{p}$ is at an angle α from the z-axis, with $\sin\alpha = \sqrt{\frac{2}{3}}$ and $\cos\alpha = \sqrt{\frac{1}{3}}$. (Why?) Next, we apply a matrix R_2 that rotates the whole space by α around the x-axis. Then $R_2R_1\boldsymbol{p}$ will lie in the z-axis and we can now multiply by a matrix R_3, which expresses rotation by the angle θ around the z-axis, and which is easy to write down. Finally, we undo the first two rotations to get the vector $\boldsymbol{p}$ back to its original position.

For the matrix R_3 we have

$$R_3 = \begin{bmatrix} \cos\theta & -\sin\theta & 0 \\ \sin\theta & \cos\theta & 0 \\ 0 & 0 & 1 \end{bmatrix}, \quad (4.3.11)$$

and setting $\theta = \frac{\pi}{4}$ we get the matrix

$$R_1 = \frac{1}{\sqrt{2}} \begin{bmatrix} 1 & -1 & 0 \\ 1 & 1 & 0 \\ 0 & 0 & \sqrt{2} \end{bmatrix}. \quad (4.3.12)$$

Similarly,

$$R_2 = \frac{1}{\sqrt{3}} \begin{bmatrix} \sqrt{3} & 0 & 0 \\ 0 & 1 & -\sqrt{2} \\ 0 & \sqrt{2} & 1 \end{bmatrix}. \quad (4.3.13)$$

Then

$$R_2R_1 = \frac{1}{\sqrt{6}}\begin{bmatrix} \sqrt{3} & -\sqrt{3} & 0 \\ 1 & 1 & -2 \\ \sqrt{2} & \sqrt{2} & \sqrt{2} \end{bmatrix}. \tag{4.3.14}$$

Because this is an orthogonal matrix,[5] its inverse equals its transpose, and so

$$R_1^{-1}R_2^{-1} = \frac{1}{\sqrt{6}}\begin{bmatrix} \sqrt{3} & 1 & \sqrt{2} \\ -\sqrt{3} & 1 & \sqrt{2} \\ 0 & -2 & \sqrt{2} \end{bmatrix}. \tag{4.3.15}$$

Thus the matrix we sought is given by

$$R_\theta = R_1^{-1}R_2^{-1}R_3R_2R_1 = \tag{4.3.16}$$

$$\frac{1}{6}\begin{bmatrix} 4\cos\theta + 2 & -2\cos\theta - 2\sqrt{3}\sin\theta + 2 & -2\cos\theta + 2\sqrt{3}\sin\theta + 2 \\ -2\cos\theta + 2\sqrt{3}\sin\theta + 2 & 4\cos\theta + 2 & -2\cos\theta - 2\sqrt{3}\sin\theta + 2 \\ -2\cos\theta - 2\sqrt{3}\sin\theta + 2 & -2\cos\theta + 2\sqrt{3}\sin\theta + 2 & 4\cos\theta + 2 \end{bmatrix}.$$

Note that for $\theta = \frac{2\pi}{3}$ this becomes the permutation matrix

$$R_{2\pi/3} = \begin{bmatrix} 0 & 0 & 1 \\ 1 & 0 & 0 \\ 0 & 1 & 0 \end{bmatrix}. \tag{4.3.17}$$

This matrix takes $\boldsymbol{i}$ into $\boldsymbol{j}$, $\boldsymbol{j}$ into $\boldsymbol{k}$ and $\boldsymbol{k}$ into $\boldsymbol{i}$, as it should.

Let us also note that we can interpret the construction above as a passive change of basis, in which nothing moves, instead of the active rotation of $\boldsymbol{p}$ into the z-axis and back. Indeed, with the notation in Theorem 3.5.3 on page 129, we may consider the ordered basis A to be the standard basis; that is, $A = I$, and the new ordered basis B as the one given by $B = R_1^{-1}R_2^{-1}$. Then Corollary 3.5.1 on page 123 gives $S = B$ and for the vector $\mathbf{x} = \boldsymbol{p} = [1, 1, 1]^T$, together with this definition of B, it gives $\mathbf{x}_B = \sqrt{3}[0, 0, 1]^T$. This vector shows that in the B basis the axis of the rotation by θ is the z'-axis. Thus we may take $M_B = R_3$. Then $M_A = R_\theta$ and Equation (3.5.40) on page 129 yields Equation (4.3.16) above.

<<>>

The last subject we take up in this section is projecting three-dimensional images onto *viewplanes*. We consider only *orthographic* (or orthog-

[5]Such matrices will be discussed in Section 5.2. For now, you may just accept the inverse in (4.3.15), which could also, of course, be computed by the usual elimination method.

onal) projections; that is, projections by rays orthogonal to the viewplane. (Perspective projections are also widely used but, because they are not linear, we do not discuss them. On the other hand, orthogonal projections in arbitrary dimensions and relative to general bases, are discussed in Chapter 5.)

The simplest orthographic projections are those onto the coordinate planes. These produce top, bottom, or side views. The components of such projections can be obtained simply by omitting one of the coordinates. For instance, the top view of the point (x, y, z) would be the point (x, y) in the xy-plane. This projection is given by the matrix

$$P(\boldsymbol{i}, \boldsymbol{j}) = \begin{bmatrix} 1 & 0 & 0 \\ 0 & 1 & 0 \end{bmatrix}. \tag{4.3.18}$$

The somewhat more difficult question is how to find the matrix that projects onto an arbitrary plane. Let the desired viewplane V be spanned by the orthogonal unit vectors $\boldsymbol{u}$, $\boldsymbol{v}$ and let $\boldsymbol{n}$ be a normal unit vector of V. We want to decompose an arbitrary vector $\boldsymbol{p} = (p_1, p_1, p_1)^T$ as

$$\boldsymbol{p} = r\boldsymbol{u} + s\boldsymbol{v} + t\boldsymbol{n} \tag{4.3.19}$$

where r, s, t are undetermined coefficients. Multiplying both sides of Equation (4.3.19) in turn by $\boldsymbol{u}$, $\boldsymbol{v}$ and $\boldsymbol{n}$, we get

$$r = \boldsymbol{p} \cdot \boldsymbol{u}, \quad s = \boldsymbol{p} \cdot \boldsymbol{v}, \quad t = \boldsymbol{p} \cdot \boldsymbol{n}. \tag{4.3.20}$$

The projection of $\boldsymbol{p}$ onto V is the vector obtained from Equation (4.3.19) by omitting the $\boldsymbol{n}$ component, and is thus

$$\boldsymbol{p}_V = (\boldsymbol{p} \cdot \boldsymbol{u})\boldsymbol{u} + (\boldsymbol{p} \cdot \boldsymbol{v})\boldsymbol{v}. \tag{4.3.21}$$

The vector $\begin{bmatrix} \boldsymbol{p} \cdot \boldsymbol{u} \\ \boldsymbol{p} \cdot \boldsymbol{v} \end{bmatrix}$ of the coefficients here is the coordinate vector (see page 121) of $\boldsymbol{p}_V$ relative to the vectors $\boldsymbol{u}$ and $\boldsymbol{v}$ that span the viewplane V. We have

$$\begin{bmatrix} r \\ s \end{bmatrix} = \begin{bmatrix} \boldsymbol{p} \cdot \boldsymbol{u} \\ \boldsymbol{p} \cdot \boldsymbol{v} \end{bmatrix} = \begin{bmatrix} u_1 & u_2 & u_3 \\ v_1 & v_2 & v_3 \end{bmatrix} \begin{bmatrix} p_1 \\ p_2 \\ p_3 \end{bmatrix} \tag{4.3.22}$$

and so the projection into the rs-coordinate system of $\boldsymbol{u}$, $\boldsymbol{v}$ is given by the matrix

$$P_V(\boldsymbol{u}, \boldsymbol{v}) = \begin{bmatrix} u_1 & u_2 & u_3 \\ v_1 & v_2 & v_3 \end{bmatrix}. \tag{4.3.23}$$

In components, Equation (4.3.19) is

$$p_{V1} = (p_1u_1 + p_2u_2 + p_3u_3)u_1 + (p_1v_1 + p_2v_2 + p_3v_3)v_1 \quad (4.3.24)$$

$$p_{V2} = (p_1u_1 + p_2u_2 + p_3u_3)u_2 + (p_1v_1 + p_2v_2 + p_3v_3)v_2 \quad (4.3.25)$$

$$p_{V3} = (p_1u_1 + p_2u_2 + p_3u_3)u_3 + (p_1v_1 + p_2v_2 + p_3v_3)v_3 \quad (4.3.26)$$

and from these equations we can read off the *projection matrix relative to the standard vectors of* $\mathbb{R}^3$ as

$$P(\boldsymbol{u}, \boldsymbol{v}) = \begin{bmatrix} u_1^2 + v_1^2 & u_1u_2 + v_1v_2 & u_1u_3 + v_1v_3 \\ u_1u_2 + v_1v_2 & u_2^2 + v_2^2 & u_2u_3 + v_2v_3 \\ u_1u_3 + v_1v_3 & u_2u_3 + v_2v_3 & u_3^2 + v_3^2 \end{bmatrix}. \quad (4.3.27)$$

EXAMPLE 4.3.3. Let us consider the house shown in front and side views in Figure 4.2 and find its view on the $x + y + z = 0$ plane.

FIGURE 4.2

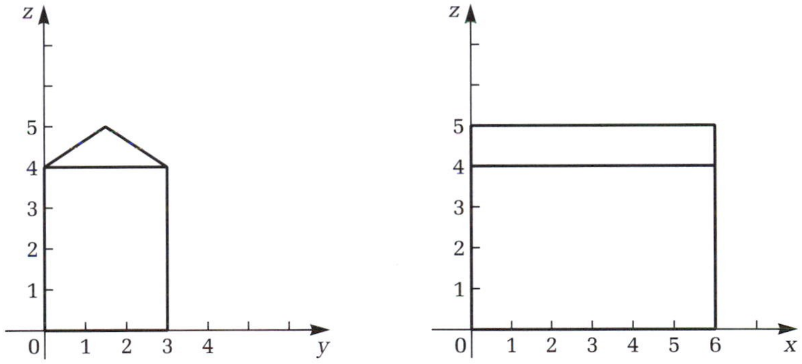

First, we must choose a coordinate system in the viewplane; that is, choose the vectors $\boldsymbol{u}$, $\boldsymbol{v}$. A good choice seems to be $\boldsymbol{u} = \frac{1}{\sqrt{2}}(-1, 1, 0)$ and $\boldsymbol{v} = \frac{1}{\sqrt{6}}(-1, -1, 2)$. (Why?) Thus

$$P_V(\boldsymbol{u}, \boldsymbol{v}) = \begin{bmatrix} -\frac{1}{\sqrt{2}} & \frac{1}{\sqrt{2}} & 0 \\ -\frac{1}{\sqrt{6}} & -\frac{1}{\sqrt{6}} & \frac{2}{\sqrt{6}} \end{bmatrix}, \quad 4.3.28)$$

and the *rs*-coordinates of each point are given, according to Equation (4.3.22), by applying this matrix to the column vector of their *xyz*-coordinates. Thus, for instance, the top-right corner in the first view corresponds to the two corners (0, 3, 4) and (6, 3, 4) in the xyz-coordinates, and so to the points with position vectors

$$P_V(\boldsymbol{u}, \boldsymbol{v})(0, 3, 4)^T = \left(\frac{3}{\sqrt{2}}, \frac{5}{\sqrt{6}}\right)^T \text{ and } P_V(\boldsymbol{u}, \boldsymbol{v})(6, 3, 4)^T = \left(\frac{-3}{\sqrt{2}}, \frac{-1}{\sqrt{6}}\right)^T$$

in the rs-system. Similarly, the bottom-right corner in the first view corresponds to the two corners (0, 3, 0) and (6, 3, 0) in the xyz-coordinates, and so to the points with position vectors

$$P_V(\boldsymbol{u}, \boldsymbol{v})(0, 3, 0)^T = \left(\frac{3}{\sqrt{2}}, \frac{-3}{\sqrt{6}}\right)^T \text{ and } P_V(\boldsymbol{u}, \boldsymbol{v})(6, 3, 0)^T = \left(\frac{-3}{\sqrt{2}}, \frac{-9}{\sqrt{6}}\right)^T$$

in the rs-system. Proceeding in a like manner for all vertices and joining those which are connected by edges, we get the picture in Figure 4.3.

FIGURE 4.3

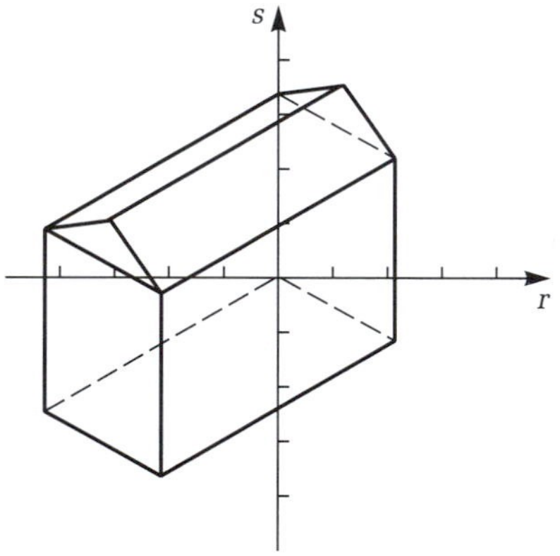

<<>>

Exercises 4.3

4.3.1. Find the matrix in homogeneous coordinates that represents a 30-degree rotation about the point (1, –2).

4.3.2. Find the matrix in homogeneous coordinates that maps the rectangle with vertices (1, –2), (1, 2), (4, 2), (4, –2) onto the unit square.

4.3.3. Find the inverse of the matrix in Exercise 4.3.2. Describe the mapping it represents.

4.3.4. Find the matrix that rotates an arbitrary vector $\boldsymbol{p}$ of $\mathbb{R}^3$ into the z-axis

a. by first rotating it about the z-axis into the yz-*plane* and then about the x-axis into the z-axis, and

b. by first rotating it about the x-axis into the xz-*plane* and then about the y-axis into the z-axis.

4.3.5. Using the result of Exercise 4.3.4, find the matrix that represents rotation by an angle θ about an arbitrary vector $\boldsymbol{p}$ of $\mathbb{R}^3$.

4.3.6. Find a 4×4 matrix, analogous to that of Equation (4.3.1), that represents translation in the $x_4 = 1$ plane by an arbitrary vector $\boldsymbol{t} = (t_1, t_2, t_3, 0)^T$.

4.3.7. Find a 4×4 matrix that represents in homogeneous coordinates the rotation by an angle θ about the $x = y = 1$, $z = 0$ line of $\mathbb{R}^3$.

4.3.8. Find a 4×4 matrix that represents in homogeneous coordinates the rotation by an angle θ about the $\boldsymbol{p} = t(1, 1, 1)^T + (1, 0, 0)^T$ line of $\mathbb{R}^3$.

4.3.9. Find the view of the house in Example 4.3.3 on the $x + 2y = 0$ plane by choosing an appropriate basis in the latter, computing the rs-coordinates of the vertices relative to it and plotting them.

4.3.10. Rederive Equation (4.3.27) by changing from the standard basis to the basis $(\boldsymbol{u}, \boldsymbol{v}, \boldsymbol{n})$, dropping the $\boldsymbol{n}$-component, and returning to the standard basis.

<<5>> ORTHOGONAL PROJECTIONS AND BASES

5.1. ORTHOGONAL PROJECTIONS AND LEAST-SQUARES APPROXIMATIONS

In this section we discuss the very practical problem of fitting a line, plane, or curve to a set of given points when this can only be done approximately. For example, we may expect some observed data to be the coordinates of points on a straight line, but they turn out to be only approximately so. Then our problem is to find a line that fits them best in some sense. The criterion generally used is the so-called least-squares principle, which we shall describe shortly. First, however, we need to discuss the following problem:

Given a point P in $\mathbb{R}^m$ and a subspace[1] V, we wish to find the point Q in V closest to P. (See Figure 5.1.) The solution has the following geometric property:

LEMMA 5.1.1. The point of V closest to P is Q if and only if $\boldsymbol{p} - \boldsymbol{q} \in V^{\perp}$.

Proof The statement is trivially true if P is in V: then $Q = P$ and $\boldsymbol{p} - \boldsymbol{q} = \boldsymbol{0}$. Otherwise, let Q be a point in V such that $\boldsymbol{p} - \boldsymbol{q} \in V^{\perp}$ (such a point does exist; why?) and let R be any point of V other than Q. Then we have $\boldsymbol{r} - \boldsymbol{q} \in V$ and so $\boldsymbol{p} - \boldsymbol{q} \perp \boldsymbol{r} - \boldsymbol{q}$. Thus the PQR triangle is a right triangle and, by the Theorem of Pythagoras (which is valid in $\mathbb{R}^m$, too), the side PQ is shorter than the hypotenuse PR. In other words, the distance $|\boldsymbol{p} - \boldsymbol{q}|$ is less than the distance $|\boldsymbol{p} - \boldsymbol{r}|$ for any $\boldsymbol{r} \neq \boldsymbol{q}$ in V.

Conversely, if R is any point in V such that $\boldsymbol{p} - \boldsymbol{r}$ is not in $V^{\perp}$, then, by the argument above, the point Q for which $\boldsymbol{p} - \boldsymbol{q} \in V^{\perp}$, is nearer to P than is the point R, and so such an R is not the point in V closest to P.

<<>>

[1] We view $\mathbb{R}^m$ and this subspace both as sets of points and as vector spaces.

FIGURE 5.1

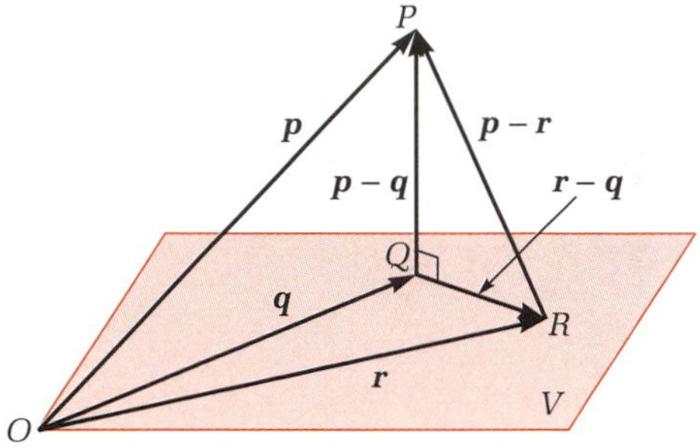

In connection with this lemma we use the following terminology:

DEFINITION 5.1.1.

If a vector $\boldsymbol{p}$ in $\mathbb{R}^m$ is decomposed into the sum of a vector $\boldsymbol{q}$ in a subspace V and a vector $\boldsymbol{p}-\boldsymbol{q} \in V^{\perp}$, then we call $\boldsymbol{q}$ and $\boldsymbol{p}-\boldsymbol{q}$ the (orthogonal) projections of $\boldsymbol{p}$ into V and $V^{\perp}$ respectively.[2]

The next question is: how do we find Q? This query is fairly easy to answer if we consider V (without any loss of generality) as the column space of an $m \times n$ matrix A with independent columns. Then we could use the theory of Section 3.4, but it is more efficient to proceed as follows.

If $\boldsymbol{q}$ and $\boldsymbol{r}$ are vectors in the column space of the matrix A, then we may write them as $\boldsymbol{q} = A\mathbf{x}$ and $\boldsymbol{r} = A\boldsymbol{y}$ with some n-vectors $\mathbf{x}$ and $\boldsymbol{y}$. The condition $\boldsymbol{p}-\boldsymbol{q} \in \mathrm{Col}(A)^{\perp}$ implies that $\boldsymbol{q}-\boldsymbol{p}$ must be orthogonal to any such $\boldsymbol{r}$, and this orthogonality can be written as

$$(A\boldsymbol{y})^T(A\mathbf{x}-\mathbf{p}) = 0 \tag{5.1.1}$$

for any $\boldsymbol{y}$. Equivalently,

$$\boldsymbol{y}^T(A^TA\mathbf{x} - A^T\mathbf{p}) = 0 \tag{5.1.2}$$

must hold for any $\boldsymbol{y}$, and that can happen only if the vector in the parentheses is the zero vector. Then

$$A^TA\mathbf{x} = A^T\mathbf{p}. \tag{5.1.3}$$

In least-squares theory the corresponding scalar equations are called the normal equations. Equation (5.1.3) is easy to remember: just multiply the usual $A\mathbf{x} = \boldsymbol{p}$ by A^T from the left on both sides. The interpretation is, how-

[2]This decomposition is unique by Theorem 3.4.4 on page 114 because V may always be considered to be the column space of a matrix.

ever, entirely different: if $\boldsymbol{p}$ is not in Col(A), then $A\mathbf{x} = \boldsymbol{p}$ has no solution, but Equation (5.1.3) always has one, as will be proved shortly. Also, $A\mathbf{x}$ and $\boldsymbol{p}$ are in $\mathbb{R}^m$, but $A^TA\mathbf{x}$ and $A^T\boldsymbol{p}$ are in $\mathbb{R}^n$.

The projection of $\boldsymbol{p}$ into the column space of A can be obtained by computing $\boldsymbol{q} = A\mathbf{x}$, once we have determined $\mathbf{x}$ from the solution of Equation (5.1.3). We shall write an explicit formula for this projection, but it is more efficient to obtain it by Gaussian elimination as in the following example.

EXAMPLE 5.1.1. Let V be the column space of the matrix

$$A = \begin{bmatrix} 1 & 0 \\ 3 & -1 \\ 1 & 1 \end{bmatrix} \tag{5.1.4}$$

and $\boldsymbol{p} = (1, 2, 3)^T$. Then Col($A$) is a plane in $\mathbb{R}^3$ just as the V in Figure 5.1 is. Furthermore,

$$A^T = \begin{bmatrix} 1 & 3 & 1 \\ 0 & -1 & 1 \end{bmatrix}, \tag{5.1.5}$$

$$A^TA = \begin{bmatrix} 11 & -2 \\ -2 & 2 \end{bmatrix} \quad \text{and} \quad A^T\boldsymbol{p} = \begin{bmatrix} 10 \\ 1 \end{bmatrix}. \tag{5.1.6}$$

Hence the normal equations are given by

$$\begin{bmatrix} 11 & -2 \\ -2 & 2 \end{bmatrix}\mathbf{x} = \begin{bmatrix} 10 \\ 1 \end{bmatrix}. \tag{5.1.7}$$

We solve this equation by Gaussian elimination as follows:

$$\left[\begin{array}{cc|c} 11 & -2 & 10 \\ -2 & 2 & 1 \end{array}\right] \to \left[\begin{array}{cc|c} 1 & 8 & 15 \\ -2 & 2 & 1 \end{array}\right] \to \left[\begin{array}{cc|c} 1 & 8 & 15 \\ 0 & 18 & 31 \end{array}\right]. \tag{5.1.8}$$

Thus,

$$\mathbf{x} = \frac{1}{18}\begin{bmatrix} 22 \\ 31 \end{bmatrix} \tag{5.1.9}$$

and

$$\boldsymbol{q} = A\mathbf{x} = \frac{1}{18}\begin{bmatrix} 1 & 0 \\ 3 & -1 \\ 1 & 1 \end{bmatrix}\begin{bmatrix} 22 \\ 31 \end{bmatrix} = \frac{1}{18}\begin{bmatrix} 22 \\ 35 \\ 53 \end{bmatrix}. \tag{5.1.10}$$

This is the orthogonal projection of $\boldsymbol{p}$ into Col(A). The vector $\boldsymbol{p} - \boldsymbol{q}$ in Col$(A)^\perp$, shown in Figure 5.1, is given by

$$\boldsymbol{p} - \boldsymbol{q} = \begin{bmatrix} 1 \\ 2 \\ 3 \end{bmatrix} - \frac{1}{18}\begin{bmatrix} 22 \\ 35 \\ 53 \end{bmatrix} = \frac{1}{18}\begin{bmatrix} -4 \\ 1 \\ 1 \end{bmatrix}. \tag{5.1.11}$$

(Notice that $\boldsymbol{q}$ and $\boldsymbol{p} - \boldsymbol{q}$ are the same as $\mathbf{x}_R$ and $\mathbf{x}_0$ in Equation (3.4.15) of Example 3.4.4 on page 112. Why?)

<<>>

We wish to show now that the normal equations always have a unique solution, as would be expected from the geometry. This end will follow from the following lemmas:

LEMMA 5.1.2. If A is any $m \times n$ matrix, then $\text{rank}(A^TA) = \text{rank}(A)$.

Proof We may prove this statement by showing that A and A^TA have the same null space, for then, by Corollary 3.4.1 on page 109, they must have the same rank as well. Let $\mathbf{x}$ be in the null space of A. Then $A\mathbf{x} = \boldsymbol{0}$ holds, and multiplying this equation by A^T from the left we obtain $A^TA\mathbf{x} = \boldsymbol{0}$, which shows that such an $\mathbf{x}$ is also in the null space of A^TA. Conversely, if $\mathbf{x}$ is in the null space of A^TA, then $A^TA\mathbf{x} = \boldsymbol{0}$ holds and, multiplying this equation by $\mathbf{x}^T$ from the left, we get

$$\mathbf{x}^TA^TA\mathbf{x} = (A\mathbf{x})^T(A\mathbf{x}) = |A\mathbf{x}|^2 = 0. \tag{5.1.12}$$

Because $\boldsymbol{0}$ is the only vector of absolute value zero, the last equality here implies $A\mathbf{x} = \boldsymbol{0}$ and so $\mathbf{x}$ is in the null space of A.

<<>>

LEMMA 5.1.3. If A is an $m \times n$ matrix with independent columns, then A^TA is invertible.

Proof If A is an $m \times n$ matrix with independent columns, then the n columns form a basis for $\text{Col}(A)$, and so $\text{rank}(A) = n$ and $n \leq m$ must hold. On the other hand, A^TA is an $n \times n$ matrix and, by Lemma 5.1.2, its rank is the same as that of A, that is, n. By Part (a) of Theorem 2.1.1 on page 45, this result shows that A^TA is invertible.

<<>>

By the last lemma, A^TA in Equation (5.1.3) is invertible, and so we may solve Equation (5.1.3) by multiplying both sides of it by $(A^TA)^{-1}$ from the left. Thus we may summarize our discussion of projections in the following theorem:

THEOREM 5.1.1. Let $\boldsymbol{p}$ be a vector in $\mathbb{R}^m$, V a subspace of $\mathbb{R}^m$, and A an $m \times n$ matrix with independent columns such that $V = \text{Col}(A)$. Then the orthogonal projection $\boldsymbol{q}$ of $\boldsymbol{p}$ into V can be obtained by solving the normal system

$$A^TA\mathbf{x} = A^T\boldsymbol{p} \tag{5.1.13}$$

for $\mathbf{x}$ and setting $\boldsymbol{q} = A\mathbf{x}$. The solution can be written explicitly as

$$\mathbf{x} = (A^TA)^{-1}A^T\boldsymbol{p} \tag{5.1.14}$$

and

$$\boldsymbol{q} = A(A^TA)^{-1}A^T\boldsymbol{p}. \tag{5.1.15}$$

The matrix $P = A(A^TA)^{-1}A^T$ in Equation 5.1.15 is called a *projection matrix* representing the projection onto Col(A) and has these important properties:

1. It is *idempotent*: $P^2 = P$, and
2. It is *symmetric*: $P^T = P$.

(The proof of these statements is left as Exercise 5.1.11.)

EXAMPLE 5.1.2. Let us compute the projection matrix that represents the projection onto the column space of the matrix A in Example 5.1.1. From Equation (5.1.6)

$$(A^TA)^{-1} = \begin{bmatrix} 11 & -2 \\ -2 & 2 \end{bmatrix}^{-1} = \frac{1}{18}\begin{bmatrix} 2 & 2 \\ 2 & 11 \end{bmatrix}, \tag{5.1.16}$$

$$(A^TA)^{-1}A^T = \frac{1}{18}\begin{bmatrix} 2 & 2 \\ 2 & 11 \end{bmatrix}\begin{bmatrix} 1 & 3 & 1 \\ 0 & -1 & 1 \end{bmatrix} = \frac{1}{18}\begin{bmatrix} 2 & 4 & 4 \\ 2 & -5 & 24 \end{bmatrix} \tag{5.1.17}$$

and so the projection matrix is

$$P = A(A^TA)^{-1}A^T = \frac{1}{18}\begin{bmatrix} 1 & 0 \\ 3 & -1 \\ 1 & 1 \end{bmatrix}\begin{bmatrix} 2 & 4 & 4 \\ 2 & -5 & 13 \end{bmatrix} = \frac{1}{18}\begin{bmatrix} 2 & 4 & 4 \\ 4 & 17 & -1 \\ 4 & -1 & 17 \end{bmatrix}, \tag{5.1.18}$$

from which we can now obtain the projection of the vector $\boldsymbol{p}$ as

$$\boldsymbol{q} = P\boldsymbol{p} = \frac{1}{18}\begin{bmatrix} 2 & 4 & 4 \\ 4 & 17 & -1 \\ 4 & -1 & 17 \end{bmatrix}\begin{bmatrix} 1 \\ 2 \\ 3 \end{bmatrix} = \frac{1}{18}\begin{bmatrix} 22 \\ 35 \\ 53 \end{bmatrix}. \tag{5.1.19}$$

<<>>

We can prove that the two properties of the matrix $A(A^TA)^{-1}A^T$ mentioned before Example 5.1.2 characterize projection matrices:

THEOREM 5.1.2. Any square matrix P that is idempotent and symmetric is a projection matrix and represents the projection onto its own column space.

Proof Let P be an $m \times m$ idempotent and symmetric matrix and $\boldsymbol{p}$ any vector in $\mathbb{R}^m$. We are going to show that $\boldsymbol{p} - P\boldsymbol{p}$ is orthogonal to the column space of P and then, $P\boldsymbol{p}$ being in Col(P), the vectors $P\boldsymbol{p}$ and $\boldsymbol{p} - P\boldsymbol{p}$ provide the decomposition of $\boldsymbol{p}$ into its projections onto Col(P) and Col$(P)^\perp$ respectively.

Since any vector of Col(P) can be written as $P\mathbf{x}$, we test the orthogonality of $\boldsymbol{p} - P\boldsymbol{p}$ to Col(P) by computing the dot product of these two vectors, using the assumed properties $P^T = P$ and $P^2 = P$:

$$(P\mathbf{x})^T(\boldsymbol{p} - P\boldsymbol{p}) = \mathbf{x}^T P^T(\boldsymbol{p} - P\boldsymbol{p}) = \mathbf{x}^T P\boldsymbol{p} - \mathbf{x}^T P^2\boldsymbol{p} = 0. \tag{5.1.20}$$

<<>>

We are now ready to discuss least-squares problems.

Suppose we are given m points (x_i, y_i) in the xy-plane and want to find the equation of a straight line, in the form $y = ax + b$, such that the sum of the squared vertical distances from the points to the line is minimized. (Hence the name least squares.) In other words, we want to minimize the function

$$f(a, b) = \sum_{i=1}^{m} d_i^2 = \sum_{i=1}^{m} (ax_i + b - y_i)^2. \tag{5.1.21}$$

(See Figure 5.2.) The solution line is called the *least-squares line* for the given points or the *line of best fit* in the least-squares sense.

FIGURE 5.2

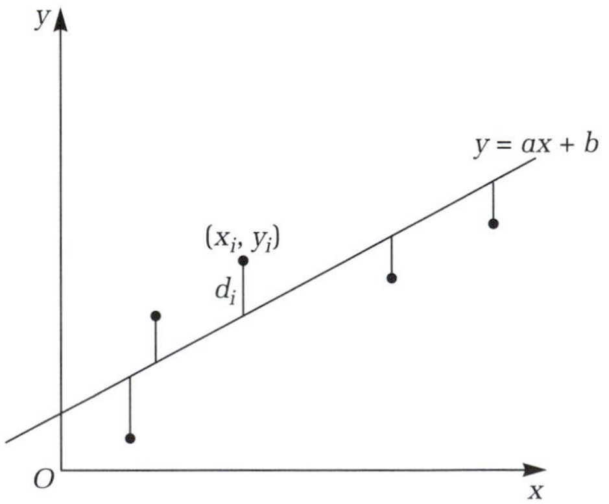

We could solve this problem by differentiating $f(a, b)$ with respect to both a and b, setting the partial derivatives equal to zero and solving the resulting equations, but we prefer to reformulate it as a projection problem in $\mathbb{R}^m$ as follows: Define

$$\mathbf{s} = \begin{bmatrix} a \\ b \end{bmatrix}, \quad A = \begin{bmatrix} x_1 & 1 \\ x_2 & 1 \\ \dots & \\ x_m & 1 \end{bmatrix}, \quad \mathbf{y} = \begin{bmatrix} y_1 \\ y_2 \\ \dots \\ y_m \end{bmatrix}. \tag{5.1.22}$$

Then we have

$$A\mathbf{s} - \mathbf{y} = \begin{bmatrix} ax_1 + b - y_1 \\ ax_2 + b - y_2 \\ \dots \\ ax_m + b - y_m \end{bmatrix} \tag{5.1.23}$$

and so

$$f(a, b) = |A\mathbf{s} - \mathbf{y}|^2. \tag{5.1.24}$$

Thus the problem of minimizing $f(a, b)$, which was originally posed in the xy-plane, can be reinterpreted as that of finding the vector $\mathbf{s}$ that minimizes the distance[3] in $\mathbb{R}^m$ from the point given by $\mathbf{y}$ to the column space of A. By Lemma 5.1.1 on page 167, this minimum is achieved by the vector $\mathbf{s}$, for which $A\mathbf{s}$ is the orthogonal projection of $\mathbf{y}$ onto the column space of A, and the solution is given, according to Theorem 5.1.1, by the solution of the normal system

$$A^T A\mathbf{s} = A^T\mathbf{y}. \tag{5.1.25}$$

Using the definitions of A and $\mathbf{y}$ from (5.1.22), in the present case we have

$$A^T A = \begin{bmatrix} x_1 & x_2 & \cdots & x_m \\ 1 & 1 & \cdots & 1 \end{bmatrix} \begin{bmatrix} x_1 & 1 \\ x_2 & 1 \\ \dots & \\ x_m & 1 \end{bmatrix} = \begin{bmatrix} \sum x_i^2 & \sum x_i \\ \sum x_i & m \end{bmatrix} \tag{5.1.26}$$

and

$$A^T\mathbf{y} = \begin{bmatrix} x_1 & x_2 & \cdots & x_m \\ 1 & 1 & \cdots & 1 \end{bmatrix} \begin{bmatrix} y_1 \\ y_2 \\ \dots \\ y_m \end{bmatrix} = \begin{bmatrix} \sum x_i y_i \\ \sum y_i \end{bmatrix}. \tag{5.1.27}$$

Let us look at an example:

[3]In Equation (5.1.24) we have the *square* of the distance and not the distance itself, but this makes no difference because the two are minimized simultaneously. (Why?)

EXAMPLE 5.1.3. Find the least-squares line for the points (–1, 0), (1, 1), (2, 1), (3, 2), and (5, 3).

First, we compute the expressions in the last two formulas:

$$\sum x_i^2 = (-1)^2 + 1^2 + 2^2 + 3^2 + 5^2 = 40, \tag{5.1.28}$$

$$\sum x_i = -1 + 1 + 2 + 3 + 5 = 10, \quad \sum y_i = 0 + 1 + 1 + 2 + 3 = 7 \tag{5.1.29}$$

and

$$\sum x_i y_i = (-1) \cdot 0 + 1 \cdot 1 + 2 \cdot 1 + 3 \cdot 2 + 5 \cdot 3 = 24. \tag{5.1.30}$$

Hence the normal system is given by

$$\begin{bmatrix} 40 & 10 \\ 10 & 5 \end{bmatrix} \begin{bmatrix} a \\ b \end{bmatrix} = \begin{bmatrix} 24 \\ 7 \end{bmatrix} \tag{5.1.31}$$

and its solution is $a = \frac{1}{2}$ and $b = \frac{2}{5}$. Thus the least-squares line is given by

$$y = \frac{1}{2}x + \frac{2}{5}. \tag{5.1.32}$$

<<>>

The problem of fitting a least-squares plane to m data points (x_i, y_i, z_i) in $\mathbb{R}^3$ is similar to the one above, except that we now have three unknowns a, b, c instead of the previous two. We are looking for the best-fitting plane with equation $z = ax + by + c$ such that the function

$$f(a, b, c) = \sum_{i=1}^{m} d_i^2 = \sum_{i=1}^{m} (ax_i + by_i + c - z_i)^2 \tag{5.1.33}$$

is minimized. This problem can again be reformulated as a projection problem in $\mathbb{R}^m$: If we define

$$\mathbf{s} = \begin{bmatrix} a \\ b \\ c \end{bmatrix}, \quad A = \begin{bmatrix} x_1 & y_1 & 1 \\ x_2 & y_2 & 1 \\ & \cdots & \\ x_m & y_m & 1 \end{bmatrix}, \quad \mathbf{z} = \begin{bmatrix} z_1 \\ z_2 \\ \cdots \\ z_m \end{bmatrix}, \tag{5.1.34}$$

then we can write

$$f(a, b, c) = |A\mathbf{s} - \mathbf{z}|^2, \tag{5.1.35}$$

which can be minimized by solving the 3×3 normal system

$$A^TA\mathbf{s} = A^T\mathbf{z}. \tag{5.1.36}$$

We leave the solution of this equation for Exercise 5.1.18.

The least-squares method is applicable to curves and curved surfaces as well, as long as we know what the form of the equation should be and if we need to find only unknown coefficients that occur linearly. For example, if we have reason to believe that some data points (x_i, y_i) fall approximately on a parabola whose equation is of the form $y = ax^2 + bx + c$, then we can set up the problem for the unknown coefficients a, b, c much as in the preceding cases and find the best-fitting parabola in the same way. In case the coefficients do not occur linearly, we may still be able to transform the problem to one in which they do: For instance, if we want to fit a curve with equation $y = ae^{bx}$, then we can take logarithms on each side to obtain an equation of the form $\ln y = \ln a + bx$ and this is linear in the unknown coefficients $\ln a$ and b. Thus, we can proceed exactly as before if we just replace a by $\ln a$ and y by $\ln y$.

Exercises 5.1

5.1.1. In $\mathbb{R}^3$ find a matrix A with independent columns whose column space is the x-axis.

5.1.2. In $\mathbb{R}^3$ find a matrix A with independent columns whose column space is the xy-plane.

• **5.1.3.** In $\mathbb{R}^3$ find a matrix A with independent columns whose column space is the $x = y$ plane.

• **5.1.4.** In $\mathbb{R}^3$ find the projection of the vector $(1, -1, 2)^T$ into the column space of the matrix

$$A = \begin{bmatrix} 1 & 2 \\ 1 & -1 \\ 2 & 1 \end{bmatrix}.$$

5.1.5. Show that, for a matrix A with independent columns, if $\boldsymbol{p}$ is in Col(A), then the equations $A\mathbf{x} = \boldsymbol{p}$ and $A^TA\mathbf{x} = A^T\boldsymbol{p}$ both have the same solution $\mathbf{x}$.

• **5.1.6.** In $\mathbb{R}^3$ find a projection matrix P that represents the projection onto the x-axis.

5.1.7. In $\mathbb{R}^3$ find a projection matrix P that represents the projection onto the xy-plane.

• **5.1.8.** In $\mathbb{R}^3$ find a projection matrix P that represents the projection onto the line through the origin given by $x = at$, $y = bt$, $z = ct$. Compare with Corollary 1.2.1 on page 19.

5.1.9. In $\mathbb{R}^3$ find a projection matrix P that represents the projection onto the $x = y$ plane and find the projections of the vectors $(1, -1, 2)^T$, $(1, 2, 3)^T$, $(2, -1, -2)^T$ into the $x = y$ plane.

5.1.10. In $\mathbb{R}^3$ what is the orthogonal complement of the $x = y$ plane? Find a projection matrix Q that represents the projection onto it.

5.1.11. Prove that, for any matrix A with independent columns, the matrix $A(A^TA)^{-1}A^T$ is idempotent and symmetric.

5.1.12. Prove that, for any projection matrix P that represents the projection onto the column space of a matrix A, the matrix $I - P$ is also a projection matrix and it represents the projection onto Left-null$(A) = \text{Col}(A)^{\perp}$.

5.1.13. What is wrong with the following "proof" of Theorem 5.1.2?

As we have seen $A(A^TA)^{-1}A^T$ represents the projection onto Col(A). Substituting P for A into this expression and, making use of Properties 1 and 2 of projection matrices, we get the projection matrix representing the projection onto Col(P) as $P(P^TP)^{-1}P^T = P(PP)^{-1}P = PP^{-1}P = P$.

***5.1.14.** Let A be a matrix with independent rows.

a. Show that AA^T is invertible.

b. Show that if $\boldsymbol{b}$ is any vector in Col(A), then the equation $A\mathbf{x} = \boldsymbol{b}$ has a solution in Row(A) given by $\mathbf{x}_r = A^T(AA^T)^{-1}\boldsymbol{b}$. (The matrix $A^T(AA^T)^{-1}$ is called the *pseudoinverse* of A, and is usually denoted A^+. It is a right inverse of A, and coincides with the two-sided inverse A^{-1} if A is square.)

c. Show that the mapping of Col(A) to Row(A) given by $\mathbf{x}_r = A^T(AA^T)^{-1}\boldsymbol{b}$ is an isomorphism.

5.1.15. Let A be a matrix with independent rows. Find a formula for the matrix of the projection onto Null(A).

5.1.16. Find the least-squares line for the points $(1, -2)$, $(-3, 1)$, $(2, 0)$, $(3, -2)$, and $(-5, 3)$.

5.1.17. Using Equations (5.1.25), (5.1.26), (5.1.27), prove that any least-squares line passes through the centroid of the given points (x_i, y_i).

5.1.18. Find formulas analogous to Equations (5.1.26), (5.1.27) for the coefficients in the normal system (5.1.36) for the least-squares plane to m data points (x_i, y_i, z_i) in $\mathbb{R}^3$.

5.1.19. Using the formulas obtained in Exercise 5.1.18, prove that any least-squares plane passes through the centroid of the given points (x_i, y_i, z_i).

5.1.20. Find formulas analogous to Equations (5.1.26), (5.1.27), for the coefficients in the normal system for the least-squares parabola $y = ax^2 + bx + c$ to m data points (x_i, y_i) in $\mathbb{R}^2$.

MATLAB Exercises 5.1

ML5.1.1. Let

$$A = \begin{bmatrix} 1 & 0 & 1 \\ 3 & 4 & 0 \\ 2 & 3 & 2 \\ 0 & -1 & 0 \end{bmatrix}.$$

a. Use MATLAB to find the matrix P of the projection into $\text{Col}(A)$.

b. Verify that P is a projection matrix.

c. The command B = **orth**(A) computes a basis-matrix for $\text{Col}(A)$. Verify that the matrix P projects into $\text{Col}(A)$ by showing that $P\mathbf{e}_i$ is a linear combination of the columns of B for each standard vector $\mathbf{e}_i$ of $\mathbb{R}^4$.

ML5.1.2. If the linear system $A\mathbf{s} = \mathbf{y}$ is overdetermined—that is, there are more equations than unknowns—then the MATLAB command $\mathbf{s} = A\backslash\mathbf{y}$ returns the solution of the corresponding normal system $A^TA\mathbf{s} = A^T\mathbf{y}$. Use this command and Equations (5.1.22) through (5.1.25) to solve the problem of Example 5.1.3 with MATLAB. Plot the result by computing also $z = s(1) * \mathbf{x} + s(2)$ or $z =$ **polyval**$(\mathbf{s}, \mathbf{x})$ and entering the command **plot**$(\mathbf{x}, \mathbf{y}, \text{'o'}, \mathbf{x}, \mathbf{z})$.

ML5.1.3. Use the MATLAB routine **polyfit** instead of $A\backslash\mathbf{y}$ to find the vector $\mathbf{s}$ of Exercise ML5.1.2.

ML5.1.4. Use the method of Exercise ML5.1.2 and Equations (5.1.34) through (5.1.36) to find the plane that best fits the points (1, 2, 3), (2, 2, 4), (−1, 0, 3), (5, −2, 2) and (7, 0, −1). (You may also want to use MATLAB to plot these points and the plane in a three-dimensional coordinate system.)

ML5.1.5. Use **polyfit** to do Exercise 5.1.16.

ML5.1.6. Use **polyfit** to find the parabola that fits best to the points (1, 2), (2, 4), (−1, 0), (−2, 5), and (4, 14), and use MATLAB to plot it, together with the given points.

5.2. ORTHOGONAL BASES

In Section 3.2 we saw how to decompose vectors into linear combinations of given-vectors and how to test the latter for independence. Both of these procedures required the solution of linear systems. If, however, the given vectors are orthogonal to each other, then their independence becomes automatic and the decomposition of other vectors can be achieved much more simply by taking dot products, as in the following example.

EXAMPLE 5.2.1. In $\mathbb{R}^2$ let us consider the basis $\{\boldsymbol{a}_1, \boldsymbol{a}_2\}$ with

$$\boldsymbol{a}_1 = \begin{bmatrix} 1 \\ 1 \end{bmatrix} \text{ and } \boldsymbol{a}_2 = \begin{bmatrix} -1 \\ 1 \end{bmatrix}.$$

We want to write an arbitrary vector $\mathbf{x} = (x_1, x_2)^T$ as

$$\mathbf{x} = x_{A1}\boldsymbol{a}_1 + x_{A2}\boldsymbol{a}_2. \tag{5.2.1}$$

Multiplying both sides of Equation (5.2.1) by $\boldsymbol{a}_1$ and considering that $\boldsymbol{a}_1 \cdot \boldsymbol{a}_2 = 0$, we get

$$\boldsymbol{a}_1 \cdot \mathbf{x} = x_{A1}\boldsymbol{a}_1 \cdot \boldsymbol{a}_1, \tag{5.2.2}$$

which reduces to

$$x_1 + x_2 = 2x_{A1} \tag{5.2.3}$$

and yields

$$x_{A1} = \frac{x_1 + x_2}{2}. \tag{5.2.4}$$

Similarly, multiplying both sides of Equation (5.2.1) by $\boldsymbol{a}_2$, we get

$$x_{A2} = \frac{x_2 - x_1}{2}. \tag{5.2.5}$$

<<>>

The procedure illustrated in the example above can be stated in general terms as follows:

THEOREM 5.2.1. Let $\boldsymbol{a}_1, \boldsymbol{a}_2, \ldots, \boldsymbol{a}_n$ be mutually orthogonal nonzero vectors in a vector space X with an inner product. Then any vector $\mathbf{x} \in \text{Span}\{\boldsymbol{a}_1, \boldsymbol{a}_2, \ldots, \boldsymbol{a}_n\}$ can be decomposed uniquely as

$$\mathbf{x} = \sum_{j=1}^{n} x_{Aj}\boldsymbol{a}_j \tag{5.2.6}$$

with the coefficients given by

$$x_{Ai} = \frac{\boldsymbol{a}_i \cdot \mathbf{x}}{\boldsymbol{a}_i \cdot \boldsymbol{a}_i}. \tag{5.2.7}$$

Furthermore $\boldsymbol{a}_1, \boldsymbol{a}_2, \ldots, \boldsymbol{a}_n$ are independent.

Proof Since $\mathbf{x} \in \text{Span}\{\boldsymbol{a}_1, \boldsymbol{a}_2, \ldots, \boldsymbol{a}_n\}$, there exists a decomposition of $\mathbf{x}$ in the form of Equation (5.2.6) with some coefficients x_{Ai}. Taking the dot product of both sides of Equation (5.2.6) with $\boldsymbol{a}_i$ and utilizing the assumed orthogonality $\boldsymbol{a}_i \cdot \boldsymbol{a}_j = 0$ for all $i \neq j$, we get Equation (5.2.7). Also, if we take $\mathbf{x} = \mathbf{0}$, then Equation (5.2.7) shows that each x_{Ai} equals zero, and so the vectors $\boldsymbol{a}_1, \boldsymbol{a}_2, \ldots, \boldsymbol{a}_n$ are independent.

<<>>

Comparing Theorem 5.2.1 with Corollary 1.2.1 on page 19, we see that each component $x_{Ai}\boldsymbol{a}_i$ above is the orthogonal projection of $\mathbf{x}$ onto $\boldsymbol{a}_i$.

Frequently the vectors $\boldsymbol{a}_1, \boldsymbol{a}_2, \ldots, \boldsymbol{a}_n$ are also "*normalized*" so that $\boldsymbol{a}_i \cdot \boldsymbol{a}_i = 1$ for each i; that is, they are taken as mutually orthogonal unit vectors. In this case they are said to constitute an *orthonormal* set, and Equation (5.2.7) takes on the especially simple form

$$x_{Ai} = \boldsymbol{a}_i \cdot \mathbf{x}. \tag{5.2.8}$$

The formula giving the projection matrix that represents the projection onto the column space of a matrix A also becomes much simpler if the columns of A are orthonormal. We have already seen a particular case of this simplification on page 163 in Section 4.3. In general we had the formula $P = A(A^TA)^{-1}A^T$ (see page 171) for such a projection matrix, and if the columns of A are orthonormal, then $A^TA = I$ holds (why?) and the formula reduces to $P = AA^T$. If we write $A = (\boldsymbol{a}_1, \boldsymbol{a}_2, \ldots, \boldsymbol{a}_n)$ in this equation, then we can write the projection of any $\mathbf{x}$ into the column space of A as

$$P\mathbf{x} = AA^T\mathbf{x} = (\boldsymbol{a}_1\ \boldsymbol{a}_2 \ldots \boldsymbol{a}_n)\begin{bmatrix} \boldsymbol{a}_1^T\mathbf{x} \\ \cdots \\ \boldsymbol{a}_n^T\mathbf{x} \end{bmatrix} = (\boldsymbol{a}_1 \cdot \mathbf{x})\boldsymbol{a}_1 + \ldots + (\boldsymbol{a}_n \cdot \mathbf{x})\boldsymbol{a}_n. \tag{5.2.9}$$

Thus, in this special case the projection onto the column space of the matrix A is the sum of the projections onto the individual columns.

Still another very important simplification results from orthonormality. To wit: the computation of the inverse of a square matrix with orthonormal columns, called an *orthogonal matrix*,[4] becomes trivial:

[4] *Orthonormal* matrix would be a better name, but we have no choice.

THEOREM 5.2.2. If Q is an orthogonal matrix, then it is invertible and

$$Q^{-1} = Q^T. \tag{5.2.10}$$

Furthermore, in this case the rows of Q are orthonormal as well as its columns.

Proof If Q is an $n \times n$ orthogonal matrix, then by Theorem 5.2.1 its n columns are independent, and so its rank is n. Thus it is invertible. Also, the assumed orthogonality can be written as $Q^TQ = I$, and therefore Q^{-1} must equal Q^T. This equality implies the relation $QQ^T = I$, which shows the orthonormality of the rows.

<<>>

Orthogonal matrices occur in many applications because they represent distance-preserving transformations:

$$|Q\mathbf{x}| = |\mathbf{x}| \tag{5.2.11}$$

for any $n \times n$ orthogonal matrix Q and any n-vector $\mathbf{x}$. (The proof of this statement is left as an exercise.) Matrices representing rotations and reflections are orthogonal.

Since orthogonality of basis vectors is such a useful property, in some applications where we have a natural basis that is not orthogonal, we often make a changeover to an orthogonal or orthonormal basis, as in the following example.

EXAMPLE 5.2.2. Find an orthonormal basis for the subspace U of $\mathbb{R}^4$ spanned by the vectors $\boldsymbol{a}_1 = (2, 0, -1, 1)^T$, $\boldsymbol{a}_2 = (1, 1, 0, 1)^T$, and $\boldsymbol{a}_3 = (1, -3, -1, 3)^T$.

To avoid dealing with unpleasant fractions, first we just find an orthogonal basis $\{\boldsymbol{b}_1, \boldsymbol{b}_2, \boldsymbol{b}_3\}$, and normalize its vectors afterward. We may take any one of the given vectors as $\boldsymbol{b}_1$; say, we take $\boldsymbol{b}_1 = \boldsymbol{a}_1$. Next, we compute the orthogonal projection $\boldsymbol{p}_2$ of $\boldsymbol{a}_2$ onto $\boldsymbol{b}_1$, and take $\boldsymbol{b}_2$ as some scalar multiple of $\boldsymbol{a}_2 - \boldsymbol{p}_2$, since then $\boldsymbol{b}_2$ will be in the plane of $\boldsymbol{a}_1$ and $\boldsymbol{a}_2$, and will be orthogonal to $\boldsymbol{b}_1$. Thus

$$\boldsymbol{a}_2 - \boldsymbol{p}_2 = \boldsymbol{a}_2 - \frac{\boldsymbol{a}_2 \cdot \boldsymbol{b}_1}{\boldsymbol{b}_1 \cdot \boldsymbol{b}_1}\boldsymbol{b}_1 = (1, 1, 0, 1)^T - \frac{3}{6}(2, 0, -1, 1)^T. \tag{5.2.12}$$

To avoid fractions, let $\boldsymbol{b}_2$ be 2 times this vector:

$$\boldsymbol{b}_2 = (2, 2, 0, 2)^T - (2, 0, -1, 1)^T = (0, 2, 1, 1)^T. \tag{5.2.13}$$

Next, take the projection $\boldsymbol{p}_3$ of $\boldsymbol{a}_3$ onto the plane of $\boldsymbol{b}_1$ and $\boldsymbol{b}_2$ and subtract it from $\boldsymbol{a}_3$. We get

$$\boldsymbol{a}_3 - \boldsymbol{p}_3 = \boldsymbol{a}_3 - \frac{\boldsymbol{a}_3 \cdot \boldsymbol{b}_1}{\boldsymbol{b}_1 \cdot \boldsymbol{b}_1}\boldsymbol{b}_1 - \frac{\boldsymbol{a}_3 \cdot \boldsymbol{b}_2}{\boldsymbol{b}_2 \cdot \boldsymbol{b}_2}\boldsymbol{b}_2$$

$$= (1, -3, -1, 3)^T - \frac{6}{6}(2, 0, -1, 1)^T - \frac{-4}{6}(0, 2, 1, 1)^T \qquad (5.2.14)$$

and taking $\boldsymbol{b}_3$ as 3 times this we obtain

$$\boldsymbol{b}_3 = (3, -9, -3, 9)^T - (6, 0, -3, 3)^T + (0, 4, 2, 2)^T = (-3, -5, 2, 8)^T. \qquad (5.2.15)$$

Normalizing the $\boldsymbol{b}_i$ vectors we have found, we get an orthonormal basis for U consisting of the vectors

$$\boldsymbol{c}_1 = \frac{1}{\sqrt{6}}(2, 0, -1, 1)^T, \boldsymbol{c}_2 = \frac{1}{\sqrt{6}}(0, 2, 1, 1)^T, \boldsymbol{c}_3 = \frac{1}{\sqrt{102}}(-3, -5, 2, 8)^T. \qquad (5.2.16)$$

<<>>

The procedure illustrated in the example above is called the *Gram-Schmidt orthogonalization procedure.* It is used mostly in function spaces such as the space of polynomials mentioned in Example 3.4.3 on page 109, in which an inner product can be defined by a suitable integral. We shall not discuss this subject here. In general, the algorithm can be described as follows:

THEOREM 5.2.3. Let U be an inner product space with basis $\{\boldsymbol{a}_1, \boldsymbol{a}_2, \ldots, \boldsymbol{a}_n\}$. Define new vectors $\boldsymbol{b}_1, \boldsymbol{b}_2, \ldots, \boldsymbol{b}_n$ successively as $\boldsymbol{b}_1 = \boldsymbol{a}_1$ and

$$\boldsymbol{b}_k = \boldsymbol{a}_k - \sum_{i=1}^{k-1} \frac{\boldsymbol{a}_k \cdot \boldsymbol{b}_i}{\boldsymbol{b}_i \cdot \boldsymbol{b}_i}\boldsymbol{b}_i \text{ for } k = 2, 3, \ldots, n. \qquad (5.2.17)$$

Then these $\boldsymbol{b}_i$ vectors form an orthogonal basis for U, and if we also normalize them, then the unit vectors $\boldsymbol{c}_i = \dfrac{\boldsymbol{b}_i}{|\boldsymbol{b}_i|}$ form an orthonormal basis for U.

Exercises 5.2

- **5.2.1.** In $\mathbb{R}^3$ find the projection of the vector $\mathbf{x} = (2, 3, 4)^T$ into the plane of the vectors $(2, 1, 2)^T$ and $(1, 0, -1)^T$.
- **5.2.2.** Use Equation (5.2.9) to rederive Equation (4.3.27) on page 163 for the matrix giving the projection into the plane spanned by the orthonormal unit vectors $\boldsymbol{u}, \boldsymbol{v}$ in $\mathbb{R}^3$.

5.2.3. In $\mathbb{R}^4$, find (a) the projection of the vector $\mathbf{x} = (1, 2, 3, -1)^T$ into the subspace U of Example 5.2.2, (b) the projection of $\mathbf{x}$ into $U^\perp$, (c) a basis for $U^\perp$, and (d) a vector $\boldsymbol{c}_4$ such that $\{\boldsymbol{c}_1, \boldsymbol{c}_2, \boldsymbol{c}_3, \boldsymbol{c}_4\}$ is an orthonormal basis for $\mathbb{R}^4$.

5.2.4. Prove Equation (5.2.11). (*Hint:* Write $|Q\mathbf{x}|^2$ as a dot product.) Is Q angle preserving as well?

5.2.5. Show that if Equation (5.2.11) holds for a given $n \times n$ matrix Q and every n-vector $\mathbf{x}$, then Q is orthogonal.

5.2.6. Prove that if P and Q are orthogonal matrices of the same size, then so too is PQ.

• **5.2.7.** Find an orthonormal basis for $\mathbb{R}^3$ that includes the vectors

$\frac{1}{3}(-1, 2, 2)^T$ and $\frac{1}{3}(2, -1, 2)^T$.

5.2.8. Find an orthonormal basis for the subspace U of $\mathbb{R}^4$ spanned by the vectors

$\boldsymbol{a}_1 = (0, 0, -1, 1)^T$, $\boldsymbol{a}_2 = (1, 0, 0, 1)^T$ and $\boldsymbol{a}_3 = (1, 0, -1, 0)^T$,

and extend it to an orthonormal basis for $\mathbb{R}^4$.

5.2.9. Show that if $\boldsymbol{q}_1, \boldsymbol{q}_2, \ldots, \boldsymbol{q}_n$ denote the columns of an orthogonal matrix, then

$$\sum_{i=1}^{n} \boldsymbol{q}_i \boldsymbol{q}_i^T = I.$$

MATLAB Exercises 5.2

In MATLAB the command C = **orth**(A) returns an orthonormal basis-matrix for Col(A) and so, by Equation (5.2.9), $P = CC^T$ is the matrix of the projection onto Col(A). Similarly, the command N = **null**(A) returns an orthonormal basis-matrix for Null(A) and so $Q = NN^T$ is the matrix of the projection onto Null(A). Use these to solve the following exercises.

ML5.2.1. In $\mathbb{R}^3$ let a plane S be given by the equation $2x_1 + 3x_2 - x_3 = 0$.

a. Find the matrix of the projection onto S and the matrix of the projection onto the normal vector of S.

b. Check that the sum of the projection matrices found in Part (a) is I.

c. Use the projection matrices found in Part (a) to decompose the vector $\mathbf{x} = (2, 3, 4)^T$ into a component in S and one orthogonal to S.

d. Find the distance of the point $A = (2, 3, 4)$ from S.

ML5.2.2. In $\mathbb{R}^3$ let a line L be given by the parametric equation

$\boldsymbol{p} = (1, 3, 5)^T t$.

a. Find the matrix of the projection onto L and the matrix of the projection onto the orthogonal complement of L.

b. Check that the sum of the projection matrices found in Part (a) is I.

c. Use the projection matrices found in Part (a) to decompose the vector $\mathbf{x} = (2, 3, 4)^T$ into a component in L and one orthogonal to L.

d. Find the distance of the point $A = (2, 3, 4)$ from L.

ML5.2.3. In $\mathbb{R}^3$ let a plane S be given by the parametric equation

$$\boldsymbol{p} = (1, 2, 0)^T s + (1, 3, 5)^T t.$$

a. Find the matrix of the projection onto S and the matrix of the projection onto the normal vector of S.

b. Check that the sum of the projection matrices found in Part (a) is I.

c. Use the projection matrices found in Part (a) to decompose the vector $\mathbf{x} = (2, 3, 4)^T$ into a component in S and one orthogonal to S.

d. Find the distance of the point $A = (2, 3, 4)$ from S.

ML5.2.4. In $\mathbb{R}^3$ let a line L be given by the equations $2x_1 + 3x_2 - x_3 = 0$ and $-x_1 + 2x_2 + 5x_3 = 0$.

a. Find the matrix of the projection onto L and the matrix of the projection onto the orthogonal complement of L.

b. Check that the sum of the projection matrices found in Part (a) is I.

c. Use the projection matrices found in Part (a) to decompose the vector $\mathbf{x} = (2, 3, 4)^T$ into a component in L and one orthogonal to L.

d. Find the distance of the point $A = (2, 3, 4)$ from L.

<<6>> DETERMINANTS

6.1. DEFINITION AND BASIC PROPERTIES

Determinants are certain complicated functions of square matrices (or, equivalently, of their column vectors or of their entries). Their usefulness follows mainly from two of their properties: first, that they can be used to compute areas and volumes, and second, that a zero determinant characterizes singular matrices. Computing areas and volumes brings determinants into the formulas for changing variables in multiple integrals, and their vanishing for singular matrices is at the heart of Chapter 7 for evaluating so-called eigenvalues of matrices, which occur in many geometrical and physical applications.

Rather than give an explicit formula right away, we define determinants by three very simple properties, from which we derive some further ones, and only then do we turn to their evaluation. The defining properties will be obtained by examining how areas and volumes of parallelograms and parallelepipeds depend on their edge vectors.

First, the area of a parallelogram is the absolute value of a linear function of each of two edge vectors if the other edge vector is fixed. More precisely, if we denote the area of the parallelogram spanned by the vectors $\boldsymbol{a}$ and $\boldsymbol{b}$ in $\mathbb{R}^2$ by $A(\boldsymbol{a}, \boldsymbol{b})$, then we have $A(\boldsymbol{a}, \lambda\boldsymbol{b}) = |\lambda| A(\boldsymbol{a}, \boldsymbol{b})$ for any real λ, and $A(\boldsymbol{a}, \boldsymbol{b}) + A(\boldsymbol{a}, \boldsymbol{c}) = A(\boldsymbol{a}, \boldsymbol{b} + \boldsymbol{c})$ for any vectors $\boldsymbol{a}$, $\boldsymbol{b}$, $\boldsymbol{c}$ such that $\boldsymbol{b}$ and $\boldsymbol{c}$ are on the same side of $\boldsymbol{a}$. These follow at a glance from Figures 6.1 and 6.2.

FIGURE 6.1

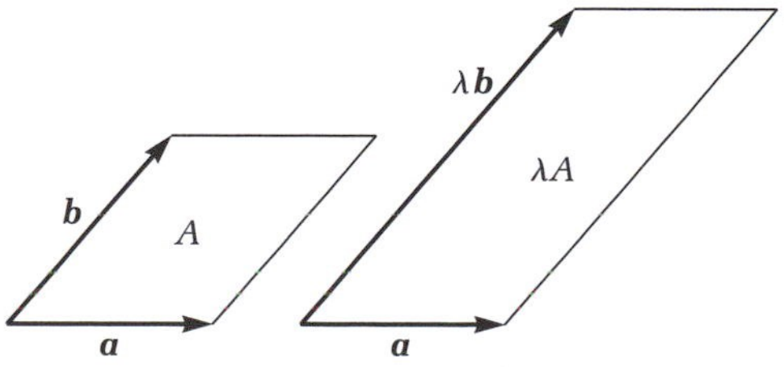

FIGURE 6.2

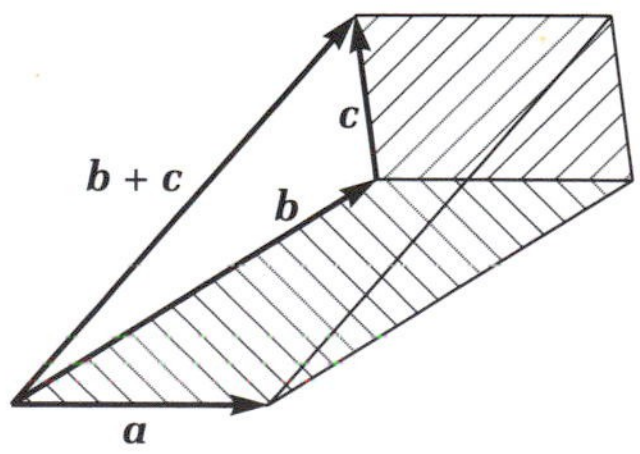

(The latter is a genuinely two-dimensional figure; try not to see it as a wedge in three dimensions.) In Figure 6.2, all three parallelograms have the same base length $|\boldsymbol{a}|$ and the heights of the shaded ones add up to the height of the unshaded one.

If, however, $\boldsymbol{b}$ and $\boldsymbol{c}$ point to opposite sides of $\boldsymbol{a}$, then we have $A(\boldsymbol{a}, \boldsymbol{b} + \boldsymbol{c}) = A(\boldsymbol{a}, \boldsymbol{b}) - A(\boldsymbol{a}, \boldsymbol{c})$, with subtraction instead of addition, as can be seen from Figure 6.3. Nevertheless, we can turn the right side of this equation into a sum for a *signed* area function $D(\boldsymbol{a}, \boldsymbol{b}) = \pm A(\boldsymbol{a}, \boldsymbol{b})$ if we choose the plus sign when $\boldsymbol{a}$ followed by $\boldsymbol{b}$ indicates a counterclockwise traversal of the parallelogram spanned by $\boldsymbol{a}$ and $\boldsymbol{b}$, and choose the minus sign otherwise. Furthermore, the function D is homogeneous, that is, $D(\boldsymbol{a}, \lambda\boldsymbol{b}) = \lambda D(\boldsymbol{a}, \boldsymbol{b})$ for any real λ, without the absolute value that was present for A. We do not go into this subject any further here, but we shall return to the relationship between areas and determinants at the end of Section 6.2. Relations similar to those above for areas hold for volumes of parallelepipeds in $\mathbb{R}^3$ as well.

FIGURE 6.3

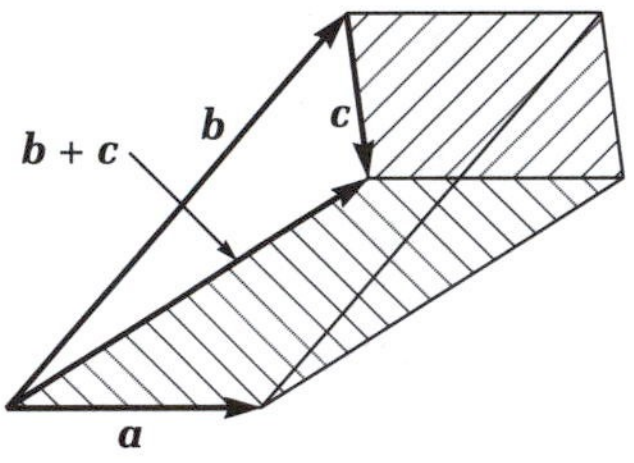

We need two more obvious properties of areas: first, that if the two spanning edges of a parallelogram coincide, then the area is zero, and second, that the area of the unit square is 1. For the function A (and also for the function D) these mean $A(\boldsymbol{a}, \boldsymbol{a}) = 0$ for any $\boldsymbol{a}$ and $A(\boldsymbol{e}_1, \boldsymbol{e}_2) = 1$. We use analogs of these two properties and linearity to define determinants:

DEFINITION 6.1.1.

The determinant of order n is a function that assigns to every $n \times n$ matrix $A = (\boldsymbol{a}_1\ \boldsymbol{a}_2 \dots \boldsymbol{a}_n)$ a number, denoted $\det(A) = \det(\boldsymbol{a}_1\ \boldsymbol{a}_2 \dots \boldsymbol{a}_n)$ or $|A| = |\boldsymbol{a}_1\ \boldsymbol{a}_2 \dots \boldsymbol{a}_n|$, such that

1. It is a *multilinear function* of the columns; that is,

$$|\boldsymbol{a}_1 \cdots \boldsymbol{a}_{i-1}\ s\boldsymbol{a}_i + t\boldsymbol{a}_i'\ \boldsymbol{a}_{i+1} \cdots \boldsymbol{a}_n|$$

$$= s|\boldsymbol{a}_1 \cdots \boldsymbol{a}_{i-1}\ \boldsymbol{a}_i\ \boldsymbol{a}_{i+1} \cdots \boldsymbol{a}_n| + t|\boldsymbol{a}_1 \cdots \boldsymbol{a}_{i-1}\ \boldsymbol{a}_i'\ \boldsymbol{a}_{i+1} \cdots \boldsymbol{a}_n| \quad (6.1.1)$$

holds for any real s, t and any i,

2. It is zero if any two columns are equal; that is,

$$|\cdots \boldsymbol{a}_i \cdots \boldsymbol{a}_j \cdots| = 0 \text{ if } \boldsymbol{a}_i = \boldsymbol{a}_j \quad (6.1.2)$$

holds for any $i \neq j$, and

3. The "volume" of the unit hypercube is 1; that is,

$$|\boldsymbol{e}_1\ \boldsymbol{e}_2 \cdots \boldsymbol{e}_n| = 1. \quad (6.1.3)$$

We are going to show shortly that these properties do indeed define a unique function for every n, by giving formulas for it in Theorems 6.1.2, 6.1.3, and 6.1.4 below. But before computing any determinant we need one more property, which follows from those above:

THEOREM 6.1.1. If the matrix A' is obtained from A by interchanging any two columns, then $|A'| = -|A|$.

Proof Suppose we have

$$|A| = |\cdots \boldsymbol{a}_i \cdots \boldsymbol{a}_j \cdots| \tag{6.1.4}$$

and

$$|A'| = |\cdots \boldsymbol{a}_j \cdots \boldsymbol{a}_i \cdots|. \tag{6.1.5}$$

Consider the determinant in which the ith and jth columns are both equal to $\boldsymbol{a}_i + \boldsymbol{a}_j$. By Axiom 2 above, its value is 0 and, by Axiom 1, it can be reduced as follows:

$$\begin{aligned}
0 &= |\cdots \boldsymbol{a}_i + \boldsymbol{a}_j \cdots \boldsymbol{a}_i + \boldsymbol{a}_j \cdots| \\
&= |\cdots \boldsymbol{a}_i \cdots \boldsymbol{a}_i + \boldsymbol{a}_j \cdots| + |\cdots \boldsymbol{a}_j \cdots \boldsymbol{a}_i + \boldsymbol{a}_j \cdots| \\
&= |\cdots \boldsymbol{a}_i \cdots \boldsymbol{a}_i \cdots| + |\cdots \boldsymbol{a}_i \cdots \boldsymbol{a}_j \cdots| \\
&\quad + |\cdots \boldsymbol{a}_j \cdots \boldsymbol{a}_i \ldots| + |\cdots \boldsymbol{a}_j \cdots \boldsymbol{a}_j \cdots|.
\end{aligned} \tag{6.1.6}$$

On the right side the first and last terms are zero by Axiom 2 and so the sum of the middle terms is also zero; that is,

$$|A| + |A'| = 0. \tag{6.1.7}$$

This equation is equivalent to the statement of the theorem.

<<>>

We now have enough information to evaluate determinants in case $n = 2$:

THEOREM 6.1.2. The determinant of order two is given by the formula

$$|A| = \begin{vmatrix} a_{11} & a_{12} \\ a_{21} & a_{22} \end{vmatrix} = a_{11}a_{22} - a_{21}a_{12}. \tag{6.1.8}$$

Proof By the linearity assumption we have

$$|A| = |\boldsymbol{a}_1\ \boldsymbol{a}_2| = |a_{11}\mathbf{e}_1 + a_{21}\mathbf{e}_2,\ a_{12}\mathbf{e}_1 + a_{22}\mathbf{e}_2|$$

$$= a_{11}|\mathbf{e}_1,\ a_{12}\mathbf{e}_1 + a_{22}\mathbf{e}_2| + a_{21}|\mathbf{e}_2,\ a_{12}\mathbf{e}_1 + a_{22}\mathbf{e}_2|$$

$$= a_{11}(a_{12}|\mathbf{e}_1\ \mathbf{e}_1| + a_{22}|\mathbf{e}_1\ \mathbf{e}_2|) + a_{21}(a_{12}|\mathbf{e}_2\ \mathbf{e}_1| + a_{22}|\mathbf{e}_2\ \mathbf{e}_2|), \qquad (6.1.9)$$

where, according to Axioms 2 and 3 and Theorem 6.1.1, the determinants of the standard vectors equal 0, 1, –1, and 0 respectively. Thus Equation (6.1.9) reduces to the statement of the theorem.

<<>>

EXAMPLE 6.1.1.

$$\begin{vmatrix} 2 & 3 \\ 4 & 5 \end{vmatrix} = 2 \cdot 5 - 4 \cdot 3 = -2. \qquad (6.1.10)$$

<<>>

Let us now find a formula for determinants of order three:

THEOREM 6.1.3. For any 3×3 matrix A we have

$$|A| = \begin{vmatrix} a_{11} & a_{12} & a_{13} \\ a_{21} & a_{22} & a_{23} \\ a_{31} & a_{32} & a_{33} \end{vmatrix}$$

$$= a_{11}a_{22}a_{33} + a_{21}a_{32}a_{13} + a_{31}a_{12}a_{23}$$

$$- a_{11}a_{32}a_{23} - a_{21}a_{12}a_{33} - a_{31}a_{22}a_{13}. \qquad (6.1.11)$$

Proof The proof of this theorem is much like that of the 2×2 case. By the linearity property of determinants we have

$$|A| = |\boldsymbol{a}_1\ \boldsymbol{a}_2\ \boldsymbol{a}_3| = \left|\sum_{i=1}^{3} a_{i1}\mathbf{e}_i,\ \sum_{j=1}^{3} a_{j2}\mathbf{e}_j,\ \sum_{k=1}^{3} a_{k3}\mathbf{e}_k\right| = \sum_{i,j,k=1}^{3} a_{i1}a_{j2}a_{k3}|\mathbf{e}_i\ \mathbf{e}_j\ \mathbf{e}_k| \quad (6.1.12)$$

Now each determinant on the right is zero whenever two of the standard vectors are equal, is +1 when $|\mathbf{e}_i\ \mathbf{e}_j\ \mathbf{e}_k|$ is $|\mathbf{e}_1\ \mathbf{e}_2\ \mathbf{e}_3|$ or can be obtained from the latter by two column exchanges, and is –1 when $|\mathbf{e}_i\ \mathbf{e}_j\ \mathbf{e}_k|$ can be obtained from $|\mathbf{e}_1\ \mathbf{e}_2\ \mathbf{e}_3|$ by just one exchange, since every column exchange changes only the sign of a determinant. (For example, in the second term on the right of Equation (6.1.11) $i = 2$, $j = 3$, $k = 1$ and, if we exchange $\mathbf{e}_1$ and $\mathbf{e}_2$ in $|\mathbf{e}_1\ \mathbf{e}_2\ \mathbf{e}_3| = 1$, we get $|\mathbf{e}_2\ \mathbf{e}_1\ \mathbf{e}_3| = -1$ and, if next we exchange $\mathbf{e}_1$ and $\mathbf{e}_3$, we obtain $|\mathbf{e}_2\ \mathbf{e}_3\ \mathbf{e}_1| = 1$ for the desired values of i, j, k.)

<<>>

EXAMPLE 6.1.2. Let us evaluate such a determinant:

$$\begin{vmatrix} 1 & 4 & 7 \\ 2 & 5 & 8 \\ 3 & 6 & 9 \end{vmatrix} = 1 \cdot 5 \cdot 9 + 2 \cdot 6 \cdot 7 + 3 \cdot 4 \cdot 8 - 1 \cdot 6 \cdot 8$$

$$- 2 \cdot 4 \cdot 9 - 3 \cdot 5 \cdot 7 = 45 + 84 + 96 - 48 - 72 - 105 = 0. \quad (6.1.13)$$

That the value turned out to be zero is a result of the three column vectors' being linearly dependent, as will be seen shortly. Soon we shall also discuss better methods for evaluating determinants.

<<>>

In the general case we proceed much as in those above. We use linearity to write any nth order determinant as a linear combination of determinants of standard vectors analogously to Equation (6.1.12), and reduce the nonzero ones among the latter to ± 1 by exchanges of adjacent $\mathbf{e}_i$ vectors.

For any determinant of the n standard vectors of $\mathbb{R}^n$, every exchange of *adjacent* $\mathbf{e}_i$ vectors, called a *transposition,* changes the sign. Consequently, if in any such determinant the number of transpositions we use to bring the standard vectors into their natural order is odd, then the result is -1, and if the number of transpositions used is even, then it is $+1$. Although the natural order can be obtained from a given arrangement through various different sequences of transpositions, nevertheless, as will be seen below, it does not matter which of these sequences we choose.

Let us consider the set of all possible arrangements of the integers $1, 2, \ldots, n$ in a row. Any such arrangement is called a *permutation* of the natural arrangement (or natural permutation) $(1, 2, \ldots, n)$.

LEMMA 6.1.1. The number of permutations of n elements is $n!$.

Proof For $n = 1$ the only permutation is (1) and so its number is $1! = 1$. For $n = 2$ we have the two permutations (1, 2) and (2, 1). These can be thought of as arising from the preceding (1) by placing the digit 2 on either side of the 1 and so their number is $2 \cdot 1! = 2 \cdot 1 = 2!$. For $n = 3$ we can obtain all permutations by placing the digit 3 in all the possible places in the previous permutations of two digits; that is, to the left of the first digit, between the two digits, or after them. Thus their number is $3 \cdot 2! = 3 \cdot 2 \cdot 1 = 3!$. This process can be continued to arbitrary n.

<<>>

An *inversion* will be said to exist between two digits of a permutation if the larger number precedes the smaller one. Thus, for example, the permutation (3, 2, 4, 5, 1) has the 5 inversions (3, 2), (3, 1), (2, 1), (4, 1), (5, 1).

A permutation is called *even* if it has an even number of inversions and *odd* otherwise. In particular, the natural permutation is even because it has zero inversions.

Any transposition changes the number of inversions by 1, because it affects only the relative order of the two transposed digits, and so it changes any even permutation into an odd one and vice versa. Hence the number of even permutations must equal the number of odd permutations, since if we consider all even permutations and transpose their first two digits, then we get odd permutations, from which a second such transposition reestablishes all the even ones. Thus the numbers of even and of odd permutations both equal $n!/2$.

Obviously any permutation can be changed into the natural permutation by a sequence of transpositions, and each transposition changes the number of inversions by one, and so any odd permutation requires an odd number of transpositions for it to be changed into the natural permutation (which is even), and any even permutation requires an even number of transpositions for it to be changed into the natural permutation, regardless of the particular transpositions used.

Let us denote any permutation of the integers $1, 2, \ldots, n$ by P; that is, let $P = (p_1, p_2, \ldots, p_n)$, where the p_i are the numbers $1, 2, \ldots, n$ permuted. Define a function ε on the set of all such permutations by

$$\varepsilon(P) = \begin{cases} 1 \text{ if } P \text{ is even} \\ -1 \text{ if } P \text{ is odd.} \end{cases} \tag{6.1.14}$$

With this notation the discussion above thus proves the following theorem:

THEOREM 6.1.4. The determinant of any $n \times n$ matrix A is given by

$$\det(A) = \sum_P \varepsilon(P) a_{p_1 1} a_{p_2 2} \cdots a_{p_n n}, \tag{6.1.15}$$

where the sum runs through all permutations of $1, 2, \ldots, n$.

This formula is hopelessly inefficient for computing determinants for large n, because the number of terms is $n!$, which grows very fast. Already for $n = 5$ or 6 we have $5! = 120$ and $6! = 720$. We shall, however, use this formula to prove some other properties that will lead to better evaluation methods. Also, it is theoretically very important to know that the defining axioms are satisfied by a unique expression.

The next three theorems will show how determinants can be evaluated by elementary operations on the rows or columns.

THEOREM 6.1.5.

1. If the matrix A' is obtained from A by adding any scalar c times one column to another, then $|A'| = |A|$.
2. If a matrix A has a zero column, then $|A| = 0$.

Proof

1. Let

$$|A| = |\cdots\ \boldsymbol{a}_i \cdots \boldsymbol{a}_j \cdots| \tag{6.1.16}$$

and assume that A' is obtained by adding c times the jth column of A to the ith one. Then by Axioms 1 and 2 we have

$$|A'| = |\cdots\ \boldsymbol{a}_i + c\boldsymbol{a}_j \cdots \boldsymbol{a}_j \cdots|$$

$$= |\cdots\ \boldsymbol{a}_i \cdots \boldsymbol{a}_j \cdots| + c|\cdots\ \boldsymbol{a}_j \cdots \boldsymbol{a}_j \cdots| = |A| + 0 = |A|. \tag{6.1.17}$$

2. Add any other column to the zero column. Then, by Part 1, the determinant does not change and, because the resulting determinant has two equal columns, it equals zero.

<<>>

THEOREM 6.1.6. For any square matrix A we have $\det(A^T) = \det(A)$.

Proof From Theorem 6.1.4 we have the formula

$$\det(A) = \sum_P \varepsilon(P) a_{p_1 1} a_{p_2 2} \cdots a_{p_n n}, \tag{6.1.18}$$

where the sum runs through all permutations of 1, 2, ..., n. Now every term in this sum contains one matrix element from each row and one from each column and they are arranged in the order of the columns. Their product in each term remains unchanged if we rearrange them in the order of the rows; that is, if we rearrange the product $a_{p_1 1} a_{p_2 2} \cdots a_{p_n n}$ to $a_{1 q_1} a_{2 q_2} \cdots a_{n q_n}$.

Now $Q = (q_1, q_2, \ldots, q_n)$ is the inverse of the permutation P; that is, it is the permutation that brings $P = (p_1, p_2, \ldots, p_n)$ back to the natural order. Thus Q can be obtained with exactly as many transpositions as P can, and so it is even when P is even and odd when P is odd. Therefore

$$\det(A) = \sum_P \varepsilon(P) a_{p_1 1} a_{p_2 2} \cdots a_{p_n n}$$

$$= \sum_Q \varepsilon(Q) a_{1 q_1} a_{2 q_2} \cdots a_{n q_n} = \det(A^T). \tag{6.1.19}$$

<<>>

COROLLARY 6.1.1. Every property of columns of determinants is also valid for the rows.

This result will enable us to compute determinants by the more familiar row reductions rather than by column operations. The next theorem tells us what the last step of the reduction is in most cases.

THEOREM 6.1.7. If A is an upper triangular matrix, then $|A|$ equals the product of its diagonal elements.

Proof The product of the diagonal elements is, of course, one of the terms in the expansion of Theorem 6.1.4. Thus we want to show that in this case all the other terms are necessarily zero.

Assume first that none of the diagonal elements is zero. Consider any one of the terms in the expansion, say T. How can we make it nonzero? Because every term in the sum contains exactly one matrix element from each row and one from each column, from the first column T must contain a_{11}, since otherwise it would be zero. Next, T cannot contain a_{12}, since it already has a factor from the first row. From the second column it must contain a_{22}, since it cannot contain a_{12} nor one of the zeroes below the diagonal. Similarly, from the third column it must contain a_{33}, since it cannot contain a_{13} or a_{23} because it already contains elements of the first and second rows. This argument could be continued for the remaining factors, and shows that the product of the diagonal elements is the only nonzero term in the expansion of Theorem 6.1.4.

If one of the diagonal elements is zero, then in the expansion even the term consisting of the product of the diagonal elements is zero, and so the product of the diagonal elements and the determinant equal each other, since both are zero.

<<>>

We can use the foregoing properties to compute determinants by row reductions, as in the following examples:

EXAMPLE 6.1.3. Let us again evaluate the determinant

$$D = \begin{vmatrix} 1 & 4 & 7 \\ 2 & 5 & 8 \\ 3 & 6 & 9 \end{vmatrix}. \tag{6.1.20}$$

If we subtract the second row from the third row, and the first row from the second row, then, by Part 1 of Theorem 6.1.5 applied to rows, the value of D remains unchanged and we get

$$D = \begin{vmatrix} 1 & 4 & 7 \\ 1 & 1 & 1 \\ 1 & 1 & 1 \end{vmatrix}. \tag{6.1.21}$$

Hence by Axiom 2 applied to rows we obtain $D = 0$.

<<>>

EXAMPLE 6.1.4. Let us evaluate the determinant

$$D = \begin{vmatrix} 1 & 1 & 2 \\ 2 & 5 & 4 \\ 3 & 6 & 9 \end{vmatrix}. \tag{6.1.22}$$

By Axiom 1 applied to rows we can factor out a 3 from the third row to get

$$D = 3 \cdot \begin{vmatrix} 1 & 1 & 2 \\ 2 & 5 & 4 \\ 1 & 2 & 3 \end{vmatrix}. \tag{6.1.23}$$

By Part 1 of Theorem 6.1.5 applied to rows we can subtract twice the first row from the second row and the first row from the third row to get

$$D = 3 \cdot \begin{vmatrix} 1 & 1 & 2 \\ 0 & 3 & 0 \\ 0 & 1 & 1 \end{vmatrix}. \tag{6.1.24}$$

Exchanging the last two rows we obtain, by Theorem 6.1.1 applied to rows, the equation

$$D = -3 \cdot \begin{vmatrix} 1 & 1 & 2 \\ 0 & 1 & 1 \\ 0 & 3 & 0 \end{vmatrix}. \tag{6.1.25}$$

Finally, if we subtract 3 times the second row from the last row and apply Theorem 6.1.7, then we find

$$D = -3 \cdot \begin{vmatrix} 1 & 1 & 2 \\ 0 & 1 & 1 \\ 0 & 0 & -3 \end{vmatrix} = -3 \cdot 1 \cdot 1 \cdot (-3) = 9. \tag{6.1.26}$$

<<>>

From the properties of determinants above a very important relation between a matrix and its determinant can be deduced:

THEOREM 6.1.8. A matrix A is singular if and only if its determinant equals zero.

Proof First, if the matrix A is singular, then it can be reduced by elementary row operations to an upper triangular matrix U with a zero last row.

If we perform the same row operations on the determinant $|A|$, then in each step the determinant either remains unchanged or we can factor out some number. We have $|U| = 0$ and so $|A| = c|U| = 0$.

If, on the other hand, the matrix A is nonsingular, then it can be reduced by elementary row operations to the unit matrix I. We have $|I| = 1$ and, if we perform the same row operations on the determinant $|A|$, then in each step the determinant either remains unchanged or we can take out a nonzero factor. Hence $|A| = c|I| = c$ for some nonzero scalar c.

<<>>

There exists another useful relation between matrices and their determinants:

THEOREM 6.1.9. If A and B are square matrices of the same size, then $|AB| = |A||B|$.

Proof If we write $C = AB$ and $\boldsymbol{c}_1, \boldsymbol{c}_2, \ldots, \boldsymbol{c}_n$ for the columns of C, then we can express C in terms of the columns of B as

$$C = (A\boldsymbol{b}_1 \; A\boldsymbol{b}_2 \cdots A\boldsymbol{b}_n) \tag{6.1.27}$$

and each $\boldsymbol{c}_k = A\boldsymbol{b}_k$ here can also be written as a linear combination of the columns of A; that is, as

$$\boldsymbol{c}_k = \sum_i b_{ik}\boldsymbol{a}_i. \tag{6.1.28}$$

Then we can write the determinant of the $\boldsymbol{c}_i$ vectors as

$$|\boldsymbol{c}_1 \; \boldsymbol{c}_2 \cdots \boldsymbol{c}_n| = \left|\sum_i b_{i1}\boldsymbol{a}_i \quad \sum_j b_{j2}\boldsymbol{a}_j \quad \cdots \quad \sum_z b_{zn}\boldsymbol{a}_z\right|$$

$$= \sum_{i,j\cdots z} b_{i1}b_{j2}\cdots b_{zn}|\boldsymbol{a}_i \; \boldsymbol{a}_j \cdots \boldsymbol{a}_z| \tag{6.1.29}$$

The determinants on the right can be evaluated in much the same way as those of the standard vectors in Theorem 6.1.3; that is, whenever two of the subscripts are the same the determinant of the $\boldsymbol{a}_i$ vectors vanishes and otherwise it reduces to $\pm|A|$, with the sign given by $\varepsilon(P)$, where P stands for the permutation $(i, j, \ldots, z)$ of $(1, 2, \ldots, n)$. This results in

$$|C| = \sum_P \varepsilon(P) b_{i1}b_{j2}\cdots b_{zn}|\boldsymbol{a}_1 \; \boldsymbol{a}_2 \cdots \boldsymbol{a}_n| = |A||B|. \tag{6.1.30}$$

<<>>

An immediate consequence of this theorem is the following result, whose proof is left as an exercise.

COROLLARY 6.1.2. If A is an invertible matrix, then

$$|A^{-1}| = \frac{1}{|A|}. \tag{6.1.31}$$

Exercises 6.1

In the first four exercises, evaluate the determinants by row reduction.

6.1.1. $\begin{vmatrix} 2 & -3 & 2 \\ 1 & 4 & 0 \\ 0 & 1 & -5 \end{vmatrix}$

• **6.1.2.** $\begin{vmatrix} 0 & 1 & 2 \\ 4 & 0 & 3 \\ 3 & 2 & 1 \end{vmatrix}$

6.1.3. $\begin{vmatrix} -1 & 1 & 2 & 3 \\ 2 & 0 & -5 & 0 \\ 0 & 0 & 0 & -1 \\ 3 & -3 & 1 & 4 \end{vmatrix}$

6.1.4. $\begin{vmatrix} -1 & -2 & 2 & 1 \\ 0 & 0 & -1 & 0 \\ 0 & 0 & 1 & -1 \\ 1 & -1 & 1 & -1 \end{vmatrix}$

6.1.5. Prove that in the trivial case of $n = 1$, Property 2 of Definition 6.1.1 does not apply and the other properties give $\det(A) = \det(a_{11}) = a_{11}$.

6.1.6. Prove that for any $n \times n$ matrix A and any scalar c we have $\det(cA) = c^n \det(A)$.

6.1.7. Show that the result of Theorem 6.1.7 holds for lower triangular matrices as well; that is, that

$$\begin{vmatrix} a_{11} & 0 & 0 \\ a_{21} & a_{22} & 0 \\ a_{31} & a_{32} & a_{33} \end{vmatrix} = a_{11}a_{22}a_{33}$$

and similar relations hold for any n.

6.1.8. Is the analog of Theorem 6.1.7 true for matrices lower triangular with respect to the secondary diagonal; that is, does

$$\begin{vmatrix} 0 & 0 & a_{13} \\ 0 & a_{22} & a_{23} \\ a_{31} & a_{32} & a_{33} \end{vmatrix} = -a_{13}a_{22}a_{31}$$

hold, and similar relations for $n \neq 3$? Prove your result!

6.1.9. Prove Corollary 6.1.2.

• **6.1.10.** Show that if A and B are similar matrices, then $\det(A) = \det(B)$.

• **6.1.11.** Use Theorems 6.1.7 and 6.1.8 to show that for any $n \times n$ matrices A and B the product AB is invertible if and only if both A and B are.

6.1.12. A matrix A is called *skew-symmetric* if $A^T = -A$. Show that for any 3×3 skew-symmetric matrix $\det(A) = 0$. Is this true for other values of n?

6.1.13. Prove that for any orthogonal matrix Q we have $\det(Q) = \pm 1$.

6.1.14. Show that for the so-called Vandermonde determinant of order three

$$\begin{vmatrix} 1 & a & a^2 \\ 1 & b & b^2 \\ 1 & c & c^2 \end{vmatrix} = (b-a)(c-b)(c-a). \qquad (6.1.32)$$

***6.1.15.** Generalize the result of Exercise 6.1.14 to $n > 3$ and prove your formula.

6.1.16. Show, using Theorem 6.1.8 and Equation (6.1.32), that the monomials $1, x, x^2$ are linearly independent in the vector space $\mathcal{P}$ of all polynomials over $\mathbb{R}$. (*Hint:* Substitute three numbers a, b, c for x into the definition of linear independence applied to these functions.)

MATLAB Exercises 6.1

In MATLAB the determinant of a matrix A is given by **det**(A). Note that it is preferable to use **rank**(A) rather than **det**(A) to determine whether A is singular or not, because MATLAB's computation of the latter is more affected by roundoff errors.

ML6.1.1. Let A = **round**(10 ∗ **rand**(5)) and **x** = **round**(10 ∗ **rand**(5, 1)).

a. Create a matrix B by entering $B = A$; $B(:, 4)$ = **x**. What does this command do?

b. Create matrices as above, in which the third column of A is replaced by multiples of random vectors, and use these to illustrate the linear dependence of the **det** function on the third column.

c. Is **det**(A) + **det**(B) equal to **det**($A + B$) in general, when A and B are the same size? Experiment with random matrices. Explain.

d. Let A be as above and compute $B = A$; $B(:, 4) = B(:, 4) + 3 * B(:, 1)$. Compare **det**($A$) and **det**($B$). Explain.

ML6.1.2. Enter the following program and explain what it does:

$\mathbf{x} = 1:6$
$y = \mathbf{randperm}(6)$
$A = \mathbf{vander}(\mathbf{x})'$
$B = \mathbf{vander}(y)'$
$P = B\backslash A$
$\mathbf{det}(P)$

ML6.1.3. The following program achieves the same result as the one above but in a much more efficient way. Enter it and explain what it does:

$n = 6,$
$\mathbf{x} = \mathbf{randperm}(n),$
$I = \mathbf{eye}(n),$
$P = I,$
for $i = 1:n$
  $P(i,:) = I(\mathbf{x}(i),:)$
end
$\mathbf{det}(P)$

6.2. FURTHER PROPERTIES OF DETERMINANTS

Frequently, determinants are evaluated by reduction formulas, called *expansions along a row or a column,* in which an nth order determinant for any $n \geq 2$ is[1] expressed in terms of certain determinants of order $n-1$. The latter have special names:

DEFINITION 6.2.1.

Given any $n \times n$ matrix A, we define its ijth minor M_{ij} as the *determinant* of the submatrix S_{ij} obtained from A by deleting the ith row and jth column. The quantity $A_{ij} = (-1)^{i+j}M_{ij}$ is called the *cofactor* of a_{ij}.

Before stating the general theorem let us examine the 3×3 case. Then

$$|A| = \begin{vmatrix} a_{11} & a_{12} & a_{13} \\ a_{21} & a_{22} & a_{23} \\ a_{31} & a_{32} & a_{33} \end{vmatrix}$$

$$\begin{aligned} &= a_{11}a_{22}a_{33} + a_{21}a_{32}a_{13} + a_{31}a_{12}a_{23} \\ &\quad - a_{11}a_{32}a_{23} - a_{21}a_{12}a_{33} - a_{31}a_{22}a_{13}. \end{aligned} \tag{6.2.1}$$

[1]We shall assume throughout this section without further mention that $n > 1$ holds.

If we factor out the elements of the first column and apply the definition of the 2×2 determinant, then we get

$$|A| = a_{11}(a_{22}a_{33} - a_{32}a_{23}) - a_{21}(a_{12}a_{33} - a_{32}a_{13}) + a_{31}(a_{12}a_{23} - a_{22}a_{13}) \tag{6.2.2}$$

and

$$|A| = a_{11}\begin{vmatrix} a_{22} & a_{23} \\ a_{32} & a_{33} \end{vmatrix} - a_{21}\begin{vmatrix} a_{12} & a_{13} \\ a_{32} & a_{33} \end{vmatrix} + a_{31}\begin{vmatrix} a_{12} & a_{13} \\ a_{22} & a_{23} \end{vmatrix}. \tag{6.2.3}$$

The determinants in this formula are the minors M_{11}, M_{21}, M_{31} respectively, and the same determinants with the signs included (in this case it makes a difference only in the second term) are the corresponding cofactors. Equation (6.2.3) is the expansion of $|A|$ along its first column. By factoring out the elements of any other row or column we would obtain analogous expansions with respect to those.

THEOREM 6.2.1. The determinant of any $n \times n$ matrix A can, for any fixed i, be evaluated as

$$|A| = a_{i1}A_{i1} + a_{i2}A_{i2} + \cdots + a_{in}A_{in} \tag{6.2.4}$$

and also, for any fixed j, as

$$|A| = a_{1j}A_{1j} + a_{2j}A_{2j} + \cdots + a_{nj}A_{nj}. \tag{6.2.5}$$

Proof Let us consider the expansion along the jth column; that is, let us start with the proof of the second formula. Write $\sum_i a_{ij}\boldsymbol{e}_i$ for $\boldsymbol{a}_j$ in $|A|$:

$$|A| = |\boldsymbol{a}_1 \cdots \boldsymbol{a}_{j-1} \quad \sum_i a_{ij}\boldsymbol{e}_i \quad \boldsymbol{a}_{j+1} \cdots \boldsymbol{a}_n| \tag{6.2.6}$$

and by linearity rewrite this equation as

$$|A| = \sum_i a_{ij}|\boldsymbol{a}_1 \cdots \boldsymbol{a}_{j-1} \quad \boldsymbol{e}_i \quad \boldsymbol{a}_{j+1} \cdots \boldsymbol{a}_n|. \tag{6.2.7}$$

We are going to show that the determinants in the sum on the right are exactly the cofactors A_{ij}. We write the determinant of the ith term with the components of the column vectors as

$$|\boldsymbol{a}_1 \cdots \boldsymbol{a}_{j-1}\ \boldsymbol{e}_i\ \boldsymbol{a}_{j+1} \cdots \boldsymbol{a}_n| = \begin{vmatrix} a_{11} & a_{12} & \dots & 0 & \dots & a_{1n} \\ a_{21} & a_{22} & \dots & 0 & \dots & a_{2n} \\ \vdots & \vdots & & \vdots & & \vdots \\ a_{i1} & a_{i2} & \dots & 1 & \dots & a_{in} \\ \vdots & \vdots & & \vdots & & \vdots \\ a_{n1} & a_{n2} & \dots & 0 & \dots & a_{nn} \end{vmatrix}. \qquad (6.2.8)$$

By subtracting appropriate multiples of the jth column from the others, we can change this to

$$|\boldsymbol{a}_1 \cdots \boldsymbol{a}_{j-1}\ \boldsymbol{e}_i\ \boldsymbol{a}_{j+1} \cdots \boldsymbol{a}_n| = \begin{vmatrix} a_{11} & a_{12} & \dots & 0 & \dots & a_{1n} \\ a_{21} & a_{22} & \dots & 0 & \dots & a_{2n} \\ \vdots & \vdots & & \vdots & & \vdots \\ 0 & 0 & \dots & 1 & \dots & 0 \\ \vdots & \vdots & & \vdots & & \vdots \\ a_{n1} & a_{n2} & \dots & 0 & \dots & a_{nn} \end{vmatrix}, \qquad (6.2.9)$$

where the 0's and the 1 replace the original ith row and jth column. We can move the ith row to the top with $i-1$ transpositions and the jth column to the left with $j-1$ transpositions. Since each transposition multiplies the determinant by -1, these moves result in

$$|\boldsymbol{a}_1 \cdots \boldsymbol{a}_{j-1}\ \boldsymbol{e}_i\ \boldsymbol{a}_{j+1} \cdots \boldsymbol{a}_n| = (-1)^{i+j} \begin{vmatrix} 1 & \boldsymbol{0} \\ \boldsymbol{0} & S_{ij} \end{vmatrix}, \qquad (6.2.10)$$

where S_{ij} is the submatrix obtained from A by deleting the ith row and jth column. The determinant on the right equals the minor $M_{ij} = \det(S_{ij})$ because reducing it to upper triangular form we can use the same elementary row operations as we would on the corresponding rows of $|S_{ij}|$, and would get the same multipliers factored out and the same products of the diagonal elements. Thus

$$|\boldsymbol{a}_1 \cdots \boldsymbol{a}_{j-1}\ \boldsymbol{e}_i\ \boldsymbol{a}_{j+1} \cdots \boldsymbol{a}_n| = (-1)^{i+j}|S_{ij}| = A_{ij}. \qquad (6.2.11)$$

Substituting this expression into Equation (6.2.7) we obtain Equation (6.2.5) of the theorem. Equation (6.2.4); that is, the expansion along any row, follows from Equation (6.2.5) and Theorem 6.1.6.

<<>>

Before giving an example let us remark that the signs $(-1)^{i+j}$ in the definition of the cofactors alternate in a checkerboard-like pattern, as shown in Figure 6.4, and are usually taken from there rather than from the formula $(-1)^{i+j}$.

FIGURE 6.4

+	−	+	−	⋯
−	+	−	+	⋯
+	−	+	−	⋯
−	+	−	+	⋯
⋯		⋯		⋯

EXAMPLE 6.2.1. Evaluate the determinant

$$D = \begin{vmatrix} 1 & 1 & 0 & 2 \\ 2 & 3 & 5 & 4 \\ 1 & 0 & 0 & 1 \\ 0 & 6 & 2 & 3 \end{vmatrix} \tag{6.2.12}$$

using Theorem 6.2.1.

Since the third row and the third column have the most zeroes, it is simplest to expand along one of those. Let us choose the third row, say. Then we get

$$D = 1 \cdot \begin{vmatrix} 1 & 0 & 2 \\ 3 & 5 & 4 \\ 6 & 2 & 3 \end{vmatrix} - 1 \cdot \begin{vmatrix} 1 & 1 & 0 \\ 2 & 3 & 5 \\ 0 & 6 & 2 \end{vmatrix}. \tag{6.2.13}$$

The 3×3 determinants here can be expanded similarly, along their first rows, say. Thus we obtain

$$D = 1 \cdot \begin{vmatrix} 5 & 4 \\ 2 & 3 \end{vmatrix} + 2 \cdot \begin{vmatrix} 3 & 5 \\ 6 & 2 \end{vmatrix} - 1 \cdot \begin{vmatrix} 3 & 5 \\ 6 & 2 \end{vmatrix} + 1 \cdot \begin{vmatrix} 2 & 5 \\ 0 & 2 \end{vmatrix}. \tag{6.2.14}$$

Finally, this expression can be evaluated as

$$\begin{aligned} D = {} & (5 \cdot 3 - 2 \cdot 4) + 2 \cdot (3 \cdot 2 - 6 \cdot 5) \\ & - (3 \cdot 2 - 6 \cdot 5) + 2 \cdot 2 = -13. \end{aligned} \tag{6.2.15}$$

<<>>

We are now in a position to be able to state the solution of any $n \times n$ system of linear equations in closed form using determinants:

THEOREM 6.2.2. Cramer's Rule:[2] The solution of $A\boldsymbol{x} = \boldsymbol{b}$ for any invertible $n \times n$ matrix A is given by

$$x_i = \frac{|\boldsymbol{a}_1 \cdots \boldsymbol{a}_{i-1}\, \boldsymbol{b}\, \boldsymbol{a}_{i+1} \cdots \boldsymbol{a}_n|}{|A|} \quad \text{for } i = 1, 2, \ldots, n. \tag{6.2.16}$$

[2]Named after Gabriel Cramer (1704–1752).

Proof Write the system to be solved in the form

$$x_1\boldsymbol{a}_1 + x_2\boldsymbol{a}_2 + \cdots + x_n\boldsymbol{a}_n = \boldsymbol{b}. \tag{6.2.17}$$

Using each side of this equation to replace the ith column of the determinant of the $\boldsymbol{a}_j$ vectors, we get

$$\begin{aligned} &|\boldsymbol{a}_1 \cdots \boldsymbol{a}_{i-1} \quad x_1\boldsymbol{a}_1 + x_2\boldsymbol{a}_2 + \cdots + x_n\boldsymbol{a}_n \quad \boldsymbol{a}_{i+1} \cdots \boldsymbol{a}_n| \\ &= |\boldsymbol{a}_1 \cdots \boldsymbol{a}_{i-1} \quad \boldsymbol{b} \quad \boldsymbol{a}_{i+1} \cdots \boldsymbol{a}_n|. \end{aligned} \tag{6.2.18}$$

On the left, subtract from the ith column x_1 times the first column, x_2 times the second column, and so on. This operation results in

$$|\boldsymbol{a}_1 \cdots \boldsymbol{a}_{i-1} \quad x_i\boldsymbol{a}_i \quad \boldsymbol{a}_{i+1} \cdots \boldsymbol{a}_n| = |\boldsymbol{a}_1 \cdots \boldsymbol{a}_{i-1} \quad \boldsymbol{b} \quad \boldsymbol{a}_{i+1} \cdots \boldsymbol{a}_n|, \tag{6.2.19}$$

which can be changed to

$$x_i|\boldsymbol{a}_1 \cdots \boldsymbol{a}_{i-1} \quad \boldsymbol{a}_i \quad \boldsymbol{a}_{i+1} \cdots \boldsymbol{a}_n| = |\boldsymbol{a}_1 \cdots \boldsymbol{a}_{i-1} \quad \boldsymbol{b} \quad \boldsymbol{a}_{i+1} \cdots \boldsymbol{a}_n|. \tag{6.2.20}$$

Dividing by $|A|$, which is not zero by the assumed invertibility of A, we obtain the formula of the theorem. Since we know that $A\mathbf{x} = \boldsymbol{b}$ has a unique solution for any invertible A, it must be the one with these values of the x_i. (A direct check is left as Exercise 6.2.5.)

<<>>

In addition to Cramer's Rule for solving linear systems, there also exists an explicit formula involving determinants for the inverse of a matrix. However, before presenting it we state an interesting intermediate result that we shall need in the proof:

LEMMA 6.2.1. For any $n \times n$ matrix A, if we add the elements of one row (column) multiplied by the cofactors of another row (column), then we get zero; that is,

$$a_{k1}A_{j1} + a_{k2}A_{j2} + \cdots + a_{kn}A_{jn} = 0 \text{ for } k \neq j \tag{6.2.21}$$

and

$$a_{1k}A_{1j} + a_{2k}A_{2j} + \cdots + a_{nk}A_{nj} = 0 \text{ for } k \neq j. \tag{6.2.22}$$

Proof The proof of this lemma is similar to that of Theorem 6.2.1. Consider the determinant of the matrix A with the jth column replaced by $\boldsymbol{a}_k$ for some $k \neq j$. Then, because two columns are equal, we have

$$|\boldsymbol{a}_1 \cdots \boldsymbol{a}_{j-1} \quad \boldsymbol{a}_k \quad \boldsymbol{a}_{j+1} \cdots \boldsymbol{a}_n| = 0 \tag{6.2.23}$$

and, expanding the $\boldsymbol{a}_k$ vector that is in the jth place, we can write this equation as

$$|\boldsymbol{a}_1 \cdots \boldsymbol{a}_{j-1} \quad \sum_i a_{ik}\boldsymbol{e}_i \quad \boldsymbol{a}_{j+1} \cdots \boldsymbol{a}_n| = 0, \tag{6.2.24}$$

which is equivalent to

$$\sum_i a_{ik} |\boldsymbol{a}_1 \cdots \boldsymbol{a}_{j-1} \quad \boldsymbol{e}_i \quad \boldsymbol{a}_{j+1} \cdots \boldsymbol{a}_n| = 0. \tag{6.2.25}$$

Equation (6.2.11) tells us that the determinant in the ith term here is A_{ij}. Thus we have

$$\sum_i a_{ik} A_{ij} = 0, \tag{6.2.26}$$

which is the same as Equation (6.2.22). Equation (6.2.21) then follows by Theorem 6.1.6.

<<>>

In Equation (6.2.21) we have a dot product of the ith row of A with the jth row of a matrix whose elements are the cofactors. We want to make use of this fact in the next theorem, but we generally prefer to write such dot products as products of a row by a column of a matrix, and so we define the second matrix as follows:

DEFINITION 6.2.2.

The transposed matrix of the cofactors of A is called the *adjoint matrix of A* and is written

$$\operatorname{adj}(A) = (A_{ij})^T. \tag{6.2.27}$$

THEOREM 6.2.3.

The inverse of any invertible matrix A is given by

$$A^{-1} = \frac{\operatorname{adj}(A)}{|A|}. \tag{6.2.28}$$

Proof If we take the matrix product of A and $\operatorname{adj}(A)$ in this order, then by Equation (6.2.21) the offdiagonal elements all vanish, and by Theorem 6.2.1 the diagonal elements are all equal to $|A|$. Thus $|A|$ can be factored out and we get

$$A \operatorname{adj}(A) = |A| I. \tag{6.2.29}$$

This implies the statement of the theorem.

<<>>

EXAMPLE 6.2.2. Let us use Theorem 6.2.3 to find the inverse of

$$A = \begin{bmatrix} 1 & 1 & 2 \\ 2 & 5 & 4 \\ 3 & 6 & 9 \end{bmatrix}. \tag{6.2.30}$$

From Example 6.1.4 we know that $|A| = 9$, and the minors are $A_{11} = 5 \cdot 9 - 6 \cdot 4 = 21$, $A_{12} = -(2 \cdot 9 - 3 \cdot 4) = -6$, $A_{13} = 2 \cdot 6 - 3 \cdot 5 = -3$, $A_{21} = -(1 \cdot 9 - 6 \cdot 2) = 3$, $A_{22} = 1 \cdot 9 - 3 \cdot 2 = 3$, etc. Thus

$$A^{-1} = \frac{1}{9}\begin{bmatrix} 21 & 3 & -6 \\ -6 & 3 & 0 \\ -3 & -3 & 3 \end{bmatrix}. \tag{6.2.31}$$

<<>>

There remains only one property of determinants to discuss, the one that we used to motivate their definition at the beginning of Section 6.1, namely their relationship to areas and volumes. The only problem that we need to clear up is how the nonnegativity of the latter can be reconciled with the linearity of determinants. We are going to show in a somewhat indirect way that the area of the parallelogram spanned by the vectors $\boldsymbol{a}_1$ and $\boldsymbol{a}_2$ in $\mathbb{R}^2$ is given by $|\det(\boldsymbol{a}_1, \boldsymbol{a}_2)|$.

The area properties that we use will be based in part on the introductory discussion, but instead of the linearity property we employ the first property of Theorem 6.1.5 on page 190; that is, the equation

$$|\boldsymbol{a}_1, \boldsymbol{a}_2| = |\boldsymbol{a}_1, \boldsymbol{a}_2 + \lambda\boldsymbol{a}_1|. \tag{6.2.32}$$

FIGURE 6.5

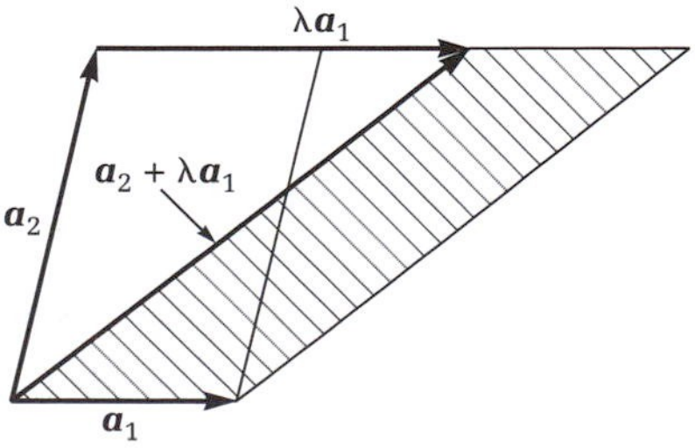

That a corresponding equation is valid for areas spanned by any vectors $\boldsymbol{a}_1$ and $\boldsymbol{a}_2$ in $\mathbb{R}^2$ and any real λ can be seen from Figure 6.5, where the area of the shaded parallelogram equals the area of the unshaded one. Thus we define the notion of an area function as follows:

DEFINITION 6.2.3.

An *area function on* $\mathbb{R}^2$ is a function F that assigns to every pair of vectors $\boldsymbol{a}_1$ and $\boldsymbol{a}_2$ in $\mathbb{R}^2$ a number $F(\boldsymbol{a}_1, \boldsymbol{a}_2)$, such that

$$F(\boldsymbol{a}_1, \boldsymbol{a}_2) = F(\boldsymbol{a}_1, \boldsymbol{a}_2 + \lambda\boldsymbol{a}_1), \tag{6.2.33}$$

$$F(\boldsymbol{a}_1, \lambda\boldsymbol{a}_2) = |\lambda| F(\boldsymbol{a}_1, \boldsymbol{a}_2), \tag{6.2.34}$$

and

$$F(\boldsymbol{a}_1, \boldsymbol{a}_2) = F(\boldsymbol{a}_2, \boldsymbol{a}_1) \tag{6.2.35}$$

hold for any real λ and any vectors $\boldsymbol{a}_1$ and $\boldsymbol{a}_2$, and

$$F(\boldsymbol{e}_1, \boldsymbol{e}_2) = 1. \tag{6.2.36}$$

It is clear that $|\det|$ is an area function, and the next theorem shows that it is the only one.

THEOREM 6.2.4.

The only function on $\mathbb{R}^2$ with the properties above is the absolute value of the determinant.

Proof Assume that F is an area function on $\mathbb{R}^2$. Then first we see that $F(\boldsymbol{a}_1, \boldsymbol{0}) = 0$, for by Equation (6.2.34) we have

$$F(\boldsymbol{a}_1, \boldsymbol{0}) = F(\boldsymbol{a}_1, 0\boldsymbol{0}) = 0F(\boldsymbol{a}_1, \boldsymbol{0}) = 0. \tag{6.2.37}$$

In this case $\det(\boldsymbol{a}_1, \boldsymbol{0}) = 0$ holds, too, and so F is the same as $|\det|$.

If, on the other hand, $\boldsymbol{a}_2 \neq \boldsymbol{0}$, then assuming $a_{22} \neq 0$ we can reduce $F(\boldsymbol{a}_1, \boldsymbol{a}_2)$ as follows:

$$\begin{aligned} F(\boldsymbol{a}_1, \boldsymbol{a}_2) &= F\left(\boldsymbol{a}_1 - \frac{a_{21}}{a_{22}}\boldsymbol{a}_2, \boldsymbol{a}_2\right) = F\left(\left(a_{11} - \frac{a_{21}}{a_{22}}a_{12}\right)\boldsymbol{e}_1, \boldsymbol{a}_2\right) = F\left(\frac{|A|}{a_{22}}\boldsymbol{e}_1, \boldsymbol{a}_2\right) \\ &= \left|\frac{|A|}{a_{22}}\right|(F(\boldsymbol{e}_1, \boldsymbol{a}_2) = \left|\frac{|A|}{a_{22}}\right| F(\boldsymbol{e}_1, \boldsymbol{a}_2 - a_{12}\boldsymbol{e}_1) = \left|\frac{|A|}{a_{22}}\right| F(\boldsymbol{e}_1, a_{22}\boldsymbol{e}_2) \\ &= |\det(A)| F(\boldsymbol{e}_1, \boldsymbol{e}_2) = |\det(A)|. \end{aligned} \tag{6.2.38}$$

Thus F is the same as $|\det|$ in this case as well.

In the remaining case of $\boldsymbol{a}_2 \neq \boldsymbol{0}$ and $a_{22} = 0$ we must have $a_{12} \neq 0$ and we may proceed much as above, but use a_{12} as we used a_{22} before.

<<>>

Theorem 6.2.4 does not explain the geometric significance of the sign of the determinant. What we have is that $\det(\boldsymbol{a}_1, \boldsymbol{a}_2)$ is positive if $\boldsymbol{a}_1$ followed by $\boldsymbol{a}_2$ indicates a counterclockwise traversal of the parallelogram spanned

by $\boldsymbol{a}_1$ and $\boldsymbol{a}_2$, and is negative otherwise. We do not prove this statement here.

In three dimensions we could define a volume function for parallelepipeds by properties similar to those in Definition 6.2.3 and prove the theorem analogous to the last one, that the only such function is the absolute value of the third-order determinant. Because all this work would be very much like the discussion above, just more involved, we do not present it. Furthermore, the sign of $\det(\boldsymbol{a}_1, \boldsymbol{a}_2, \boldsymbol{a}_3)$ is positive if $\boldsymbol{a}_1, \boldsymbol{a}_2, \boldsymbol{a}_3$ form a right-handed triple in this order, and negative otherwise. Again, this subject will not be discussed here any further.

Exercises 6.2

In the first four exercises, evaluate the determinants of the corresponding exercises in Section 6.1 by expansion along any row or column:

6.2.1. $$\begin{vmatrix} 2 & -3 & 2 \\ 1 & 4 & 0 \\ 0 & 1 & -5 \end{vmatrix}$$

• **6.2.2.** $$\begin{vmatrix} 0 & 1 & 2 \\ 4 & 0 & 3 \\ 3 & 2 & 1 \end{vmatrix}$$

6.2.3. $$\begin{vmatrix} -1 & 1 & 2 & 3 \\ 2 & 0 & -5 & 0 \\ 0 & 0 & 0 & -1 \\ 3 & -3 & 1 & 4 \end{vmatrix}$$

6.2.4. $$\begin{vmatrix} -1 & -2 & 2 & 1 \\ 0 & 0 & -1 & 0 \\ 0 & 0 & 1 & -1 \\ 1 & -1 & 1 & -1 \end{vmatrix}$$

6.2.5. Check Cramer's Rule by direct substitution from Equation (6.2.16) into Equation (6.2.17).

In the next three exercises, solve $A\boldsymbol{x} = \boldsymbol{b}$ by Cramer's Rule with the given data, if possible.

6.2.6. $A = \begin{bmatrix} 1 & 3 \\ 2 & 1 \end{bmatrix}$ and $\boldsymbol{b} = \begin{bmatrix} -1 \\ 3 \end{bmatrix}$.

• **6.2.7.** $A = \begin{bmatrix} 1 & 1 & 2 \\ 2 & 0 & 4 \\ 0 & 3 & 1 \end{bmatrix}$ and $\boldsymbol{b} = \begin{bmatrix} 1 \\ 2 \\ 3 \end{bmatrix}$.

6.2.8. $A = \begin{bmatrix} 1 & 1 & 2 & 2 \\ 2 & 0 & 4 & 1 \\ 1 & 3 & 4 & 0 \\ 0 & 3 & 1 & 1 \end{bmatrix}$ and $\boldsymbol{b} = \begin{bmatrix} 1 \\ 0 \\ 4 \\ 3 \end{bmatrix}$.

• **6.2.9.** Use Theorem 6.2.3 to find the inverse of the matrix A in Exercise 6.2.6.

6.2.10. Use Theorem 6.2.3 to find the inverse of the matrix A in Exercise 6.2.7.

6.2.11. Show that for any invertible $n \times n$ matrix A we have

$$\det(\operatorname{adj}(A)) = (\det(A))^{n-1}. \tag{6.2.39}$$

• **6.2.12.** Show that for any invertible $n \times n$ matrix A the matrix $\operatorname{adj}(A)$ is also invertible and satisfies

$$(\operatorname{adj}(A))^{-1} = \operatorname{adj}(A^{-1}). \tag{6.2.40}$$

6.2.13. Show that in the plane the area of the triangle with vertices (a_1, a_2), (b_1, b_2), (c_1, c_2) is given by the absolute value of

$$\frac{1}{2}\begin{vmatrix} a_1 & a_2 & 1 \\ b_1 & b_2 & 1 \\ c_1 & c_2 & 1 \end{vmatrix}. \tag{6.2.41}$$

6.2.14. Show that in the plane an equation of the line through the points (x_1, y_1), (x_2, y_2) is given by

$$\begin{vmatrix} x & y & 1 \\ x_1 & y_1 & 1 \\ x_2 & y_2 & 1 \end{vmatrix} = 0. \tag{6.2.42}$$

• **6.2.15.** Show that in the plane an equation of the form $y = ax^2 + bx + c$ for a parabola through the distinct points (x_1, y_1), (x_2, y_2), (x_3, y_3) is equivalent to

$$\begin{vmatrix} y & x^2 & x & 1 \\ y_1 & x_1^2 & x_1 & 1 \\ y_2 & x_2^2 & x_2 & 1 \\ y_3 & x_3^2 & x_3 & 1 \end{vmatrix} = 0. \tag{6.2.43}$$

6.2.16. Show that in the plane an equation of the form $x^2 + y^2 + ax + by + c = 0$ for a circle through the points (x_1, y_1), (x_2, y_2), (x_3, y_3) is equivalent to

$$\begin{vmatrix} x^2 + y^2 & x & y & 1 \\ x_1^2 + y_1^2 & x_1 & y_1 & 1 \\ x_2^2 + y_2^2 & x_2 & y_2 & 1 \\ x_3^2 + y_3^2 & x_3 & y_3 & 1 \end{vmatrix} = 0. \tag{6.2.44}$$

6.2.17. Use the result of Exercise 6.2.16 to find an equation of the circle through the points (0, 0), (2, −1), (4, 0) and its center and radius.

***6.2.18.** Show that the volume of a tetrahedron with vertices (a_1, a_2, a_3), (b_1, b_2, b_3), (c_1, c_2, c_3), (d_1, d_2, d_3) is given by the absolute value of

$$\frac{1}{6}\begin{vmatrix} a_1 & a_2 & a_3 & 1 \\ b_1 & b_2 & b_3 & 1 \\ c_1 & c_2 & c_3 & 1 \\ d_1 & d_2 & d_3 & 1 \end{vmatrix}. \tag{6.2.45}$$

6.2.19. Show that in $\mathbb{R}^3$ an equation of the plane through the points (x_1, y_1, z_1), (x_2, y_2, z_2), (x_3, y_3, z_3) is given by

$$\begin{vmatrix} x & y & z & 1 \\ x_1 & y_1 & z_1 & 1 \\ x_2 & y_2 & z_2 & 1 \\ x_3 & y_3 & z_3 & 1 \end{vmatrix} = 0. \tag{6.2.46}$$

6.2.20. Show that if $\boldsymbol{T} : \mathbb{R}^2 \to \mathbb{R}^2$ is a linear transformation, then it maps the unit square to a parallelogram of area $|\det([T])|$.

6.3. THE CROSS PRODUCT OF VECTORS IN $\mathbb{R}^3$

Consider two arbitrary vectors $\boldsymbol{u}$ and $\boldsymbol{v}$ in $\mathbb{R}^3$. Project the corresponding parallelogram onto the coordinate planes and define a vector $\boldsymbol{u} \times \boldsymbol{v}$ with components equal to the areas of these projections with appropriate signs. (See Figure 6.6 for noncollinear $\boldsymbol{u}$ and $\boldsymbol{v}$. If they are collinear, the areas are zero and $\boldsymbol{u} \times \boldsymbol{v} = \boldsymbol{0}$.)

FIGURE 6.6

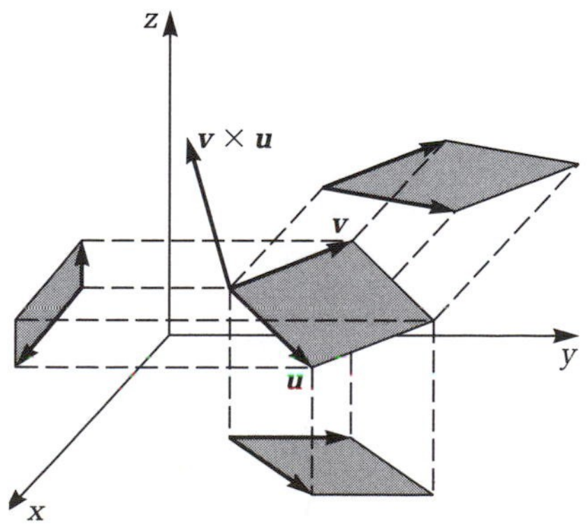

The formal definition is given below. As we shall see shortly, this definition makes $\boldsymbol{u} \times \boldsymbol{v}$ perpendicular to $\boldsymbol{u}$ and $\boldsymbol{v}$ and its length equal to the area of the parallelogram spanned by $\boldsymbol{u}$ and $\boldsymbol{v}$. It is, however, easier to start with the components.

DEFINITION 6.3.1.

We define the *vector product* or *cross product* for any vectors $\boldsymbol{u}$ and $\boldsymbol{v}$ of $\mathbb{R}^3$ as the vector of $\mathbb{R}^3$ given by

$$\boldsymbol{u} \times \boldsymbol{v} = (u_2v_3 - u_3v_2)\boldsymbol{i} - (u_1v_3 - u_3v_1)\boldsymbol{j} + (u_1v_2 - u_2v_1)\boldsymbol{k}. \quad (6.3.1)$$

This is usually abbreviated as

$$\boldsymbol{u} \times \boldsymbol{v} = \begin{vmatrix} \boldsymbol{i} & \boldsymbol{j} & \boldsymbol{k} \\ u_1 & u_2 & u_3 \\ v_1 & v_2 & v_3 \end{vmatrix}. \quad (6.3.2)$$

This construction works only in three dimensions, because in other dimensions the number of coordinate planes is different from the number of coordinate axes. In four dimensions, for example, we have six coordinate planes. (Why?) But, because our world is three-dimensional, this is a very important special case. Also, in this context it is customary to write $\boldsymbol{i}, \boldsymbol{j}, \boldsymbol{k}$ for the standard vectors.

EXAMPLE 6.3.1. The cross product of $\boldsymbol{u} = (1, 2, 3)^T$ and $\boldsymbol{v} = (4, 5, 6)^T$ is given by

$$\boldsymbol{u} \times \boldsymbol{v} = \begin{vmatrix} \boldsymbol{i} & \boldsymbol{j} & \boldsymbol{k} \\ 1 & 2 & 3 \\ 4 & 5 & 6 \end{vmatrix}$$

$$= (2 \cdot 6 - 3 \cdot 5)\boldsymbol{i} - (1 \cdot 6 - 3 \cdot 4)\boldsymbol{j} + (1 \cdot 5 - 2 \cdot 4)\boldsymbol{k} = -3\boldsymbol{i} + 6\boldsymbol{j} - 3\boldsymbol{k}. \quad (6.3.3)$$

<<>>

EXAMPLE 6.3.2. The cross products of the standard vectors can easily be computed from the definition. For instance, since $\boldsymbol{i} = (1, 0, 0)^T$ and $\boldsymbol{j} = (0, 1, 0)^T$, we have

$$\boldsymbol{i} \times \boldsymbol{j} = \begin{vmatrix} \boldsymbol{i} & \boldsymbol{j} & \boldsymbol{k} \\ 1 & 0 & 0 \\ 0 & 1 & 0 \end{vmatrix} = \boldsymbol{k}. \quad (6.3.4)$$

Similarly,

$$\boldsymbol{i} \times \boldsymbol{j} = -\boldsymbol{j} \times \boldsymbol{i} = \boldsymbol{k}, \boldsymbol{j} \times \boldsymbol{k} = -\boldsymbol{k} \times \boldsymbol{j} = \boldsymbol{i}, \boldsymbol{k} \times \boldsymbol{i} = -\boldsymbol{i} \times \boldsymbol{k} = \boldsymbol{j}, \quad (6.3.5)$$

and

$$\boldsymbol{i} \times \boldsymbol{i} = \boldsymbol{j} \times \boldsymbol{j} = \boldsymbol{k} \times \boldsymbol{k} = \boldsymbol{0}. \tag{6.3.6}$$

<<>>

In Theorem 6.3.1 we list the most useful properties of the cross product.

THEOREM 6.3.1. For any vectors $\boldsymbol{u}$, $\boldsymbol{v}$, $\boldsymbol{w}$ of $\mathbb{R}^3$ and any scalar c we have

1. $(c\boldsymbol{u}) \times \boldsymbol{v} = \boldsymbol{u} \times (c\boldsymbol{v}) = c(\boldsymbol{u} \times \boldsymbol{v})$
2. $\boldsymbol{v} \times \boldsymbol{u} = -\boldsymbol{u} \times \boldsymbol{v}$
3. $\boldsymbol{u} \times \boldsymbol{u} = \boldsymbol{0}$
4. $\boldsymbol{u} \cdot (\boldsymbol{u} \times \boldsymbol{v}) = 0$ and $\boldsymbol{v} \cdot (\boldsymbol{u} \times \boldsymbol{v}) = 0$
5. $\boldsymbol{u}$, $\boldsymbol{v}$, and $\boldsymbol{u} \times \boldsymbol{v}$, in this order, form a right-handed triple
6. $\boldsymbol{u} \times (\boldsymbol{v} + \boldsymbol{w}) = \boldsymbol{u} \times \boldsymbol{v} + \boldsymbol{u} \times \boldsymbol{w}$
7. $(\boldsymbol{u} + \boldsymbol{v}) \times \boldsymbol{w} = \boldsymbol{u} \times \boldsymbol{w} + \boldsymbol{v} \times \boldsymbol{w}$
8. $|\boldsymbol{u} \times \boldsymbol{v}|^2 = |\boldsymbol{u}|^2 |\boldsymbol{v}|^2 - (\boldsymbol{u} \cdot \boldsymbol{v})^2$
9. $|\boldsymbol{u} \times \boldsymbol{v}| = |\boldsymbol{u}||\boldsymbol{v}| \sin \theta$, where $\theta \in [0, \pi]$ denotes the angle between $\boldsymbol{u}$ and $\boldsymbol{v}$, and the right-hand side gives the area of the parallelogram spanned by $\boldsymbol{u}$ and $\boldsymbol{v}$
10. $\boldsymbol{u} \times \boldsymbol{v} = \boldsymbol{0}$ if and only if $\boldsymbol{u}$ and $\boldsymbol{v}$ are linearly dependent
11. $\boldsymbol{u} \cdot (\boldsymbol{v} \times \boldsymbol{w}) = \boldsymbol{v} \cdot (\boldsymbol{w} \times \boldsymbol{u}) = \boldsymbol{w} \cdot (\boldsymbol{u} \times \boldsymbol{v}) = \det(\boldsymbol{u} \;\; \boldsymbol{v} \;\; \boldsymbol{w})$
12. $(\boldsymbol{u} \times \boldsymbol{v}) \times \boldsymbol{w} = (\boldsymbol{u} \cdot \boldsymbol{w})\boldsymbol{v} - (\boldsymbol{v} \cdot \boldsymbol{w})\boldsymbol{u}$
13. $\boldsymbol{u} \times (\boldsymbol{v} \times \boldsymbol{w}) = (\boldsymbol{u} \cdot \boldsymbol{w})\boldsymbol{v} - (\boldsymbol{u} \cdot \boldsymbol{v})\boldsymbol{w}$.

Proof Statements 1, 2, and 3 follow by straightforward substitution from Definition 6.3.1. Notice that Statement 2 says that the cross product is not commutative but, as we say, *anticommutative*. Furthermore, Statement 3 coupled with Statement 1 for $\boldsymbol{v} = \boldsymbol{u}$ says that the cross product of collinear vectors is zero, as mentioned earlier.

The first part of Statement 4 can be proved as follows: From Equation (6.3.1) we have

$$\begin{aligned}\boldsymbol{u} \cdot (\boldsymbol{u} \times \boldsymbol{v}) &= (u_2v_3 - u_3v_2)\boldsymbol{u} \cdot \boldsymbol{i} - (u_1v_3 - u_3v_1)\boldsymbol{u} \cdot \boldsymbol{j} + (u_1v_2 - u_2v_1)\boldsymbol{u} \cdot \boldsymbol{k} \\ &= (u_2v_3 - u_3v_2)u_1 - (u_1v_3 - u_3v_1)u_2 + (u_1v_2 - u_2v_1)u_3 \\ &= \begin{vmatrix} u_1 & u_2 & u_3 \\ u_1 & u_2 & u_3 \\ v_1 & v_2 & v_3 \end{vmatrix} = 0. \end{aligned} \tag{6.3.7}$$

The second part of Statement 4 can be established similarly. Geometrically these two equations mean that the cross product is orthogonal to its factors. This is the most important property of the cross product, the one that makes it so useful in many applications.

Statement 5 is true for the standard vectors, as can be seen from the formulas in Example 6.3.2. For other vectors it could be proved by rotating and stretching or shrinking them into the standard vectors, because these operations do not change the "handedness" of a triple. We omit the details.

Statements 6 and 7 follow again by straightforward substitution from Definition 6.3.1.

Statement 8 can be proved as follows:

$$|\boldsymbol{u} \times \boldsymbol{v}|^2 = \sum_{i<j}(u_iv_j - u_jv_i)^2 = \frac{1}{2}\sum_{i\neq j}(u_iv_j - u_jv_i)^2 = \frac{1}{2}\sum_{i,\,j=1}^{3}(u_iv_j - u_jv_i)^2$$

$$= \frac{1}{2}\sum_{i,\,j=1}^{3}\left[(u_iv_j)^2 - 2u_iv_ju_jv_i + (u_jv_i)^2\right]$$

$$= \frac{1}{2}\sum u_i{}^2 \sum v_j{}^2 - \sum u_iv_i \sum u_jv_j + \frac{1}{2}\sum u_j{}^2 \sum v_i{}^2$$

$$= |\boldsymbol{u}|^2|\boldsymbol{v}|^2 - (\boldsymbol{u}\cdot\boldsymbol{v})^2. \qquad (6.3.8)$$

Statement 9 follows from Statement 8, because

$$|\boldsymbol{u}|^2|\boldsymbol{v}|^2 - (\boldsymbol{u}\cdot\boldsymbol{v})^2 = |\boldsymbol{u}|^2|\boldsymbol{v}|^2 - (|\boldsymbol{u}||\boldsymbol{v}|\cos\theta)^2 = (|\boldsymbol{u}||\boldsymbol{v}|\sin\theta)^2, \qquad (6.3.9)$$

and, both $|\boldsymbol{u} \times \boldsymbol{v}|$ and $|\boldsymbol{u}||\boldsymbol{v}|\sin\theta$ being nonnegative, we can take square roots in Equations (6.3.8) and (6.3.9). The expression $|\boldsymbol{u}||\boldsymbol{v}|\sin\theta$ gives the area of the parallelogram spanned by $\boldsymbol{u}$ and $\boldsymbol{v}$ because, as Figure 6.7 shows, $|\boldsymbol{u}|$ is the parallelogram's base length and $|\boldsymbol{v}|\sin\theta$ its height.

FIGURE 6.7

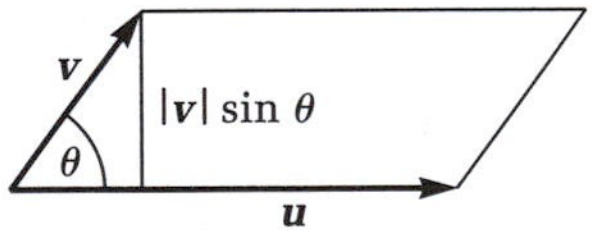

The "if" part of Statement 10 follows from the fact that two vectors are linearly dependent if and only if they are collinear (the zero vector is collinear with every vector by definition), and for collinear vectors we already know that the cross product is zero. The "only if" part follows from Statement 9, since if $\boldsymbol{u}$ and $\boldsymbol{v}$ are not collinear, then none of the factors in $|\boldsymbol{u}||\boldsymbol{v}|\sin\theta$ is zero, and so $|\boldsymbol{u} \times \boldsymbol{v}| \neq 0$.

The proof of Statement 11 is similar to Equation (6.3.7) above. We leave it for Exercise 6.3.4.

Statements 12 and 13 could be proved simply by writing out each expression in terms of the components u_i, v_i, w_i, but we gain a little more insight if we proceed as follows. First, if $\boldsymbol{u}$ and $\boldsymbol{v}$ are linearly dependent, then both sides of Statement 12 are $\boldsymbol{0}$. Otherwise, $\boldsymbol{u} \times \boldsymbol{v}$ being orthogonal to $\boldsymbol{u}$ and $\boldsymbol{v}$, all vectors orthogonal to $\boldsymbol{u} \times \boldsymbol{v}$ must lie in the plane of $\boldsymbol{u}$ and $\boldsymbol{v}$; that is, must be linear combinations of $\boldsymbol{u}$ and $\boldsymbol{v}$. Now, $(\boldsymbol{u} \times \boldsymbol{v}) \times \boldsymbol{w}$ is orthogonal to $\boldsymbol{u} \times \boldsymbol{v}$, and so it must satisfy

$$(\boldsymbol{u} \times \boldsymbol{v}) \times \boldsymbol{w} = a\boldsymbol{u} + b\boldsymbol{v} \tag{6.3.10}$$

for some scalars a and b. Taking the dot product of both sides with $\boldsymbol{w}$, and considering that the left side is orthogonal to it, we obtain

$$a(\boldsymbol{u} \cdot \boldsymbol{w}) + b(\boldsymbol{v} \cdot \boldsymbol{w}) = 0. \tag{6.3.11}$$

The solutions of this equation for the unknown a and b can be written in the form

$$a = c(\boldsymbol{v} \cdot \boldsymbol{w}) \text{ and } b = -c(\boldsymbol{u} \cdot \boldsymbol{w}) \tag{6.3.12}$$

with another scalar c still to be determined. Substituting these values into Equation (6.3.10), we get

$$(\boldsymbol{u} \times \boldsymbol{v}) \times \boldsymbol{w} = c(\boldsymbol{v} \cdot \boldsymbol{w})\boldsymbol{u} - c(\boldsymbol{u} \cdot \boldsymbol{w})\boldsymbol{v}. \tag{6.3.13}$$

Since both sides are homogeneous linear expressions of the u_i, v_i, w_i components and this is an identity, which must hold for any choice of the vectors $\boldsymbol{u}$, $\boldsymbol{v}$, $\boldsymbol{w}$, the scalar c must be independent of this choice. Thus we can evaluate it conveniently by setting $\boldsymbol{u} = \boldsymbol{w} = \boldsymbol{i}$ and $\boldsymbol{v} = \boldsymbol{j}$ in Equation (6.3.13). This leads to $c = -1$, which proves Statement 12. Statement 13 could be proved similarly.

Incidentally, Statements 12 and 13 show that the cross product is not associative.

<<>>

Let us now look at some examples of applications of the cross product:

EXAMPLE 6.3.3. Find a nonparametric equation of the plane S through the three points $A = (1, 0, 2)$, $B = (3, 1, 2)$, $C = (2, 1, 4)$.

We can find a normal vector of S by taking the cross product of any two independent vectors lying in it, say, $\overrightarrow{AB} = (2, 1, 0)^T$ and $\overrightarrow{AC} = (1, 1, 2)^T$. Thus we may take

$$\boldsymbol{n} = \begin{vmatrix} \boldsymbol{i} & \boldsymbol{j} & \boldsymbol{k} \\ 2 & 1 & 0 \\ 1 & 1 & 2 \end{vmatrix} = 2\boldsymbol{i} - 4\boldsymbol{j} + \boldsymbol{k}. \tag{6.3.14}$$

If we denote the position vector of a general point of S by $\boldsymbol{p} = (x, y, z)^T$ and the position vector of A by $\boldsymbol{a}$, then the desired equation can be written in general as

$$\boldsymbol{n} \cdot (\boldsymbol{p} - \boldsymbol{a}) = 0, \tag{6.3.15}$$

and for this plane as

$$2(x-1) - 4(y-0) + 1(z-2) = 0. \tag{6.3.16}$$

<<>>

In general, the parametric vector equation of a plane S was written on page 28 as

$$\boldsymbol{p} = \boldsymbol{p}_0 + s\boldsymbol{u} + t\boldsymbol{v}, \tag{6.3.17}$$

where $\boldsymbol{u}$ and $\boldsymbol{v}$ denote two vectors lying in S, $\boldsymbol{p}$ and $\boldsymbol{p}_0$ the position vectors of a variable and a fixed point of S, and s and t the parameters. Taking the dot product of both sides with the vector $\boldsymbol{n} = \boldsymbol{u} \times \boldsymbol{v}$, which is orthogonal to $\boldsymbol{u}$ and $\boldsymbol{v}$, we obtain the general nonparametric equation of a plane as

$$\boldsymbol{n} \cdot \boldsymbol{p} = \boldsymbol{n} \cdot \boldsymbol{p}_0. \tag{6.3.18}$$

This is the simple way of eliminating s and t that we alluded to in footnote 7 on page 29.

EXAMPLE 6.3.4. Find a vector $\boldsymbol{n}$ normal to the triangle T with vertices $A = (0, -2, 2)$, $B = (0, 2, 3)$, $C = (2, 0, 2)$, and whose length equals the area of T.

We can find a normal vector of T by taking the cross product of any two of its edge-vectors, say, $\overrightarrow{AB} = (0, 4, 1)^T$ and $\overrightarrow{AC} = (2, 2, 0)^T$. Thus we may take

$$\boldsymbol{n} = \begin{vmatrix} \boldsymbol{i} & \boldsymbol{j} & \boldsymbol{k} \\ 0 & 4 & 1 \\ 2 & 2 & 0 \end{vmatrix} = -2\boldsymbol{i} + 2\boldsymbol{j} - 8\boldsymbol{k}. \tag{6.3.19}$$

The area of T is given, according to Statement 9, by $\frac{1}{2}|\boldsymbol{n}|$. Thus

$$\text{Area}(T) = \sqrt{1^2 + 1^2 + 4^2} = \sqrt{18}. \tag{6.3.20}$$

<<>>

A vector normal to a plane figure S and having length equal to the area of S, like the vector $1/2\mathbf{n}$ above, is sometimes called an *area-vector of S*.

The cross product has many applications in physics. We discuss just two of these briefly:

EXAMPLE 6.3.5. Any object in a rotating frame of reference experiences some forces due to its inertia; that is, its tendency to move uniformly in a straight line. A well-known example is the centrifugal force pushing us outward in a turning car. If the object is also moving relative to the rotating frame, then there is an additional such force acting on it, called the *Coriolis force*, which is shown in physics courses to be given by a cross product:

$$\boldsymbol{F} = 2m\boldsymbol{v} \times \boldsymbol{\omega}, \tag{6.3.21}$$

where m denotes the mass of the object, $\boldsymbol{v}$ its velocity vector relative to the rotating frame, and ω the angular velocity vector of the frame's rotation relative to the universe. In the case of the Earth $\boldsymbol{\omega}$ is a vector parallel to the Earth's axis pointing from the South Pole to the North Pole and having length $2\pi/(24 \text{ hrs})$.

Notice that the right-hand rule of Statement 5 of Theorem 6.3.1 reflects an essential property of this force, in contrast to the geometrical examples above, where it played no role.

The Coriolis force has powerful effects in meteorology. For example, the hot climate near the equator makes the air rise, which then cools off and descends at moderate latitudes. In Figure 6.8, the vector $\boldsymbol{v}$ represents the velocity of this descending air somewhere in the middle of the northern hemisphere and the vector $\boldsymbol{F}$ is the Coriolis force, which, according to the right-hand rule, points to the east. This is the reason for the prevailing westerly winds there.

FIGURE 6.8

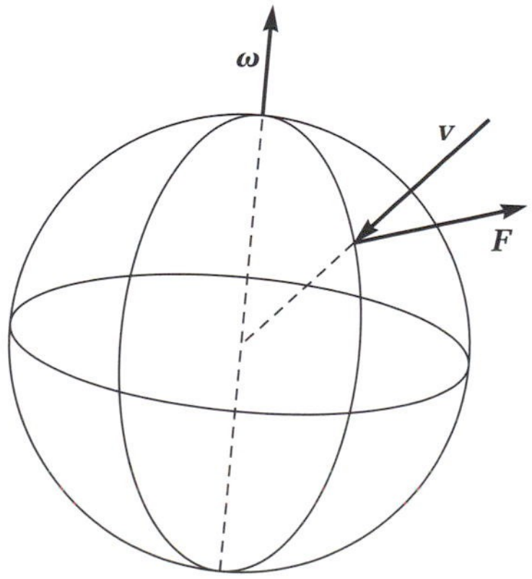

Similarly, a hurricane starts around a region where the barometric pressure is very low, consequently the outside air starts moving toward it. This movement creates a Coriolis force, which is perpendicular to this radial flow and starts the circulation of the hurricane. There are other forces involved as well, and with the changed wind direction the direction of the Coriolis force changes too, but it is still the major cause of the hurricane. The right-hand rule shows that in the northern hemisphere the circulation must be counterclockwise (we leave the explanation of this effect for Exercise 6.3.10).

<<>>

EXAMPLE 6.3.6. Another instance of a cross product is in the formula for the so-called Lorentz force. This is the force exerted by a magnetic field $\boldsymbol{B}$ on a particle with an electric charge q and moving with velocity $\boldsymbol{v}$. It is given by

$$\boldsymbol{F} = q\boldsymbol{v} \times \boldsymbol{B}. \tag{6.3.22}$$

The effect of this force can be seen, for example, in cloud-chamber photographs of the tracks of elementary particles. In a transverse magnetic field this force changes the particles' straight-line paths to circles. This is also the force that drives electric motors by acting on the electrons comprising the current in the motor's coils.

<<>>

Exercises 6.3

6.3.1. Find the cross product of the vectors $\boldsymbol{u} = (1, -1, 0)^T$ and $\boldsymbol{v} = (1, 2, 0)^T$, and verify by elementary geometry that $|\boldsymbol{u} \times \boldsymbol{v}|$ equals the area of the parallelogram spanned by $\boldsymbol{u}$ and $\boldsymbol{v}$.

• **6.3.2.** Use the cross product to find an equation of the plane S through the three points $A = (1, -1, 2)$, $B = (0, -1, 3)$, $C = (3, 0, 2)$.

6.3.3. Verify Statements 11, 12, and 13 of Theorem 6.3.1 for the vectors $\boldsymbol{u} = (1, -1, 0)^T$, $\boldsymbol{v} = (1, 2, 0)^T$, and $\boldsymbol{w} = (1, 0, 3)^T$.

6.3.4. Prove Statement 11 of Theorem 6.3.1.

6.3.5. The expression $\boldsymbol{u} \cdot (\boldsymbol{v} \times \boldsymbol{w})$ of Statement 11 of Theorem 6.3.1 is called the *triple product* of these vectors. Show geometrically that its absolute value equals the volume of the parallelepiped spanned by the vectors $\boldsymbol{u}$, $\boldsymbol{v}$, and $\boldsymbol{w}$.

***6.3.6.** Let $\boldsymbol{n}_1, \boldsymbol{n}_2, \boldsymbol{n}_3, \boldsymbol{n}_4$ denote the outward-pointing area-vectors of a tetrahedron. Prove that their sum equals $\boldsymbol{0}$. (*Hint:* Let $\boldsymbol{a}_1, \ldots \boldsymbol{a}_6$ denote the edge-vectors, write each area-vector as a cross product of these, and apply the appropriate properties from Theorem 6.3.1 to the sum.)

***6.3.7.** Prove that for any vectors $\boldsymbol{a}$, $\boldsymbol{b}$, $\boldsymbol{c}$, $\boldsymbol{d}$ of $\mathbb{R}^3$ we have

$$(\boldsymbol{a} \times \boldsymbol{b}) \cdot (\boldsymbol{c} \times \boldsymbol{d}) = (\boldsymbol{a} \cdot \boldsymbol{c})(\boldsymbol{b} \cdot \boldsymbol{d}) - (\boldsymbol{a} \cdot \boldsymbol{d})(\boldsymbol{b} \cdot \boldsymbol{c}). \qquad (6.3.23)$$

***6.3.8.** Prove that for any vectors $\boldsymbol{a}$, $\boldsymbol{b}$, $\boldsymbol{c}$, $\boldsymbol{d}$ of $\mathbb{R}^3$ we have

$$(\boldsymbol{a} \times \boldsymbol{b}) \times (\boldsymbol{c} \times \boldsymbol{d}) = [\det(\boldsymbol{a}\ \boldsymbol{c}\ \boldsymbol{d})]\boldsymbol{b} - [\det(\boldsymbol{b}\ \boldsymbol{c}\ \boldsymbol{d})]\boldsymbol{a} \qquad (6.3.24)$$

and

$$(\boldsymbol{a} \times \boldsymbol{b}) \times (\boldsymbol{c} \times \boldsymbol{d}) = [\det(\boldsymbol{a}\ \boldsymbol{b}\ \boldsymbol{d})]\boldsymbol{c} - [\det(\boldsymbol{a}\ \boldsymbol{b}\ \boldsymbol{c})]\boldsymbol{d}. \qquad (6.3.25)$$

6.3.9. Let $\boldsymbol{a}$ be any fixed nonzero vector of $\mathbb{R}^3$. Define the transformation $\boldsymbol{T} : \mathbb{R}^3 \to \mathbb{R}^3$ by $\boldsymbol{T}(\boldsymbol{x}) = \boldsymbol{a} \times \boldsymbol{x}$.

a. Show that $\boldsymbol{T}$ is linear.

b. Find the matrix $[\boldsymbol{T}]$ that represents $\boldsymbol{T}$ relative to the standard basis.

c. Find the nullspace of the matrix $[\boldsymbol{T}]$, and describe it geometrically.

d. What is the rank of $[\boldsymbol{T}]$?

6.3.10. Explain, using Equation (6.3.21), why in the northern hemisphere the circulation of hurricanes must be counterclockwise. What is it in the southern hemisphere and why?

<<7>> EIGENVALUES AND EIGENVECTORS

7.1. BASIC PROPERTIES

In this chapter we study another major branch of linear algebra, very different from what we have seen so far. The problems in this area arise in many applications in physics, economics, statistics, and so on. The main reason can be explained roughly as follows:

Frequently the states of a physical system can be described by an n-dimensional vector $\mathbf{x}$ and the latter's change in time by an $n \times n$ matrix A, so that the state of the system at some later time will be given by the vector $\mathbf{y} = A\mathbf{x}$. Often such changes are described by differential equations; that is, the vector $\mathbf{x}$ is an unknown differentiable vector function of time satisfying an equation of the form $\mathbf{x}' = A\mathbf{x}$. Such differential equations also lead to the same basic situation that we want to explain for the simpler case of $\mathbf{y} = A\mathbf{x}$.

Let us write the last equation in two dimensions as the pair of scalar equations

$$y_1 = a_{11}x_1 + a_{12}x_2 \tag{7.1.1}$$

and

$$y_2 = a_{21}x_1 + a_{22}x_2. \tag{7.1.2}$$

As these equations show, generally the matrix A mixes up the components of $\mathbf{x}$ and, over many such time steps and in higher dimensions, this mixing can be quite involved. The question we ask therefore is whether it is possible to find a new basis $\{\mathbf{s}_1, \mathbf{s}_2\}$ for $\mathbb{R}^2$ in which such mixing does not occur. That is, a basis in which the components are decoupled and develop separately as

$$y_{S1} = \lambda_1 x_{S1} \tag{7.1.3}$$

and

$$y_{S2} = \lambda_2 x_{S2} \tag{7.1.4}$$

where λ_1 and λ_2 are appropriate scalars depending on the matrix A. The answer is frequently yes, and this leads to greatly simplified computations when many such steps follow each other. Let us illustrate this decoupling in an example:

EXAMPLE 7.1.1. Consider the matrix

$$A = \begin{bmatrix} 1/2 & 3/2 \\ 3/2 & 1/2 \end{bmatrix} \tag{7.1.5}$$

and its action on an arbitrary vector $\mathbf{x}$. Let $\mathbf{s}_1 = \frac{1}{\sqrt{2}}(1, 1)^T$ and $\mathbf{s}_2 = \frac{1}{\sqrt{2}}(-1, 1)^T$. Then

$$A\mathbf{s}_1 = \frac{1}{\sqrt{2}} \begin{bmatrix} 1/2 & 3/2 \\ 3/2 & 1/2 \end{bmatrix} \begin{bmatrix} 1 \\ 1 \end{bmatrix} = \frac{1}{\sqrt{2}} \begin{bmatrix} 2 \\ 2 \end{bmatrix} = 2\mathbf{s}_1 \tag{7.1.6}$$

and

$$A\mathbf{s}_2 = \frac{1}{\sqrt{2}} \begin{bmatrix} 1/2 & 3/2 \\ 3/2 & 1/2 \end{bmatrix} \begin{bmatrix} -1 \\ 1 \end{bmatrix} = \frac{1}{\sqrt{2}} \begin{bmatrix} 1 \\ -1 \end{bmatrix} = -\mathbf{s}_2. \tag{7.1.7}$$

Thus if we write $\mathbf{x}$ and $\boldsymbol{y} = A\mathbf{x}$ in terms of the basis $\{\mathbf{s}_1, \mathbf{s}_2\}$ as

$$\mathbf{x} = x_{S1}\mathbf{s}_1 + x_{S2}\mathbf{s}_2 \text{ and } \boldsymbol{y} = y_{S1}\mathbf{s}_1 + y_{S2}\mathbf{s}_2 \tag{7.1.8}$$

then we get

$$\boldsymbol{y} = A\mathbf{x} = x_{S1}A\mathbf{s}_1 + x_{S2}A\mathbf{s}_2 = 2x_{S1}\mathbf{s}_1 - x_{S2}\mathbf{s}_2, \tag{7.1.9}$$

and from this

$$y_{S1} = 2x_{S1} \tag{7.1.10}$$

and

$$y_{S2} = -x_{S2}. \tag{7.1.11}$$

Thus the action of A on the S-components of $\mathbf{x}$ is simple multiplication (see Figure 7.1), but in the standard basis it would be linear combination. Similarly, the S-components of $A^2\mathbf{x}$ would be $2^2 x_{S1}$ and $(-1)^2 x_{S1}$, and so on for higher powers; much simpler than in the standard basis.

FIGURE 7.1

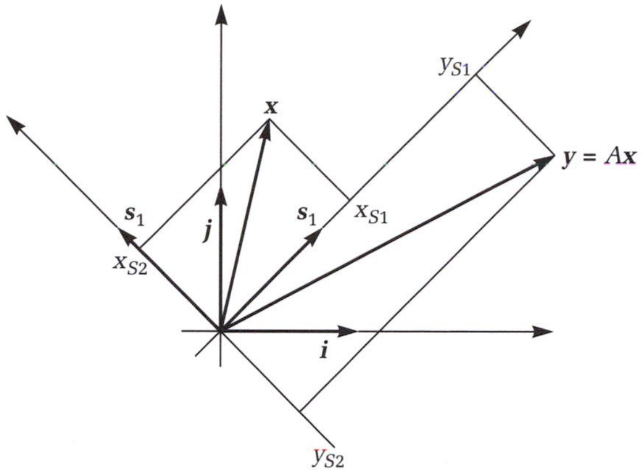

<<>>

How can we find such a decoupling basis in general as the one in the example above? We must find nontrivial $\{\mathbf{s}_1, \mathbf{s}_2\}$ and corresponding λ_1 and λ_2 such that

$$A\mathbf{s}_1 = \lambda_1\mathbf{s}_1 \text{ and } A\mathbf{s}_2 = \lambda_2\mathbf{s}_2 \tag{7.1.12}$$

hold. Indeed, in that case, substituting

$$\mathbf{x} = x_{S1}\mathbf{s}_1 + x_{S2}\mathbf{s}_2 \text{ and } \mathbf{y} = y_{S1}\mathbf{s}_1 + y_{S2}\mathbf{s}_2 \tag{7.1.13}$$

into the equation $\mathbf{y} = A\mathbf{x}$, we get

$$y_{S1}\mathbf{s}_1 + y_{S2}\mathbf{s}_2 = A(x_{S1}\mathbf{s}_1 + x_{S2}\mathbf{s}_2)$$

$$= x_{S1}A\mathbf{s}_1 + x_{S2}A\mathbf{s}_2 = x_{S1}\lambda_1\mathbf{s}_1 + x_{S2}\lambda_2\mathbf{s}_2, \tag{7.1.14}$$

from which

$$y_{S1} = \lambda_1 x_{S1} \tag{7.1.15}$$

and

$$y_{S2} = \lambda_2 x_{S2} \tag{7.1.16}$$

follow. We shall see more detailed examples later. For now we base our fundamental definition on Equations (7.1.12), which lie at the heart of this whole theory:

DEFINITION 7.1.1.

Given any $n \times n$ matrix A, if for a scalar λ and a *nonzero* vector $\mathbf{s}$ the equation

$$A\mathbf{s} = \lambda\mathbf{s} \tag{7.1.17}$$

holds, then λ is called an *eigenvalue* of the matrix A and $\mathbf{s}$ an *eigenvector* of A corresponding or belonging to λ. For any eigenvalue λ the zero vector is always a solution of Equation (7.1.17) and is called the trivial eigenvector of A belonging to λ.

The word *eigen* is German for "own" and was taken over into English, although some authors use *proper value* and *proper vector* or *characteristic value* and *characteristic vector* instead. The Greek letter *lambda* is traditional in this context.

Geometrically, Equation (7.1.17) expresses the fact that an eigenvector of a matrix A is a vector whose direction is preserved by A, and the corresponding eigenvalue is the magnification factor by which the matrix alters this eigenvector. (Magnification may, of course, mean diminution or no change and may also involve reflection, depending on the size and sign of λ.)

How then do we solve Equation (7.1.17)? This is the usual procedure: We rewrite it as

$$(A - \lambda I)\mathbf{s} = \mathbf{0}, \tag{7.1.18}$$

since in this form we have an equation closely resembling the familiar $A\mathbf{s} = \mathbf{0}$, except that the matrix A is replaced by the matrix $A - \lambda I$. Recall that, for any fixed λ, a homogeneous equation like this has nontrivial solutions if and only if its matrix is singular. By Theorem 6.1.8 this condition is equivalent to

$$\det(A - \lambda I) = 0. \tag{7.1.19}$$

This is called the *characteristic equation* of the matrix A and the left side of the equation the *characteristic polynomial* of A. It does not contain the unknown vector $\mathbf{s}$ and is therefore used to find the eigenvalues.[1] For an $n \times n$ matrix it is an algebraic equation of degree n. Once a value of λ is known, we substitute it into Equation (7.1.18), and we solve the latter by Gaussian elimination. Let us see how all this is done:

[1]Though of paramount theoretical importance, for large values of n this method is hopelessly inefficient. There exist fast, approximate methods for finding eigenvalues; we shall discuss some of them in Chapter 8.

EXAMPLE 7.1.2. Find the eigenvalues and eigenvectors of

$$A = \begin{bmatrix} 1 & 2 \\ 2 & 1 \end{bmatrix}. \tag{7.1.20}$$

The corresponding characteristic equation is

$$|A - \lambda I| = \det\left(\begin{bmatrix} 1 & 2 \\ 2 & 1 \end{bmatrix} - \lambda \begin{bmatrix} 1 & 0 \\ 0 & 1 \end{bmatrix}\right) = \begin{vmatrix} 1-\lambda & 2 \\ 2 & 1-\lambda \end{vmatrix} = 0. \tag{7.1.21}$$

Expanding the determinant, we get

$$(1-\lambda)^2 - 2^2 = 0, \tag{7.1.22}$$

$$(1-\lambda-2)(1-\lambda+2) = 0, \tag{7.1.23}$$

and so

$$(\lambda+1)(\lambda-3) = 0. \tag{7.1.24}$$

The solutions of this equation are $\lambda_1 = -1$ and $\lambda_2 = 3$.

To find the eigenvectors we substitute these eigenvalues, one after the other, into Equation (7.1.18), and solve for the unknown vector $\boldsymbol{s}$. Let us start with $\lambda_1 = -1$. Then Equation (7.1.18) becomes

$$\begin{bmatrix} 1-(-1) & 2 \\ 2 & 1-(-1) \end{bmatrix}\boldsymbol{s} = \begin{bmatrix} 2 & 2 \\ 2 & 2 \end{bmatrix}\boldsymbol{s} = \boldsymbol{0}. \tag{7.1.25}$$

The solutions of this equation are of the form $\boldsymbol{s} = s(1, -1)^T$. Thus the eigenvectors belonging to the eigenvalue $\lambda_1 = -1$ form a one-dimensional subspace with basis vector $\boldsymbol{s}_1 = (1, -1)^T$.

Substituting $\lambda_2 = 3$ into Equation (7.1.18) we obtain

$$\begin{bmatrix} 1-3 & 2 \\ 2 & 1-3 \end{bmatrix}\boldsymbol{s} = \begin{bmatrix} -2 & 2 \\ 2 & -2 \end{bmatrix}\boldsymbol{s} = 0. \tag{7.1.26}$$

The solutions of this equation are all the multiples of $\boldsymbol{s}_2 = (1, 1)^T$.

<<>>

As we have seen in this example, the eigenvectors belonging to a fixed eigenvalue form a subspace, the so-called *eigenspace* corresponding to the given eigenvalue. This is true in general, since these vectors constitute the null space of a fixed matrix. In the example above we had two one-dimensional eigenspaces, but in others they can be of higher dimensions, as in the following example:

EXAMPLE 7.1.3. Find the eigenvalues and eigenvectors of

$$A = \begin{bmatrix} 3 & 0 & 1 \\ 0 & 3 & 0 \\ 0 & 0 & 1 \end{bmatrix}. \tag{7.1.27}$$

The corresponding characteristic equation is

$$|A - \lambda I| = \begin{vmatrix} 3-\lambda & 0 & 1 \\ 0 & 3-\lambda & 0 \\ 0 & 0 & 1-\lambda \end{vmatrix} = 0. \tag{7.1.28}$$

Expanding the determinant we get

$$(\lambda - 1)(\lambda - 3)^2 = 0. \tag{7.1.29}$$

The solutions are $\lambda_1 = 1$ and $\lambda_2 = 3$. Since $\lambda - 3$ occurs squared in Equation (7.1.29) we call $\lambda_2 = 3$ a *double eigenvalue* and we say that its (*algebraic*) *multiplicity* is 2.

Substituting $\lambda_1 = 1$ into Equation (7.1.18), we obtain

$$\begin{bmatrix} 3-1 & 0 & 1 \\ 0 & 3-1 & 0 \\ 0 & 0 & 1-1 \end{bmatrix} \mathbf{s} = \begin{bmatrix} 2 & 0 & 1 \\ 0 & 2 & 0 \\ 0 & 0 & 0 \end{bmatrix} \mathbf{s} = 0. \tag{7.1.30}$$

The solutions of this equation are all the multiples of $\mathbf{s}_1 = (-1, 0, 2)^T$.

Substituting $\lambda_2 = 3$ into Equation (7.1.18) we obtain

$$\begin{bmatrix} 3-3 & 0 & 1 \\ 0 & 3-3 & 0 \\ 0 & 0 & 1-3 \end{bmatrix} \mathbf{s} = \begin{bmatrix} 0 & 0 & 1 \\ 0 & 0 & 0 \\ 0 & 0 & -2 \end{bmatrix} \mathbf{s} = 0. \tag{7.1.31}$$

The solutions of this equation are all the linear combinations of $\mathbf{s}_2 = (1, 0, 0)^T$ and $\mathbf{s}_3 = (0, 1, 0)^T$.

<<>>

As we have seen, for an $n \times n$ matrix the characteristic equation is of degree n, and as such, according to a theorem known as the Fundamental Theorem of Algebra, it always has n roots, provided we count multiplicities and allow complex numbers. This means that, in principle at least, the characteristic equation can be reformulated as

$$(\lambda - \lambda_1)^{k_1} (\lambda - \lambda_2)^{k_2} \cdots (\lambda - \lambda_r)^{k_r} = 0. \tag{7.1.32}$$

where r is the number of distinct roots, k_i is the multiplicity of the root λ_i, and

$$k_1 + k_2 + \cdots + k_r = n. \tag{7.1.33}$$

The case of complex solutions is very important in many applications—for instance rotation matrices have complex eigenvalues—and we shall discuss this case in Section 7.3. Also, the solution of higher-degree equations is generally very difficult, and for $n > 4$ it cannot even be done in a finite number of algebraic steps except in some special cases. We want to avoid such difficulties and consider only examples in which factoring of the characteristic polynomial is easy.

The foregoing examples seem to suggest that not only do we always have n eigenvalues, but that the sum of the eigenspaces has dimension n. Unfortunately this is not true in general, as the next example shows.

EXAMPLE 7.1.4. Find the eigenvalues and eigenvectors of

$$A = \begin{bmatrix} 4 & 1 \\ -1 & 2 \end{bmatrix}. \tag{7.1.34}$$

The corresponding characteristic equation is

$$|A - \lambda I| = \det\left(\begin{bmatrix} 4 & 1 \\ -1 & 2 \end{bmatrix} - \lambda \begin{bmatrix} 1 & 0 \\ 0 & 1 \end{bmatrix}\right) = \begin{vmatrix} 4-\lambda & 1 \\ -1 & 2-\lambda \end{vmatrix} = 0. \tag{7.1.35}$$

Expanding the determinant we get

$$(4 - \lambda)(2 - \lambda) + 1 = 0, \tag{7.1.36}$$

$$\lambda^2 - 6\lambda + 9 = 0, \tag{7.1.37}$$

and so

$$(\lambda - 3)^2 = 0. \tag{7.1.38}$$

The only solution of this equation is $\lambda_1 = 3$, which is thus a double eigenvalue.

Substituting $\lambda_1 = 3$ into Equation (7.1.18) we obtain

$$\begin{bmatrix} 4-3 & 1 \\ -1 & 2-3 \end{bmatrix} \boldsymbol{s} = \begin{bmatrix} 1 & 1 \\ -1 & -1 \end{bmatrix} \boldsymbol{s} = \boldsymbol{0}. \tag{7.1.39}$$

The solutions of this equation are all the multiples of $\boldsymbol{s}_1 = (1, -1)^T$, and so the sole eigenspace is one-dimensional, although $n = 2$. We say that the matrix A is *defective,* and that the *geometric multiplicity* of λ_1 is 1 while its algebraic multiplicity is 2.

<<>>

In general there is no way to predict whether a matrix is defective or not. We have to compute the eigenvalues and eigenvectors and see, although there are some important special cases in which we can be assured of a "full set" of eigenvectors. We now take up two of these.

THEOREM 7.1.1. If the $n \times n$ matrix A has n eigenvalues of multiplicity 1 each, then there is a one-dimensional eigenspace for each eigenvalue. If $\mathbf{s}_i$ is any nonzero eigenvector from the ith eigenspace, for $i = 1, 2, \ldots, n$, then the vectors $\mathbf{s}_1, \mathbf{s}_2, \ldots, \mathbf{s}_n$ are linearly independent and form a basis for $\mathbb{R}^n$.

Proof If λ_i is any eigenvalue, then by definition the equation

$$(A - \lambda_i I)\mathbf{s}_i = \mathbf{0} \tag{7.1.40}$$

has a nontrivial solution $\mathbf{s}_i$ for each i. Thus there is associated with each eigenvalue an eigenspace of dimension at least 1.

To see that the $\mathbf{s}_i$ vectors are independent, let $c_1, c_2, \ldots, c_n$ be scalars such that

$$c_1\mathbf{s}_1 + c_2\mathbf{s}_2 + \cdots + c_n\mathbf{s}_n = \mathbf{0}. \tag{7.1.41}$$

Multiply both sides of this equation by $A - \lambda_i I$. Then, by Equation (7.1.40), the ith term will vanish. If we then multiply by $A - \lambda_j I$, then the jth term will go away, because

$$(A - \lambda_j I)(A - \lambda_i I)\mathbf{s}_j = (A - \lambda_i I)(A - \lambda_j I)\mathbf{s}_j = (A - \lambda_i I)\mathbf{0} = \mathbf{0}. \tag{7.1.42}$$

Continuing in this fashion we may annihilate all terms of Equation (7.1.41) but one. Say we keep the kth term. Then we are left with

$$\prod_{\substack{i=1 \\ i\neq k}}^{n} (A - \lambda_i I)c_k\mathbf{s}_k = \mathbf{0}. \tag{7.1.43}$$

Using the fact that $\mathbf{s}_k$ is an eigenvector of A with corresponding eigenvalue λ_k, we can reduce the last equation to

$$c_k \prod_{\substack{i=1 \\ i\neq k}}^{n} (\lambda_k - \lambda_i)\mathbf{s}_k = \mathbf{0}. \tag{7.1.44}$$

The assumption that each eigenvalue has multiplicity one implies that $\lambda_k - \lambda_i \neq 0$ for any $i \neq k$, and so $c_k = 0$ must hold for every k. This result proves the independence of the $\mathbf{s}_k$ vectors.

The only statement left to prove is that each eigenspace is one-dimensional. Indeed, the n vectors $\mathbf{s}_k$, being independent, form a basis for $\mathbb{R}^n$, consequently if any eigenspace were higher than one-dimensional, then it would overlap another eigenspace in more than just the zero vector. This is clearly impossible, because no nonzero eigenvector can belong to two different eigenvalues.

<<>>

In case the matrix A is symmetric, we can say more:

THEOREM 7.1.2. Any two eigenvectors of a symmetric matrix that correspond to different eigenvalues are orthogonal.

Proof Let $\mathbf{s}_1$ and $\mathbf{s}_2$ be two eigenvectors of the symmetric matrix A that correspond to the eigenvalues λ_1 and λ_2 respectively, with $\lambda_1 \neq \lambda_2$. Then, using the fact that $\mathbf{s}_2^T A \mathbf{s}_1$ is a scalar and that $A^T = A$, we have

$$\mathbf{s}_2^T A \mathbf{s}_1 = (\mathbf{s}_2^T A \mathbf{s}_1)^T = \mathbf{s}_1^T A^T \mathbf{s}_2 = \mathbf{s}_1^T A \mathbf{s}_2, \tag{7.1.45}$$

hence

$$\lambda_1 \mathbf{s}_2^T \mathbf{s}_1 = \lambda_2 \mathbf{s}_1^T \mathbf{s}_2. \tag{7.1.46}$$

Because $\mathbf{s}_2^T \mathbf{s}_1 = \mathbf{s}_1^T \mathbf{s}_2$ (each being just a different expression of the dot product), and since $\lambda_1 \neq \lambda_2$, Equation (7.1.46) implies that $\mathbf{s}_2^T \mathbf{s}_1 = 0$, that is, that the vectors $\mathbf{s}_1$ and $\mathbf{s}_2$ are orthogonal.

<<>>

In Section 7.3 we are going to prove that for any symmetric matrix there actually exists an orthonormal set of eigenvectors spanning $\mathbb{R}^n$.

Exercises 7.1

In the first eight exercises, find all eigenvalues and associated eigenvectors for the given matrices.

7.1.1. $A = \begin{bmatrix} 2 & 3 \\ 3 & 2 \end{bmatrix}$.

7.1.2. $A = \begin{bmatrix} 2 & 0 \\ 0 & 2 \end{bmatrix}$.

7.1.3. $A = \begin{bmatrix} 0 & 0 \\ 0 & 0 \end{bmatrix}$.

• **7.1.4.** $A = \begin{bmatrix} 0 & 0 \\ 1 & 0 \end{bmatrix}$.

• **7.1.5.** $A = \begin{bmatrix} 2 & 0 & 1 \\ 0 & 2 & 0 \\ 1 & 0 & 2 \end{bmatrix}$.

7.1.6. $A = \begin{bmatrix} 2 & 0 & 1 \\ 0 & 2 & 0 \\ 0 & 0 & 2 \end{bmatrix}$.

• **7.1.7.** $A = \begin{bmatrix} 1 & 0 & -1 \\ -2 & 3 & -1 \\ -6 & 6 & 0 \end{bmatrix}$.

• **7.1.8.** $A = \begin{bmatrix} 1 & 0 & 0 & 1 \\ 0 & 1 & 1 & 1 \\ 0 & 0 & 2 & 0 \\ 0 & 0 & 0 & 2 \end{bmatrix}$.

7.1.9. Can you find a relationship between the eigenvalues and eigenvectors of a matrix A and those of the matrix cA, where c is any scalar?

• **7.1.10.** Can you find a relationship between the eigenvalues and eigenvectors of a matrix A and those of the matrix $A + cI$, where c is any scalar?

7.1.11. Prove that a matrix is singular if and only if one of its eigenvalues is zero.

7.1.12. Prove that if $\boldsymbol{s}$ is an eigenvector of a matrix A, then it is also an eigenvector of the matrix A^2. How are the associated eigenvalues related?

7.1.13. Prove that if $\boldsymbol{s}$ is an eigenvector of a nonsingular matrix A, then it is also an eigenvector of the matrix A^{-1}. How are the associated eigenvalues related?

7.1.14. Show that a matrix A and its transpose A^T have the same eigenvalues.

7.1.15. Let $\boldsymbol{u}$ be any unit vector in $\mathbb{R}^n$. Show that the matrix $A = \boldsymbol{u}\boldsymbol{u}^T$ represents the projection onto the line of $\boldsymbol{u}$ and find its eigenvalues and eigenspaces.

***7.1.16.** Find the eigenvalues and eigenspaces of any projection matrix P.

7.1.17. A row vector $\boldsymbol{s}^T$ is called a *left eigenvector* of a matrix A belonging to the eigenvalue λ if the equation $\boldsymbol{s}^T A = \lambda \boldsymbol{s}^T$ holds. Show that $\boldsymbol{s}^T$ is a left eigenvector of a matrix A belonging to the eigenvalue λ if and only if $\boldsymbol{s}$ is an eigenvector of A^T belonging to the eigenvalue λ.

7.1.18. Show that if $\boldsymbol{u}$ and $\boldsymbol{v}$ are eigenvectors, belonging to different eigenvalues, of a matrix A and its transpose A^T respectively, then they are orthogonal to each other.

MATLAB Exercises 7.1

The MATLAB command $\boldsymbol{c}$ = **poly**(A) returns the coefficients in descending order of the characteristic polynomial of the matrix A, and the command $\boldsymbol{d}$ = **roots**($\boldsymbol{c}$) returns the roots of this polynomial; that is, the eigenvalues of the matrix A. In the following exercises use these and the command $A\backslash \boldsymbol{b}$ to solve the appropriate linear systems to find the eigenvalues and eigenvectors of each matrix A. Also, compare these results to the solutions obtained by using the MATLAB routine **eig**(A).

ML7.1.1. Let A be the matrix of Exercise 7.1.5.

ML7.1.2. Let A be the matrix of Exercise 7.1.8.

ML7.1.3. A = **hilb**(3).

ML7.1.4. A = **hilb**(4).

ML7.1.5. A = **ones**(3).

ML7.1.6. A = **ones**(4).

ML7.1.7. What conjectures can you make about the eigenvalues and eigenvectors of matrices consisting of ones, on the basis of the last two exercises? Can you prove these? (Notice that any such matrix is n times a projection matrix.)

ML7.1.8. A = **hadamard**(4).

ML7.1.9. A = **hadamard**(8).

ML7.1.10. Orthogonal matrices with all entries equal to ± 1 are called *Hadamard matrices.* What conjecture can you make about their eigenvalues and eigenvectors on the basis of the last two exercises? Can you prove any of these?

7.2. DIAGONALIZATION OF MATRICES

If an $n \times n$ matrix A has n linearly independent eigenvectors $\boldsymbol{s}_1, \boldsymbol{s}_2, \ldots, \boldsymbol{s}_n$, then we may use these as a basis for $\mathbb{R}^n$. Writing S for the matrix with these vectors as columns in the order indicated; that is, $S = (\boldsymbol{s}_1\ \boldsymbol{s}_2 \cdots \boldsymbol{s}_n)$, we find

$$AS = (A\boldsymbol{s}_1\ A\boldsymbol{s}_2 \cdots A\boldsymbol{s}_n) = (\lambda_1\boldsymbol{s}_1\ \lambda_2\boldsymbol{s}_2 \cdots \lambda_n\boldsymbol{s}_n) = S\Lambda, \qquad (7.2.1)$$

where

$$\Lambda = \begin{bmatrix} \lambda_1 & 0 & \cdots & 0 \\ 0 & \lambda_2 & \cdots & 0 \\ & & \cdots & \\ 0 & 0 & \cdots & \lambda_n \end{bmatrix} \qquad (7.2.2)$$

is the diagonal matrix whose nonzero entries are the eigenvalues corresponding to the columns of S in the same order. Since the columns of S form a basis for $\mathbb{R}^n$, the matrix S must be invertible and, multiplying Equation (7.2.1) by S^{-1}, we obtain

$$S^{-1}AS = \Lambda. \tag{7.2.3}$$

Now the left side of this is, by Theorem 3.5.3 on page 129, the matrix A_S that represents A in the basis S. In the terminology of Section 3.5, the matrices A and Λ are similar and Equation (7.2.3) specifies the similarity transformation connecting them.

We call the above process *diagonalization of* A, and S a *diagonalizing matrix for* A. As is obvious from the foregoing discussion, the diagonalizing matrix S for a given diagonalizable A is not unique: the eigenvectors may be permuted, multiplied by arbitrary nonzero scalars, and linearly combined within higher-dimensional eigenspaces if there are any. It can be shown, however, that this is all the latitude we are permitted: the columns of any diagonalizing S must be independent eigenvectors of A (Exercise 7.2.5); consequently, A is diagonalizable if and only if it has a full set of n linearly independent eigenvectors. Note that this condition places no restriction on the multiplicities of the eigenvalues but, by Theorem 7.1.1, every matrix with distinct eigenvalues is diagonalizable.

Equation (7.2.3) has the useful consequence that A^k and Λ^k, for any positive integer k, are related by the same similarity transformation as were A and Λ:

$$\Lambda^k = S^{-1}ASS^{-1}AS \cdots S^{-1}AS = S^{-1}AIAIA \cdots AS = S^{-1}A^kS. \tag{7.2.4}$$

The powers of Λ are very easy to compute:

$$\Lambda^k = \begin{bmatrix} \lambda_1^k & 0 & \cdots & 0 \\ 0 & \lambda_2^k & \cdots & 0 \\ & & \cdots & \\ 0 & 0 & \cdots & \lambda_n^k \end{bmatrix}, \tag{7.2.5}$$

and then A^k can be recovered as

$$A^k = S\Lambda^kS^{-1}. \tag{7.2.6}$$

Let us summarize the foregoing discussion in a theorem:

THEOREM 7.2.1. If an $n \times n$ matrix A has n linearly independent eigenvectors, then, writing S for the matrix with these vectors as columns, we find that A is similar with transformation matrix S to a diagonal matrix Λ, whose diagonal entries are the eigenvalues of A corresponding to the columns of S. The same similarity transforms A^k to Λ^k for any positive integer k.

As mentioned in the introduction to Section 7.1, the simplification resulting from the change to a basis of eigenvectors is the main reason for their usefulness, and in matrix form this simplification means that instead of multiplying by the general matrix A^k, which is usually very difficult to compute, in the new basis we may multiply by the very simple matrix Λ^k. Let usnow look at a concrete example:

EXAMPLE 7.2.1. Let us assume that a town was originally settled by 100 people, all less than fifty years old. We want to investigate how the age distribution changes over a long time between two age groups: less than fifty, and fifty or older. We must, of course, make some quantitative assumptions about the changes from one time period to the next. So let us say that over any decade, on the one hand, there is a net increase of 10% in the under-fifty population, and on the other hand, 20% of the under-fifty population becomes fifty or older while 40% of the initial over-fifty population dies. Denoting in the nth decade the number of people under fifty by $x_1(n)$ and the number of those fifty or over by $x_2(n)$, for $n = 0, 1, 2, \ldots$ we can write the following equations:

$$x_1(0) = 100, \tag{7.2.7}$$

$$x_2(0) = 0, \tag{7.2.8}$$

$$x_1(n+1) = 1.1x_1(n), \tag{7.2.9}$$

$$x_2(n+1) = 0.2x_1(n) + 0.6x_2(n). \tag{7.2.10}$$

In matrix form these equations become

$$\mathbf{x}(0) = \begin{bmatrix} 100 \\ 0 \end{bmatrix} \tag{7.2.11}$$

and

$$\mathbf{x}(n+1) = A\mathbf{x}(n), \tag{7.2.12}$$

where

$$A = \begin{bmatrix} 1.1 & 0 \\ 0.2 & 0.6 \end{bmatrix}. \tag{7.2.13}$$

From Equation (7.2.12) we find

$$\mathbf{x}(n) = A^n\mathbf{x}(0) \tag{7.2.14}$$

for any positive integer n. We want to diagonalize A to compute A^n here.

The characteristic equation for this A is

$$|A - \lambda I| = \begin{vmatrix} 1.1 - \lambda & 0 \\ 0.2 & 0.6 - \lambda \end{vmatrix} = (1.1 - \lambda)(0.6 - \lambda) = 0. \tag{7.2.15}$$

The solutions are $\lambda_1 = 1.1$ and $\lambda_2 = 0.6$. The corresponding eigenvectors can be found by substituting these into $(A - \lambda I)\boldsymbol{s} = \boldsymbol{0}$:

$$(A - \lambda_1 I) = \begin{bmatrix} 1.1 - 1.1 & 0 \\ 0.2 & 0.6 - 1.1 \end{bmatrix} \boldsymbol{s}_1 = \begin{bmatrix} 0 & 0 \\ 0.2 & -0.5 \end{bmatrix} \boldsymbol{s}_1 = \boldsymbol{0}. \tag{7.2.16}$$

A solution is $\boldsymbol{s}_1 = (5, 2)^T$. For the other eigenvector we have the equation

$$(A - \lambda_2 I)\boldsymbol{s}_2 = \begin{bmatrix} 1.1 - 0.6 & 0 \\ 0.2 & 0.6 - 0.6 \end{bmatrix} \boldsymbol{s}_2 = \begin{bmatrix} 0.5 & 0 \\ 0.2 & 0 \end{bmatrix} \boldsymbol{s}_2 = \boldsymbol{0}. \tag{7.2.17}$$

A solution of this equation is $\boldsymbol{s}_2 = (0, 1)^T$. Thus

$$S = \begin{bmatrix} 5 & 0 \\ 2 & 1 \end{bmatrix}, \tag{7.2.18}$$

$$\Lambda = \begin{bmatrix} 1.1 & 0 \\ 0 & 0.6 \end{bmatrix}, \tag{7.2.19}$$

and

$$S^{-1} = \frac{1}{5} \begin{bmatrix} 1 & 0 \\ -2 & 5 \end{bmatrix}. \tag{7.2.20}$$

According to Corollary 3.5.1 on page 123, the coordinate vectors of each $\mathbf{x}(n)$ for $n = 0, 1, \ldots$, relative to the basis S, are given by

$$\mathbf{x}_S(0) = S^{-1}\mathbf{x}(0) = \frac{1}{5} \begin{bmatrix} 1 & 0 \\ -2 & 5 \end{bmatrix} \begin{bmatrix} 100 \\ 0 \end{bmatrix} = \begin{bmatrix} 20 \\ -40 \end{bmatrix} \tag{7.2.21}$$

and

$$\mathbf{x}_S(n) = S^{-1}\mathbf{x}(n) = S^{-1}A^n\mathbf{x}(0) = S^{-1}A^nS\mathbf{x}_S(0)$$

$$= \Lambda^n \mathbf{x}_S(0) = \begin{bmatrix} 1.1^n & 0 \\ 0 & 0.6^n \end{bmatrix} \begin{bmatrix} 20 \\ -40 \end{bmatrix} = \begin{bmatrix} 1.1^n \cdot 20 \\ -0.6^n \cdot 40 \end{bmatrix} \tag{7.2.22}$$

Hence the solution in the standard basis is given by

$$\mathbf{x}(n) = S\mathbf{x}_S(n) = \begin{bmatrix} 5 & 0 \\ 2 & 1 \end{bmatrix} \begin{bmatrix} 1.1^n \cdot 20 \\ -0.6^n \cdot 40 \end{bmatrix} = \begin{bmatrix} 1.1^n \cdot 100 \\ (1.1^n - 0.6^n) \cdot 40 \end{bmatrix}. \tag{7.2.23}$$

For large values of n the term 0.6^n can be neglected and we get

$$\mathbf{x}(n) \approx 1.1^n \cdot \begin{bmatrix} 100 \\ 40 \end{bmatrix}. \tag{7.2.24}$$

Thus, in the long run both the under-fifty and the over-fifty population will increase 10% per decade and there will be 40 people over fifty for every 100 under fifty.

<<>>

A system of equations like the one in the example above is called a system of *first-order linear difference equations.* Equation (7.2.12) shows their general vector form. The word "difference" indicates that the equation involves an unknown function that occurs at different values of an integer-valued variable n, and "first order" refers to the fact that only n and $n + 1$ occur and no additional $n + 2$ and such. Such equations can always be solved by the method shown, provided the matrix A is diagonalizable.

Systems of *first-order linear differential equations,* as defined below, with diagonalizable matrices can also be solved in a similar manner:

Suppose we are to find a vector-valued function $\mathbf{u}$ of a scalar variable t such that

$$\frac{d\mathbf{u}}{dt} = A\mathbf{u}. \tag{7.2.25}$$

Here $\mathbf{u}(t)$ is in $\mathbb{R}^n$ for every real t, and A is a constant $n \times n$ matrix. The derivative is defined componentwise; that is, $\mathbf{u}'(t) = (\mathbf{u}_1'(t), \ldots, \mathbf{u}_n'(t))$. Equation (7.2.25) is equivalent to the following system of so-called first-order linear differential equations:

$$\frac{du_i}{dt} = \sum_{i=1}^{n} a_{ij}u_j \quad \text{for } i = 1, 2, \ldots, n. \tag{7.2.26}$$

Let us consider a change of basis such that $\mathbf{u} = S\mathbf{v}$ for some constant, invertible $n \times n$ matrix S and $\mathbf{v}(t) \in \mathbb{R}^n$ for every real t. Then

$$\frac{dS\mathbf{v}}{dt} = AS\mathbf{v} \tag{7.2.27}$$

and

$$\frac{d\mathbf{v}}{dt} = S^{-1}AS\mathbf{v}. \tag{7.2.28}$$

If S diagonalizes A, so that $S^{-1}AS = \Lambda$ is a diagonal matrix, then this equation becomes

$$\frac{d\mathbf{v}}{dt} = \Lambda\mathbf{v}, \tag{7.2.29}$$

which can be written in components as

$$\frac{dv_i}{dt} = \lambda_i v_i \quad \text{for } i = 1, 2, \ldots, n. \tag{7.2.30}$$

In contrast to Equations (7.2.26), *each equation here contains only one of the unknown functions*, and can therefore easily be solved as follows:

For each i and all t such that $v_i(t) \neq 0$ we rewrite Equations (7.2.30) as

$$\frac{1}{v_i}\frac{dv_i}{dt} = \lambda_i, \tag{7.2.31}$$

$$\frac{d \ln |v_i|}{dt} = \lambda_i \tag{7.2.32}$$

and so

$$\ln |v_i| = \lambda_i t + c_i, \tag{7.2.33}$$

where each c_i is an arbitrary constant. Thus the general solution of (7.2.31) is

$$v_i = C_i e^{\lambda_i t}, \tag{7.2.34}$$

where $C_i = \pm e^{c_i}$. Allowing $C_i = 0$ too, we have the general solution of (7.2.30) as well. From (7.2.34) we obtain the general solution of the original Equations (7.2.25) by applying $\boldsymbol{u} = S\boldsymbol{v}$.

EXAMPLE 7.2.2. In physics it is shown that the electric circuit in Figure 7.2 is governed by the equation

$$Ri + \frac{1}{C}q + L\frac{di}{dt} = E, \tag{7.2.35}$$

where R, L, C are positive numbers denoting the resistance, the inductance, and the capacitance of the indicated elements, E is the applied electromotive force or voltage, $q(t)$ is the charge on the capacitor at time t, and $i(t) = q'(t)$ is the current at time t.

FIGURE 7.2

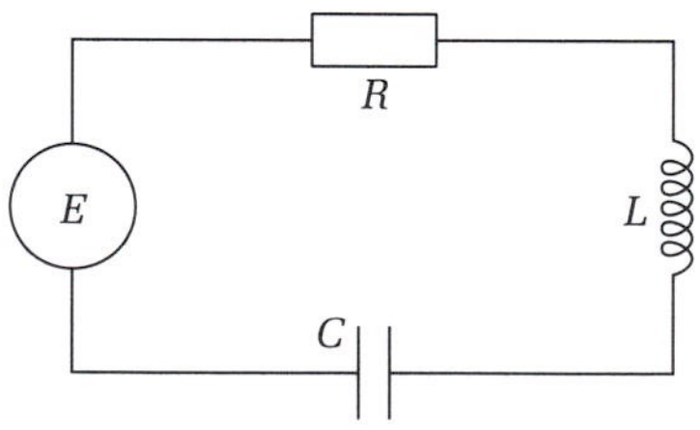

Equation (7.2.35) is an inhomogeneous equation, which means that it contains a term not involving the unknown functions i and q. The general solution of an inhomogeneous linear differential equation or system can be obtained, just as for ordinary equations (see Theorem 2.1.2 on page 47), by adding any particular solution of it to the general solution of the corresponding homogeneous equation or system. Thus, in any case, it is preferable to consider the latter first, meaning that we set $E = 0$, and proceed in that way. This procedure has a physical meaning as well: for instance, $E(t)$ can be a pulse, which is nonzero for some time, but becomes zero afterward while the circuit is closed. During the latter period the homogeneous equation rules, and because of the initial pulse it may very well have nonzero solutions.

We have two unknown functions i and q and two equations: one is Equation (7.2.35) and the other the equation $i(t) = q'(t)$. We can write the homogeneous system in the form corresponding to the general case of Equations (7.2.26) as

$$\frac{di}{dt} = -\frac{R}{L}i - \frac{1}{LC}q \tag{7.2.36}$$

and

$$\frac{dq}{dt} = i. \tag{7.2.37}$$

Thus, in this case, the matrix A is of the form

$$A = \begin{bmatrix} -\frac{R}{L} & -\frac{1}{LC} \\ 1 & 0 \end{bmatrix}. \tag{7.2.38}$$

The characteristic equation is

$$\left(-\frac{R}{L} - \lambda\right)(-\lambda) + \frac{1}{LC} = 0 \tag{7.2.39}$$

or equivalently,

$$\lambda^2 + \frac{R}{L}\lambda + \frac{1}{LC} = 0. \tag{7.2.40}$$

The discriminant of the last equation is

$$D = \frac{R^2}{L^2} - \frac{4}{LC}, \tag{7.2.41}$$

and if $D \neq 0$, then Equation (7.2.40) has the two solutions

$$\lambda_1 = \frac{1}{2}\left(-\frac{R}{L} - \sqrt{D}\right) \tag{7.2.42}$$

and

$$\lambda_2 = \frac{1}{2}\left(-\frac{R}{L} + \sqrt{D}\right). \tag{7.2.43}$$

The next step would be to compute the corresponding eigenvectors, but it is easier to avoid it by proceeding as follows:

First, observe that in the general formulation each u_k is a linear combination of the v_k, which are given by Equation (7.2.34). In the present case $u_1 = i$ and $u_2 = q$, and so the general solution must be of the form

$$i(t) = c_{11}e^{\lambda_1 t} + c_{12}e^{\lambda_2 t} \tag{7.2.44}$$

and

$$q(t) = c_{21}e^{\lambda_1 t} + c_{22}e^{\lambda_2 t} \tag{7.2.45}$$

with unknown coefficients c_{jk}. Next, substitute these expressions of $i(t)$ and $q(t)$ into Equations (7.2.36) and (7.2.37) to get two equations for the unknown c_{jk}, and prescribe two initial conditions; that is, values for $i(0)$ and $q(0)$, to have the necessary four equations for the four unknowns. For example, if there is no current at $t = 0$ but there is a charge Q on the capacitor, then the initial conditions are

$$i(0) = 0 \tag{7.2.46}$$

and

$$q(0) = Q, \tag{7.2.47}$$

and the coefficients turn out to be (Exercise 7.2.10)

$$c_{11} = -c_{12} = \frac{Q}{LC\sqrt{D}} \tag{7.2.48}$$

and

$$c_{21} = Q - c_{22} = \frac{Q\lambda_2}{\sqrt{D}}. \tag{7.2.49}$$

Notice that, if $D > 0$ holds, then both λ_1 and λ_2 are negative, and so both $i(t)$ and $q(t)$ decay to zero as $t \to \infty$. However, since $|\lambda_1| > |\lambda_2|$ holds, for large t the $e^{\lambda_1 t}$ terms will be negligible next to the $e^{\lambda_2 t}$ terms, and so the approach to zero will be like the latter.

The case $D = 0$ has to be treated differently, because then the matrix A is defective. We suggest a possible approach in Exercise 7.2.12(d).

The remaining case of $D < 0$ will be treated in Section 7.4. This case is very important, since it occurs in many circuits and leads to oscillating solutions that are entirely different from the foregoing ones.

The discussion of the inhomogeneous case is left to other courses.

<<>>

Exercises 7.2

7.2.1. Prove that a diagonalizable matrix A is invertible if and only if all its eigenvalues are different from zero, and in that case Equations (7.2.5) and (7.2.6) are valid for all negative integers k as well.

7.2.2. Define A^x for any real x and any diagonalizable A with positive eigenvalues.

7.2.3. Show that if A is symmetric with nonnegative eigenvalues, then $\sqrt{A}$ exists and is also symmetric.

7.2.4. Find A^{100} and $A^{1/2}$ for the matrix

$$A = \begin{bmatrix} 3 & 2 \\ 2 & 3 \end{bmatrix}.$$

• **7.2.5.** Find A^{100} for the matrix A of Exercise 7.1.5:

$$A = \begin{bmatrix} 2 & 0 & 1 \\ 0 & 2 & 0 \\ 1 & 0 & 2 \end{bmatrix}.$$

• **7.2.6.** Find A^4 for the matrix

$$A = \begin{bmatrix} 1 & 0 & 0 & 1 \\ 0 & 1 & 1 & 1 \\ 0 & 0 & 2 & 0 \\ 0 & 0 & 0 & 2 \end{bmatrix}$$

of Exercise 7.1.8, using the eigenvalues and eigenvectors of A.

• **7.2.7.** Prove that similar matrices have the same characteristic polynomial; that is, if A and B are similar, then $\det(A - \lambda I) = \det(B - \lambda I)$ for any λ.

7.2.8. Let A and B be similar matrices with $B = TAT^{-1}$. Prove that $\mathbf{s}$ is an eigenvector of A belonging to the eigenvalue λ if and only if $T\mathbf{s}$ is an eigenvector of B belonging to the same eigenvalue λ.

7.2.9. Prove the converse of Theorem 7.2.2: If A is orthogonally similar to a diagonal matrix Λ, that is, $S^{-1}AS = \Lambda$ for some orthogonal S, then A must be symmetric.

7.2.10. Prove Equations (7.2.48) and (7.2.49).

7.2.11. In biology the following type of simplified models for predator–prey populations are sometimes considered: Assume that in a certain area the number of animals of a predator species is $x_1(k)$ in month k, and the number of its prey is $x_2(k)$. Furthermore, the number of predators in the next month decreases in proportion to $x_1(k)$ and increases in proportion to the available food $x_2(k)$, while the number of prey animals decreases in proportion to the number of predators and increases in proportion to their own numbers. Thus, if we ignore other factors, we may for instance assume

$$x_1(k+1) = 0.8x_1(k) + 0.4x_2(k), \tag{7.2.50}$$

and

$$x_2(k+1) = -0.8x_1(k) + 2.0x_2(k). \tag{7.2.51}$$

Solve these equations for all k, assuming also that initially there were 1000 animals of each kind. What happens as $k \to \infty$?

***7.2.12.** In this exercise we outline an alternative approach to the solution of differential equations like

$$\frac{d\boldsymbol{u}}{dt} = A\boldsymbol{u}. \tag{7.2.52}$$

Define, for any square matrix A and any real t,

$$e^{At} = I + At + \frac{(At)^2}{2!} + \frac{(At)^3}{3!} + \cdots \tag{7.2.53}$$

assuming convergence and term-by-term differentiability.

a. Show that

$$\boldsymbol{u}(t) = e^{At}\boldsymbol{u}_0 \tag{7.2.54}$$

is the solution of Equation (7.2.52) satisfying the initial condition

$$\boldsymbol{u}(0) = \boldsymbol{u}_0. \tag{7.2.55}$$

b. Show that if A is diagonalizable so that $A = S\Lambda S^{-1}$, then

$$e^{At} = Se^{\Lambda t}S^{-1}, \tag{7.2.56}$$

$$e^{\Lambda t} = \begin{bmatrix} e^{\lambda_1 t} & 0 & 0 & \cdots & 0 \\ 0 & e^{\lambda_2 t} & 0 & \cdots & 0 \\ 0 & 0 & e^{\lambda_3 t} & \cdots & 0 \\ \vdots & \vdots & \vdots & & \vdots \\ 0 & 0 & 0 & \cdots & e^{\lambda_n t} \end{bmatrix}, \tag{7.2.57}$$

and Equation (7.2.54) becomes

$$\boldsymbol{u}(t) = Se^{\Lambda t}S^{-1} = \sum_{k=1}^{n} \boldsymbol{c}_k e^{\lambda_k t}, \tag{7.2.58}$$

where the $\boldsymbol{c}_k$ are appropriate constant vectors.

c. Use the formalism above to solve Equations (7.2.36) and (7.2.37) with $R = 5$, $L = 1$, $C = 1/4$, and $\boldsymbol{u}_0 = (0, 10)^T$. Plot the graphs of the solution for i and q.

d. Use Equations (7.2.53) and (7.2.54) to solve Equations (7.2.36) and (7.2.37) with $R = 2$, $L = 1$, $C = 1$, and $\boldsymbol{u}_0 = (0, 10)^T$. Plot the graphs of the solution for i and q.

MATLAB Exercises 7.2

ML7.2.1. Consider the problem of Example 7.2.1 again.

a. Enter the matrix A from Equation (7.2.13) and use MATLAB to verify that Equation (7.2.14) leads to Equation (7.2.24).

b. Experiment with different death rates for the over-fifty population, in place of the given 40%, to see what rates would lead to eventual extinction and what rate would lead to a steady population in the long run.

c. For what death rate would one of the eigenvalues equal 1? Compare this A to those examined in Part (b) and explain.

ML7.2.2. In Exercise 7.2.11 the coefficient $r = -0.8$ represents the predation rate; that is, the number of prey caught per predator per month. Experiment with different values of r to find one for which there is a stable limiting population. What is the split between the two kinds of animals in the limit for this r?

ML7.2.3. If A is an $m \times r$ matrix and has rank r, then AA^T is an $m \times m$ symmetric matrix of rank r. (Why?)

a. Use the fact above to generate random symmetric matrices of ranks 1, 2, and 3 for $m = 4, 5, 6$.

b. Use **eig** to find the eigenvalues of each matrix generated in Part (a) and note the multiplicity of the eigenvalue 0.

c. Make a conjecture about the dependence of the multiplicity of the eigenvalue 0 on m and r, and prove it.

7.3. PRINCIPAL AXES

We return now to theoretical considerations and discuss the diagonalization of symmetric matrices mentioned at the end of Section 7.1. We consider this subject because it is important in many applications and fairly easy to prove, while the general case lies beyond our scope.

First, however, we state a theorem whose proof is relegated to Section 7.4 because it requires complex numbers:

THEOREM 7.3.1. The eigenvalues of a symmetric matrix with real entries are real.

The next theorem contains the main result and is variously called the *Principal Axis Theorem* and the *Spectral Theorem for Symmetric Matrices.* These names come from applications of the theorem to determine the principal axes of ellipsoids and the color spectra of light sources. For the same reason the set of eigenvalues of any matrix is called its *spectrum.*

THEOREM 7.3.2. For any symmetric matrix A there exists an orthogonal matrix S such that $S^{-1}AS = \Lambda$ is a diagonal matrix. The columns of S are eigenvectors of A, and the diagonal entries of Λ the corresponding eigenvalues.

Proof Every $n \times n$ matrix A has at least one eigenvalue because its characteristic equation must have at least one solution according to the Fundamental Theorem of Algebra. Call such an eigenvalue λ_1. By Theorem 7.3.1, λ_1 is real and so there must exist a corresponding real unit eigenvector $\boldsymbol{s}_1$. The Gram-Schmidt algorithm guarantees that we can construct an orthogonal matrix S_1 whose first column is $\boldsymbol{s}_1$. For such an S_1 the first column of $S_1^{-1}AS_1$ is given by

$$S_1^{-1}A\boldsymbol{s}_1 = \lambda_1 S_1^{-1}\boldsymbol{s}_1 = \lambda_1 S_1^T\boldsymbol{s}_1 = \lambda_1\boldsymbol{e}_1. \tag{7.3.1}$$

Furthermore, $S_1^{-1}AS_1$ is also symmetric because

$$(S_1^{-1}AS_1)^T = (S_1^TAS_1)^T = S_1^TA^TS_1^{TT} = S_1^{-1}AS_1. \tag{7.3.2}$$

Thus $S_1^{-1}AS_1$ has the form

$$S_1^{-1}AS_1 = \left[\begin{array}{c|ccc} \lambda_1 & 0 & \cdots & 0 \\ \hline 0 & & & \\ \vdots & & A_1 & \\ 0 & & & \end{array}\right], \tag{7.3.3}$$

where A_1 is an $(n-1) \times (n-1)$ symmetric matrix.

Now we can repeat the argument above with A_1 in place of A: Then A_1 has an eigenvalue λ_2 and a corresponding unit eigenvector $\boldsymbol{s}_2 \in \mathbb{R}^{n-1}$, and there exists an $(n-1) \times (n-1)$ orthogonal matrix S_2' with $\mathbf{s}_2$ as its first column. If we set

$$S_2 = \left[\begin{array}{c|ccc} 1 & 0 & \cdots & 0 \\ \hline 0 & & & \\ \vdots & & S_2' & \\ 0 & & & \end{array}\right], \tag{7.3.4}$$

then this matrix is easily seen to be orthogonal as well, and we obtain

$$S_2^{-1}S_1^{-1}AS_1S_2 = \left[\begin{array}{cc|ccc} \lambda_1 & 0 & 0 & \cdots & 0 \\ 0 & \lambda_2 & 0 & \cdots & 0 \\ \hline 0 & 0 & & & \\ \vdots & \vdots & & A_2 & \\ 0 & 0 & & & \end{array}\right], \tag{7.3.5}$$

where the size of A_2 is $(n-2) \times (n-2)$.

Continuing in the same fashion we can reduce A to a diagonal matrix Λ by applying n similarity transformations like these. Writing

$$S = S_1S_2 \cdots S_n \tag{7.3.6}$$

we thus have

$$S^{-1}AS = \begin{bmatrix} \lambda_1 & 0 & 0 & \cdots & 0 \\ 0 & \lambda_2 & 0 & \cdots & 0 \\ 0 & 0 & \lambda_3 & \cdots & 0 \\ \vdots & \vdots & \vdots & & \vdots \\ 0 & 0 & 0 & \cdots & \lambda_n \end{bmatrix} = \Lambda. \tag{7.3.7}$$

The matrix S is orthogonal because it is the product of orthogonal matrices. Furthermore, from Equation (7.2.1) we can see that if there exists a nonsingular S that transforms A to a diagonal matrix Λ as above, then, conversely to Theorem 7.2.1, the columns of S must be eigenvectors of A corresponding to the diagonal entries of Λ as eigenvalues.

<<>>

Before giving an example of the use of this theorem, we need some terminology:

DEFINITION 7.3.1. A function from $\mathbb{R}^n$ to $\mathbb{R}$, for any $n > 1$, is called a *form.* A form Q given by the formula $Q(\mathbf{x}) = \mathbf{x}^TA\mathbf{x}$, where A is an arbitrary, symmetric $n \times n$ matrix, is called a *quadratic form.*

Note that using a symmetric matrix A in defining a quadratic form involves no loss of generality: Assume that A is not necessarily symmetric. Then, since $\mathbf{x}^TA\mathbf{x}$ is a scalar, it equals its own transpose, and so, on the one hand,

$$Q(\mathbf{x}) = \mathbf{x}^T A\mathbf{x} \tag{7.3.8}$$

and on the other,

$$Q(\mathbf{x}) = (\mathbf{x}^T A\mathbf{x})^T = \mathbf{x}^T A^T (\mathbf{x}^T)^T = \mathbf{x}^T A^T \mathbf{x}. \tag{7.3.9}$$

Adding Equations (7.3.8) and (7.3.9), we get

$$2Q(\mathbf{x}) = \mathbf{x}^T A\mathbf{x} + \mathbf{x}^T A^T \mathbf{x}, \tag{7.3.10}$$

and so

$$Q(\mathbf{x}) = \mathbf{x}^T \frac{1}{2}(A + A^T)\mathbf{x}. \tag{7.3.11}$$

Thus $Q(\mathbf{x}) = \mathbf{x}^T A\mathbf{x}$ can be written with the symmetric matrix $\frac{1}{2}(A + A^T)$ in place of the possibly nonsymmetric A, and therefore it is no restriction on Q to assume that A is symmetric to begin with.

If we make a change of basis with the orthogonal matrix S made up of eigenvectors of A, whose existence is guaranteed by Theorem 7.3.2, and write $\mathbf{x} = S\boldsymbol{y}$ (to make the notation simpler, we write $\boldsymbol{y}$ for the coordinate vector $\mathbf{x}_S$ of $\mathbf{x}$ relative to the basis S), then we get

$$\mathbf{x}^T A\mathbf{x} = \boldsymbol{y}^T S^T A S\boldsymbol{y} = \boldsymbol{y}^T \Lambda \boldsymbol{y}. \tag{7.3.12}$$

In component form, this equation becomes

$$\sum_{i=1}^{n}\sum_{j=1}^{n} a_{ij}x_i x_j = \sum_{i=1}^{n} \lambda_i y_i^2, \tag{7.3.13}$$

and so we have transformed the general quadratic form to a sum of squares weighted with the eigenvalues. The eigenvectors of the matrix Λ are the orthonormal standard vectors $\mathbf{e}_i$, since $\Lambda\mathbf{e}_i = \lambda_i \mathbf{e}_i$. These correspond to the orthonormal eigenvectors $\mathbf{s}_i$ of the matrix A, because if $\boldsymbol{y} = \mathbf{e}_i$, then $\mathbf{x} = S\boldsymbol{y} = S\mathbf{e}_i = \mathbf{s}_i$.

We know that in $\mathbb{R}^2$ the equation

$$\sum_{i=1}^{2} \lambda_i y_i^2 = 1$$

describes a conic section, for any values of the λ_i, and that the transformation by an orthogonal matrix is a rotation or a reflection. Therefore the equation

$$a_{11}x_1^2 + 2a_{12}x_1x_2 + a_{22}x_2^2 = \sum_{i=1}^{2}\sum_{j=1}^{2} a_{ij}x_ix_j = 1$$

represents a conic section in a rotated position. The type of this conic section is determined by the eigenvalues. For instance, if $0 < \lambda_1 < \lambda_2$, then we have an ellipse. Setting $\lambda_1 = 1/a^2$ and $\lambda_2 = 1/b^2$, its equation becomes the standard

$$\frac{y_1^2}{a^2} + \frac{y_2^2}{b^2} = 1. \tag{7.3.14}$$

The major and minor axes point in the directions of the vectors $\boldsymbol{y}_1 = (1, 0)^T$ and $\boldsymbol{y}_2 = (0, 1)^T$, and have half-lengths $a = 1/\sqrt{\lambda_1}$ and $b = 1/\sqrt{\lambda_2}$, respectively. (Just set successively $y_2 = 0$ and $y_1 = 0$ in Equation (7.3.14).) The vectors $\boldsymbol{y}_1$ and $\boldsymbol{y}_2$ correspond to the vectors $\mathbf{x}_1 = S\boldsymbol{y}_1 = (\boldsymbol{s}_1 \ \boldsymbol{s}_2)(1, 0)^T = \boldsymbol{s}_1$ and $\mathbf{x}_2 = S\boldsymbol{y}_2 = \boldsymbol{s}_2$ in the original basis. This result shows that the principal axes point in the directions of the eigenvectors, with the major axis corresponding to the smaller eigenvalue. Other conic sections and quadric surfaces in higher dimensions can be analyzed similarly. We consider some of these in the exercises.

EXAMPLE 7.3.1. Discuss the conic section given by

$$8x_1^2 - 12x_1x_2 + 17x_2^2 = 20. \tag{7.3.15}$$

This equation can be written in the standard form

$$\mathbf{x}^T A\mathbf{x} = 1 \tag{7.3.16}$$

with

$$A = \frac{1}{20}\begin{bmatrix} 8 & -6 \\ -6 & 17 \end{bmatrix}. \tag{7.3.17}$$

The eigenvalues and corresponding unit eigenvectors of this matrix are

$$\lambda_1 = 1/4,\ \lambda_2 = 1,\ \ \boldsymbol{s}_1 = \frac{1}{\sqrt{5}}(2, 1)^T,\ \ \boldsymbol{s}_2 = \frac{1}{\sqrt{5}}(-1, 2)^T.$$

Hence, according to the preceding discussion, Equation (7.3.15) represents the ellipse in Figure 7.3, which is centered at the origin, whose major axis has half-length $1/\sqrt{\lambda_1} = 2$ and points in the direction of the eigenvector $\boldsymbol{s}_1$, and whose minor axis has half-length $1/\sqrt{\lambda_2} = 1$ and points in the direction of the eigenvector $\boldsymbol{s}_2$.

<<>>

FIGURE 7.3

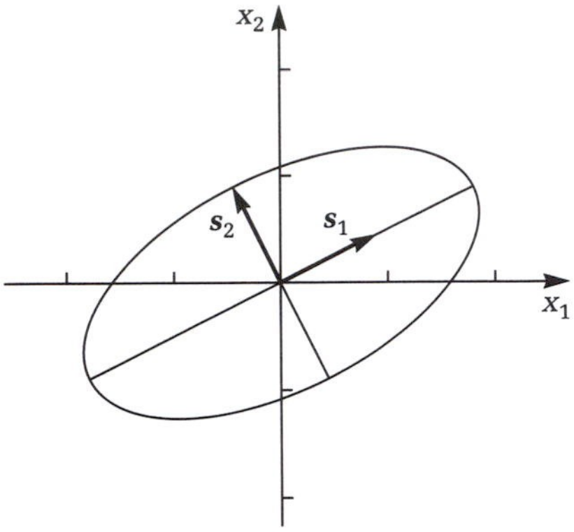

Exercises 7.3

7.3.1. Find the direction and length of each principal axis of the ellipse given by the equation below, and sketch its graph.

$$13x_1^2 - 8x_1x_2 + 7x_2^2 = 30.$$

7.3.2. Find the direction and length of each principal axis of the hyperbola given by the equation below, and sketch its graph.

$$7x_1^2 + 48x_1x_2 - 7x_2^2 = 25.$$

7.3.3. Find the direction and length of each principal axis of the hyperbola given by the equation below, and sketch its graph.

$$2x_1^2 + 4x_1x_2 - x_2^2 = 12.$$

• **7.3.4.** Find principal axes of the ellipsoid given by the equation below, change its equation to standard form so that the left side becomes a sum of squares weighted with the eigenvalues as in Equation (7.3.13), and describe its position and shape.

$$3x_1^2 + 3x_2^2 + 3x_3^2 - 2x_1x_2 - 2x_1x_3 - 2x_2x_3 = 4.$$

***7.3.5.** **a.** Show that if A is a symmetric matrix, then $\nabla(\mathbf{x}^TA\mathbf{x}) = 2(A\mathbf{x})^T$. (We have the transpose on the right because ∇f, for any f, is usually considered to be a row vector.)

b. Use the method of Lagrange multipliers to show that the extreme values of $\mathbf{x}^T A\mathbf{x}$ subject to the constraint $\mathbf{x}^T\mathbf{x} = 1$ are eigenvalues of A. (This property of A can be developed into a practical method for computing eigenvalues.)

***7.3.6.** Prove the following theorem called Schur's Lemma: For any real square matrix A with only real eigenvalues there exists an orthogonal matrix S such that $S^{-1}AS = T$ is upper triangular. (*Hint:* Modify the proof of Theorem 7.3.2 to account for the possible lack of symmetry.)

MATLAB Exercises 7.3

ML7.3.1. The ellipse of Example 7.3.1, together with the corresponding one in standard position, can be plotted in MATLAB by using polar coordinates.

a. Enter the following program and explain the steps.

```
t = 0 : .1 : 2 * pi;
c = cos(t); s = sin(t);
ra = 1./sqrt(.4 * c .^2 – .6 * c . * s + .85 * s .^2);
rb = 1./sqrt(c .^2/4 + s .^2);
polar(t, ra)
hold
polar(t, rb)
```

b. An alternative program to plot the original ellipse directly from the matrix A is the following. Enter it and explain the steps.

```
A = [8, –6; –6, 17]/20; t = 0 : .2 : 2 * pi;
x = [cos(t); sin(t)];
q = diag(x' * A * x)';
r = 1./sqrt(q);
polar(t, r)
```

ML7.3.2. Use MATLAB as in Exercise ML7.3.1 to solve Exercise 7.3.2.

ML7.3.3. Use MATLAB to plot the conic section $\mathbf{x}^T A\mathbf{x} = 1$ with

$$A = \begin{bmatrix} 4 & 1 \\ -1 & 2 \end{bmatrix}.$$

As we have seen in Example 7.1.4 on page 223, this is a defective matrix. How can you square this fact with your result?

7.4. COMPLEX EIGENVALUES AND EIGENVECTORS

As we have seen in the preceding sections, complex eigenvalues may well occur even for matrices with real entries, and in some applications, as in quantum physics, for instance, we must deal with complex-valued matrix components too. Consequently, we devote this section to such matters. We assume that the reader is familiar with complex numbers and exponential functions, but in an appendix at the back of the book we briefly review these for those who need it.

DEFINITION 7.4.1.

The complex vector space $\mathbb{C}^n$ is defined, for any positive integer n, as the set of ordered n-tuples $\mathbf{z} = (z_1, z_2, \ldots, z_n)^T$ of complex numbers written as columns, and with addition of vectors and multiplication of vectors by scalars defined so that for any such vectors and any complex scalars c we have

$$\begin{aligned}&(x_1, x_2, \ldots, x_n)^T + (y_1, y_2, \ldots, y_n)^T \\ &= (x_1 + y_1, x_2 + y_2, \ldots, x_n + y_n)^T \qquad (7.4.1)\end{aligned}$$

and

$$c(z_1, z_2, \ldots, z_n)^T = (cz_1, cz_2, \ldots, cz_n)^T. \qquad (7.4.2)$$

EXAMPLE 7.4.1. Let

$$\mathbf{x} = \begin{bmatrix} 2 + 3i \\ 1 - 4i \end{bmatrix} \text{ and } \mathbf{y} = \begin{bmatrix} 2 - 5i \\ 4 + 6i \end{bmatrix} \qquad (7.4.3)$$

be vectors of $\mathbb{C}^2$, and $c = 5 - 6i$ a scalar. Then

$$\mathbf{x} + \mathbf{y} = \begin{bmatrix} 4 - 2i \\ 5 + 2i \end{bmatrix} \text{ and } c\mathbf{x} = \begin{bmatrix} 28 + 3i \\ -19 - 26i \end{bmatrix}. \qquad (7.4.4)$$

<<>>

The length of such a vector cannot be defined in terms of the sum of the squares of the components, because such squares are generally complex, and we want a length to be real. This situation is, however, easy to remedy—we square the *absolute values* of the components:

DEFINITION 7.4.2.

The *length* or *norm* of any $\mathbf{z} = (z_1, z_2, \ldots, z_n)^T \in \mathbb{C}^n$ is defined as

$$|\mathbf{z}| = \sqrt{|z_1|^2 + |z_2|^2 + \cdots + |z_n|^2}. \tag{7.4.5}$$

We can put this formula in much simpler form, analogous to $|x|^2 = \mathbf{x}^T\mathbf{x}$ in the real case, by replacing each absolute value on the right, as follows:

$$|\mathbf{z}|^2 = \overline{z}_1 z_1 + \overline{z}_2 z_2 + \cdots + \overline{z}_n z_n. \tag{7.3.6}$$

Writing $\overline{\mathbf{z}}^T = (\overline{z}_1, \overline{z}_2, \ldots, \overline{z}_n)$ and defining matrix multiplication exactly as in the real case, we may write this as

$$|\mathbf{z}|^2 = \overline{\mathbf{z}}^T\mathbf{z}. \tag{7.4.7}$$

We could have written $|\mathbf{z}|^2 = \mathbf{z}^T\overline{\mathbf{z}}$ as well, but this form is never used. In fact there is a special name and notation for $\overline{\mathbf{z}}^T$:

DEFINITION 7.4.3.

For any $\mathbf{z} \in \mathbb{C}^n$ the row vector $\overline{\mathbf{z}}^T$ is called the *Hermitian conjugate*[2] *of* $\mathbf{z}$ and is denoted by $\mathbf{z}^H$. Similarly, for any matrix A with complex entries we define its Hermitian conjugate as $A^H = \overline{A}^T$ (read: "A–Hermitian"), where $\overline{A} = (\overline{a}_{ij})$.

Equation (7.4.7) suggests the following generalization of the dot product:

DEFINITION 7.4.4.

For any $\mathbf{x}, \mathbf{y} \in \mathbb{C}^n$ the scalar $\mathbf{x}^H\mathbf{y}$ is called their *inner product.*

Note that this definition is not commutative:

$$\mathbf{x}^H\mathbf{y} = \overline{x}_1 y_1 + \overline{x}_2 y_2 + \cdots + \overline{x}_n y_n, \tag{7.4.8}$$

but

$$\mathbf{y}^H\mathbf{x} = \overline{y}_1 x_1 + \overline{y}_2 x_2 + \cdots + \overline{y}_n x_n. \tag{7.4.9}$$

Thus,

$$\mathbf{y}^H\mathbf{x} = \overline{\mathbf{x}^H\mathbf{y}}, \tag{7.4.10}$$

and the two sides are equal if and only if $\mathbf{x}^H\mathbf{y}$ is real.

[2]After Charles Hermite (1822–1901).

EXAMPLE 7.4.2. For the vectors in Example 7.4.1 we have

$$\mathbf{x}^H = (2 - 3i, 1 + 4i) \text{ and } \mathbf{y}^H = (2 + 5i, 4 - 6i), \tag{7.4.11}$$

$$|\mathbf{x}|^2 = \mathbf{x}^H\mathbf{x} = (2 - 3i, 1 + 4i)\begin{bmatrix} 2 + 3i \\ 1 - 4i \end{bmatrix}$$

$$= (2 - 3i)(2 + 3i) + (1 + 4i)(1 - 4i) = 2^2 + 3^2 + 1^2 + 4^2 = 30, \tag{7.4.12}$$

and

$$|\mathbf{y}|^2 = \mathbf{y}^H\mathbf{y} = (2 + 5i, 4 - 6i)\begin{bmatrix} 2 - 5i \\ 4 + 6i \end{bmatrix}$$

$$= (2 + 5i)(2 - 5i) + (4 - 6i)(4 + 6i) = 2^2 + 5^2 + 4^2 + 6^2 = 81. \tag{7.4.13}$$

Similarly, the inner products can be computed as

$$\mathbf{x}^H\mathbf{y} = (2 - 3i, 1 + 4i)\begin{bmatrix} 2 - 5i \\ 4 + 6i \end{bmatrix} = (2 - 3i)(2 - 5i) + (1 + 4i)(4 + 6i)$$

$$= (-11 - 16i) + (-20 + 22i) = -31 + 6i, \tag{7.4.14}$$

and

$$\mathbf{y}^H\mathbf{x} = (2 + 5i, 4 - 6i)\begin{bmatrix} 2 + 3i \\ 1 - 4i \end{bmatrix} = (2 + 5i)(2 + 3i) + (4 - 6i)(1 - 4i)$$

$$= (-11 + 16i) + (-20 - 22i) = -31 - 6i. \tag{7.4.15}$$

<<>>

Two vectors of $\mathbb{C}^n$ are still called *orthogonal* if their inner product is zero, although the geometric meaning is lost.

For the Hermitian of a matrix product the expansion involves reversing the factors, just as for transposes:

THEOREM 7.4.1. For any matrices A and B for which AB is defined we have

$$(AB)^H = B^H A^H. \tag{7.4.16}$$

The Hermitian conjugate is used in place of the transpose to generalize the notions of symmetric and orthogonal matrices:

DEFINITION 7.4.5.

A matrix A is called *Hermitian* if

$$A^H = A, \tag{7.4.17}$$

and U is called *unitary* if

$$U^H U = I. \tag{7.4.18}$$

EXAMPLE 7.4.3. The matrix

$$A = \begin{bmatrix} 1 & 3+i \\ 3-i & 4 \end{bmatrix} \tag{7.4.19}$$

is Hermitian, for

$$A^H = \begin{bmatrix} \overline{1} & \overline{3-i} \\ \overline{3+i} & \overline{4} \end{bmatrix} = \begin{bmatrix} 1 & 3+i \\ 3-i & 4 \end{bmatrix} = A. \tag{7.4.20}$$

Notice that the diagonal entries are real, as they must be in any Hermitian matrix.

<<>>

EXAMPLE 7.4.4. The matrix

$$U = \frac{1}{\sqrt{2}}\begin{bmatrix} 1 & 1 \\ i & -i \end{bmatrix} \tag{7.4.21}$$

is unitary, since

$$U^H = \frac{1}{\sqrt{2}}\begin{bmatrix} 1 & -i \\ 1 & i \end{bmatrix} \tag{7.4.22}$$

and

$$U^H U = \frac{1}{2}\begin{bmatrix} 1 & -i \\ 1 & i \end{bmatrix}\begin{bmatrix} 1 & 1 \\ i & -i \end{bmatrix} = \begin{bmatrix} 1 & 0 \\ 0 & 1 \end{bmatrix} = I. \tag{7.4.23}$$

<<>>

The usefulness of Hermitian matrices rests on their following property:

THEOREM 7.4.2.

A Hermitian matrix has only real eigenvalues.

Proof Suppose A is Hermitian and λ is an eigenvalue of A corresponding to a nonzero eigenvector $\mathbf{s}$. Then

$$A\mathbf{s} = \lambda\mathbf{s} \tag{7.4.24}$$

and

$$\mathbf{s}^H A\mathbf{s} = \lambda\mathbf{s}^H\mathbf{s}. \tag{7.4.25}$$

Here $\mathbf{s}^H\mathbf{s}$ is real because it equals $|\mathbf{s}|^2$. The left side is real as well, because for scalars, complex conjugation and Hermitian conjugation are the same, consequently

$$\overline{\mathbf{s}^H A\mathbf{s}} = (\mathbf{s}^H A\mathbf{s})^H = \mathbf{s}^H A^H \mathbf{s}^{HH} = \mathbf{s}^H A\mathbf{s}. \tag{7.4.26}$$

Thus λ must be real.

<<>>

EXAMPLE 7.4.5. For the matrix of Example 7.4.3 we have the characteristic equation

$$|A - \lambda I| = \begin{vmatrix} 1-\lambda & 3+i \\ 3-i & 4-\lambda \end{vmatrix} = (1-\lambda)(4-\lambda) - 4 = 0, \tag{7.4.27}$$

which gives the eigenvalues $\lambda_1 = 5$ and $\lambda_2 = 0$.

<<>>

Since for a real matrix, Hermitian conjugation is the same as transposition, we have now proved Theorem 7.3.1 of page 238:

COROLLARY 7.4.1. The eigenvalues of a real symmetric matrix are real.

For the eigenvalues of a unitary matrix we have an analogous theorem:

THEOREM 7.4.3. The eigenvalues of a unitary matrix have absolute value 1.

Proof Let U be a unitary matrix and λ one of its eigenvalues with a nonzero eigenvector $\boldsymbol{0}$. Then, taking the Hermitian conjugate of both sides of

$$U\mathbf{s} = \lambda\mathbf{s}, \tag{7.4.28}$$

we get

$$\mathbf{s}^H U^H = \overline{\lambda}\mathbf{s}^H, \tag{7.4.29}$$

and multiplying corresponding sides:

$$\mathbf{s}^H U^H U \mathbf{s} = \overline{\lambda}\lambda \mathbf{s}^H \mathbf{s}. \tag{7.4.30}$$

Because U is unitary, we have $U^H U = I$, and so the left side above reduces to $\mathbf{s}^H\mathbf{s}$, which can then be canceled, leaving $\overline{\lambda}\lambda = 1$. This equation can be written as $|\lambda|^2 = 1$ and, because $|\lambda| \geq 0$, we must have $|\lambda| = 1$.

<<>>

Just as for real symmetric matrices, we have the following analogous theorems for Hermitian matrices:

THEOREM 7.4.4. Any two eigenvectors of a Hermitian matrix that belong to different eigenvalues are orthogonal to each other.

We leave the proof as Exercise 7.4.13.

THEOREM 7.4.5. (The Spectral Theorem.) For any Hermitian matrix A there exists a unitary matrix U such that $U^H A U = \Lambda$ is a real diagonal matrix. The columns of U are eigenvectors of A, and the diagonal entries of Λ the corresponding eigenvalues.

The proof is similar to the one in the real case and is omitted. We just illustrate the procedure with some examples.

EXAMPLE 7.4.6. In Example 7.4.5, we have found the eigenvalues $\lambda_1 = 5$ and $\lambda_2 = 0$ for the Hermitian matrix

$$A = \begin{bmatrix} 1 & 3+i \\ 3-i & 4 \end{bmatrix}. \tag{7.4.31}$$

The corresponding eigenvectors can be obtained by solving $(A - \lambda I)\mathbf{s} = \mathbf{0}$ with the values above for λ. For $\lambda_1 = 5$, the last equation becomes

$$\begin{bmatrix} -4 & 3+i \\ 3-i & -1 \end{bmatrix}\begin{bmatrix} s_{11} \\ s_{12} \end{bmatrix} = \begin{bmatrix} 0 \\ 0 \end{bmatrix}. \tag{7.4.32}$$

The two rows are dependent, as they should be, because the first one equals $-(3+i)$ times the second one. A solution is obviously $\mathbf{s} = (3+i, 4)^T$. Hence $\mathbf{s}^H = (3-i, 4)$, and so $\mathbf{s}^H\mathbf{s} = (3-i)(3+i) + 4^2 = 26$. Thus a normalized solution is given by

$$\mathbf{s}_1 = \frac{1}{\sqrt{26}}\begin{bmatrix} 3+i \\ 4 \end{bmatrix}. \tag{7.4.33}$$

For $\lambda_2 = 0$ the equation $(A - \lambda I)\mathbf{s} = \mathbf{0}$ becomes

$$\begin{bmatrix} 1 & 3+i \\ 3-i & 4 \end{bmatrix}\begin{bmatrix} s_{21} \\ s_{22} \end{bmatrix} = \begin{bmatrix} 0 \\ 0 \end{bmatrix}, \tag{7.4.34}$$

and a normalized solution of this equation is

$$\mathbf{s}_2 = \frac{1}{\sqrt{26}}\begin{bmatrix} -4 \\ 3-i \end{bmatrix}. \tag{7.4.35}$$

We combine the two eigenvectors above into the matrix

$$S = \frac{1}{\sqrt{26}}\begin{bmatrix} 3+i & -4 \\ 4 & 3-i \end{bmatrix}. \tag{7.4.36}$$

This S is unitary, since

$$S^H = \frac{1}{\sqrt{26}}\begin{bmatrix} 3-i & 4 \\ -4 & 3+i \end{bmatrix}, \tag{7.4.37}$$

and $S^H S = I$, as can be checked easily. Here we have denoted the unitary matrix U of Theorem 7.4.5 by S, in keeping with our earlier notation of $\mathbf{s}$ for eigenvectors. We leave it to the reader to check that $S^H A S = \Lambda$ holds; that is, that

$$S^H A S = \frac{1}{26}\begin{bmatrix} 3-i & 4 \\ -4 & 3+i \end{bmatrix}\begin{bmatrix} 1 & 3+i \\ 3-i & 4 \end{bmatrix}\begin{bmatrix} 3+i & -4 \\ 4 & 3-i \end{bmatrix} = \begin{bmatrix} 5 & 0 \\ 0 & 0 \end{bmatrix}, \tag{7.4.38}$$

as required by Theorem 7.4.5.

<<>>

EXAMPLE 7.4.7. This is a continuation of the electric circuit problem of Example 7.2.2 on page 232 for the case of negative D, which we had to omit previously. This case does occur in many real-life electric circuits and needs to be solved just as much as the earlier cases did.

The whole formalism as presented in Section 7.2 remains valid; we just have to carry it somewhat further to obtain the solutions in real rather than complex form. We can do so whenever the initial conditions are real, although the matrix is not Hermitian and the eigenvalues are complex.

Before proceeding let us mention an unfortunate notational collision between the traditional uses of the letter i for $\sqrt{-1}$ by mathematicians and for electric currents (from *i*ntensity) by physicists and engineers. The latter avoid this difficulty generally by using j for $\sqrt{-1}$. We shall stay with $i = \sqrt{-1}$, and in this section use only $i(t)$ for currents, not i as in Section 7.2.

Thus, let $D < 0$. Then the eigenvalues from Equations (7.2.42) and (7.2.43) on page 233 may be written as

$$\lambda_1 = \frac{1}{2}\left(-\frac{R}{L} - i\sqrt{|D|}\right) \tag{7.4.39}$$

and

$$\lambda_2 = \frac{1}{2}\left(-\frac{R}{L} + i\sqrt{|D|}\right). \tag{7.4.40}$$

Since these are complex conjugates of each other, we drop the subscripts and write $\lambda_2 = \lambda$ and $\lambda_1 = \overline{\lambda}$. Also, we write

$$\lambda = -\alpha + i\omega, \tag{7.4.41}$$

where

$$\alpha = \frac{R}{2L} \text{ and } \omega = \frac{\sqrt{|D|}}{2} \tag{7.4.42}$$

are nonnegative real numbers. With these notations the general solutions (7.2.44) and (7.2.45) on page 234 become

$$i(t) = e^{-\alpha t}(c_{11}e^{-i\omega t} + c_{12}e^{i\omega t}) \tag{7.4.43}$$

and

$$q(t) = e^{-\alpha t}(c_{21}e^{-i\omega t} + c_{22}e^{i\omega t}). \tag{7.4.44}$$

In view of Euler's Formula these equations represent damped oscillations; that is, oscillations with angular frequency ω and exponentially decaying amplitudes.

With the initial conditions (7.2.46) and (7.2.47) representing an initial charge Q and no initial current, we obtain from Equations (7.2.48) and (7.2.49)

$$c_{11} = -c_{12} = \frac{Q}{LCi\sqrt{|D|}} \tag{7.4.45}$$

and

$$c_{21} = Q - c_{22} = \frac{Q\lambda_2}{i\sqrt{|D|}}. \tag{7.4.46}$$

Substituting these into the general solution above, we get the corresponding particular solution as

$$i(t) = \frac{Q}{LC\sqrt{|D|}}e^{-\alpha t}\frac{e^{-i\omega t} - e^{i\omega t}}{i} = \frac{-2Q}{LC\sqrt{|D|}}e^{-\alpha t}\sin\omega t \tag{7.4.47}$$

and

$$q(t) = e^{-\alpha t}\left(Q\cos\omega t + \frac{QR}{L\sqrt{|D|}}\sin\omega t\right). \tag{7.4.48}$$

Notice that the imaginary parts of these numbers have disappeared as they ought to.

<<>>

Exercises 7.4

In the first four exercises find, for the given vectors, (a) their Hermitian conjugates, (b) their lengths, (c) their inner products in both orders.

7.4.1. $\mathbf{x} = \begin{bmatrix} 2 \\ 2i \end{bmatrix}$ and $\mathbf{y} = \begin{bmatrix} 5i \\ 4+i \end{bmatrix}$.

7.4.2. $\mathbf{x} = \begin{bmatrix} 2+4i \\ 1-2i \end{bmatrix}$ and $\mathbf{y} = \begin{bmatrix} 1-5i \\ 4+2i \end{bmatrix}$.

7.4.3. $\mathbf{x} = \begin{bmatrix} 2e^{i\pi/4} \\ 2i \end{bmatrix}$ and $\mathbf{y} = \begin{bmatrix} e^{i\pi/4} \\ e^{-i\pi/4} \end{bmatrix}$.

7.4.4. $\mathbf{x} = \begin{bmatrix} 2 \\ 2i \\ 1+i \end{bmatrix}$ and $\mathbf{y} = \begin{bmatrix} 5i \\ 4+i \\ 4-i \end{bmatrix}$.

7.4.5. Let $\boldsymbol{u}_1 = \frac{1}{\sqrt{2}}\begin{bmatrix} 1 \\ i \end{bmatrix}$.

a. Find another vector $\boldsymbol{u}_2 \in \mathbb{C}^2$ so that $\boldsymbol{u}_1, \boldsymbol{u}_2$ form an orthonormal basis for $\mathbb{C}^2$.

b. Find the coordinates of an arbitrary vector $\mathbf{x}$ with respect to this basis; that is, the coefficients x_{U1}, x_{U2} in the decomposition $\mathbf{x} = x_{U1}\boldsymbol{u}_1 + x_{U2}\boldsymbol{u}_2$.

c. Find the coordinates of the vector

$$\mathbf{x} = \begin{bmatrix} 2+4i \\ 1-2i \end{bmatrix}$$

with respect to the above basis.

7.4.6. Find the eigenvalues and eigenvectors of the rotation matrix

$$R_\theta = \begin{bmatrix} \cos\theta & -\sin\theta \\ \sin\theta & \cos\theta \end{bmatrix}.$$

7.4.7. Show that for any matrix A we have $A^{HH} = A$.

7.4.8. Show that for any matrix A the product $A^H A$ is Hermitian.

7.4.9. Show that a matrix A is Hermitian if and only if $\mathbf{x}^H A\mathbf{x}$ is real for every vector $\mathbf{x}$ (of the right size, of course).

7.4.10. Show that if U is unitary, then $|U\mathbf{x}| = |\mathbf{x}|$ for every $\mathbf{x}$ (of the right size).

7.4.11. Show that if the matrix A is Hermitian, then $U = e^{iAt}$ is unitary for every real t. (This fact is important in physics, since U provides the solutions to Schrödinger's differential equation

$$\frac{d\mathbf{u}}{dt} = iA\mathbf{u}.$$

Cf. Exercise 7.2.12 on page 236.)

7.4.12. Verify that Equations (7.4.47) and (7.4.48) give the particular solutions of Example 7.4.7 for the initial conditions $q(0) = Q$ and $i(0) = 0$.

7.4.13. Prove Theorem 7.4.4.

7.4.14. Prove that the determinant of any Hermitian matrix is real.

7.4.15. Find the eigenvalues and eigenvectors of the matrix

$$A = \begin{bmatrix} 1 & 1 \\ -1 & 1 \end{bmatrix}.$$

7.4.16. Find the eigenvalues and eigenvectors of the matrix

$$A = \begin{bmatrix} 1 & 1 & 1 \\ 0 & 1 & 1 \\ 0 & -1 & 1 \end{bmatrix}.$$

MATLAB Exercises 7.4

In MATLAB we can create random orthogonal matrices as follows: For any real matrix A, the command $[Q, R] = \mathbf{qr}(A)$ returns an orthogonal matrix Q and an upper triangular matrix R such that $A = QR$. (Here we use this command only to obtain Q and discard the matrices A and R.) In the next exercise we want to show that any 3×3 orthogonal matrix represents a rotation or the product of a rotation and a reflection, and find the axis and angle of the rotation. (Cf. also Example 4.3.2 on page 160.)

ML7.4.1. **a.** Enter the matrix

$$A = \begin{bmatrix} 1 & 1 & 0 \\ 2 & 1 & 0 \\ 3 & 0 & 1 \end{bmatrix}.$$

and find the corresponding orthogonal matrix Q. Let $[X, D] = \mathbf{eig}(Q)$. This command returns the eigenvectors of Q in X and the eigenvalues in D. Notice that one of the eigenvalues is 1 and check that the other two have absolute value 1. Let $t = \mathbf{angle}(D(2, 2))$ and show that the matrix Q represents a rotation by the angle t around the first eigenvector, as follows. Let **s1** = $X(:, 1)$. Then the command $[S, T]$ = **qr(s1)** creates an orthogonal matrix S whose first column is **s1**. Thus the columns of S are mutually orthogonal unit vectors, and so S represents a rotation or –1 times a rotation of the standard basis to the columns of S. The command $R = S' * Q * S$ transforms the matrix Q to the basis S. Show that the matrix R represents a rotation by angle t around the first vector of the new basis. (Compare R with the matrix in Exercise 7.4.6.)

b. Prove that in general, if Q is an orthogonal matrix with eigenvalues ± 1, $e^{i\theta}$, $e^{-i\theta}$, then Q can be written as $Q = \pm S * R * S'$, where S is an orthogonal matrix whose first column is the first eigenvector of Q, and

$$R = \begin{bmatrix} 1 & 0 & 0 \\ 0 & \cos\theta & -\sin\theta \\ 0 & \sin\theta & \cos\theta \end{bmatrix}.$$

<<8>> NUMERICAL METHODS

8.1. *LU* FACTORIZATION

In this section we consider a variant of Gaussian elimination in which the coefficient matrix A is written as a product of a lower triangular matrix L and an upper triangular or echelon matrix U. The main advantage of this method over the straightforward algorithm is that it is considerably more economical when we need to solve several systems of the form $A\mathbf{x} = \boldsymbol{b}$ with the same A but different right-hand sides $\boldsymbol{b}$. An additional, though less practical, advantage is that we gain some insight into the structure of Gaussian elimination in terms of matrix products.

The idea behind the new procedure is very simple: As we have seen in Chapter 2, forward elimination changes $A\mathbf{x} = \boldsymbol{b}$ into an equivalent system $U\mathbf{x} = \boldsymbol{c}$, where U is an echelon matrix, or an upper triangular matrix in case A is square. If we can write A as a product LU, then $A\mathbf{x} = \boldsymbol{b}$ becomes $LU\mathbf{x} = \boldsymbol{b}$, and multiplying $U\mathbf{x} = \boldsymbol{c}$ by L on both sides, we get $LU\mathbf{x} = L\boldsymbol{c}$. Hence we must have

$$L\boldsymbol{c} = \boldsymbol{b}. \tag{8.1.1}$$

Since L turns out to be lower triangular, it is very easy to solve this equation for $\boldsymbol{c}$ by "forward substitution," once L is known. Thus if we know L and U, then the solution of the system $A\mathbf{x} = \boldsymbol{b}$ is reduced to that of the two extremely simple systems $L\boldsymbol{c} = \boldsymbol{b}$ and $U\mathbf{x} = \boldsymbol{c}$, which express the forward-elimination and the back-substitution phases of Gaussian elimination, respectively.

To see how to find L and how to use this method, let us first consider some examples:

EXAMPLE 8.1.1. Let us find L and U for

$$A = \begin{bmatrix} 1 & 3 \\ 2 & 4 \end{bmatrix}, \tag{8.1.2}$$

and use those to solve $A\mathbf{x} = \boldsymbol{b}$ with $\boldsymbol{b} = (5, 6)^T$.

Here the first step in Gaussian elimination is that of subtracting twice the first row of A from the second row. This move is equivalent to multiplying A by the elementary matrix[1]

$$E = \begin{bmatrix} 1 & 0 \\ -2 & 1 \end{bmatrix}. \tag{8.1.3}$$

Indeed, if we write $A = (\boldsymbol{a}^1, \boldsymbol{a}^2)^T$, then

$$U = EA = \begin{bmatrix} 1\boldsymbol{a}^1 + 0\boldsymbol{a}^2 \\ -2\boldsymbol{a}^1 + 1\boldsymbol{a}^2 \end{bmatrix} = \begin{bmatrix} 1 & 3 \\ 0 & -2 \end{bmatrix}, \tag{8.1.4}$$

and so the first row of this matrix is the same as the first row of A, and the second row is (−2) times the first row of A plus the second row of A; just what we needed.

We can now proceed in two ways to obtain the vector $\boldsymbol{c}$ of the reduced system $U\mathbf{x} = \boldsymbol{c}$. First, we can simply compute it in the old way as

$$\boldsymbol{c} = E\boldsymbol{b} = \begin{bmatrix} 1 & 0 \\ -2 & 1 \end{bmatrix}\begin{bmatrix} 5 \\ 6 \end{bmatrix} = \begin{bmatrix} 5 \\ -4 \end{bmatrix}. \tag{8.1.5}$$

Second, we can compute $\boldsymbol{c}$ in the new way by finding the matrix L for which $A = LU$ holds, and solving Equation (8.1.1). In this simple example the two methods are equally easy, but for larger systems with various right sides the second one is preferable. So let us see how the new method works in this case.

From the equation $U = EA$ we obtain $A = E^{-1}U$, since E is invertible. Thus

$$L = E^{-1} = \begin{bmatrix} 1 & 0 \\ 2 & 1 \end{bmatrix}. \tag{8.1.6}$$

(Notice that for this L we have

$$LU = L(EA) = \begin{bmatrix} 1 & 0 \\ 2 & 1 \end{bmatrix}\begin{bmatrix} \boldsymbol{a}^1 \\ -2\boldsymbol{a}^1 + \boldsymbol{a}^2 \end{bmatrix} = \begin{bmatrix} \boldsymbol{a}^1 \\ 2\boldsymbol{a}^1 + (-2\boldsymbol{a}^1 + \boldsymbol{a}^2) \end{bmatrix} = A; \tag{8.1.7}$$

that is, L has the desired effect of adding back the $2\boldsymbol{a}_1$ subtracted by E in the second row of A.) Hence the equation $L\boldsymbol{c} = \boldsymbol{b}$ is now

$$\begin{bmatrix} 1 & 0 \\ 2 & 1 \end{bmatrix}\begin{bmatrix} c_1 \\ c_2 \end{bmatrix} = \begin{bmatrix} 5 \\ 6 \end{bmatrix}. \tag{8.1.8}$$

We solve this system from the top down as $c_1 = 5$ and $2 \cdot 5 + c_2 = 6$, $c_2 = -4$. Thus we get the same $\boldsymbol{c}$, of course, as before.

[1]An elementary matrix is a matrix that corresponds to an elementary row operation.

There is nothing new in the rest of the computation: We solve $U\mathbf{x} = \boldsymbol{c}$ by back substitution; that is, from

$$\begin{bmatrix} 1 & 3 \\ 0 & -2 \end{bmatrix} \begin{bmatrix} x_1 \\ x_2 \end{bmatrix} = \begin{bmatrix} 5 \\ -4 \end{bmatrix} \tag{8.1.9}$$

we compute $-2x_2 = -4$, $x_2 = 2$ and $x_1 + 3 \cdot 2 = 5$, $x_1 = -1$.

<<>>

EXAMPLE 8.1.2. Let

$$A = \begin{bmatrix} 1 & 2 & 0 \\ 3 & 6 & -1 \\ 1 & 2 & 1 \end{bmatrix} \text{ and } \boldsymbol{b} = \begin{bmatrix} 2 \\ 8 \\ 0 \end{bmatrix} \tag{8.1.10}$$

as in Example 2.1.5 on page 41.

Multiplying A by

$$E_{21} = \begin{bmatrix} 1 & 0 & 0 \\ -3 & 1 & 0 \\ 0 & 0 & 1 \end{bmatrix} \tag{8.1.11}$$

we annihilate the $a_{21} = 3$ entry and obtain[2]

$$E_{21}A = \begin{bmatrix} 1 & 2 & 0 \\ 0 & 0 & -1 \\ 1 & 2 & 1 \end{bmatrix}. \tag{8.1.12}$$

Next, we multiply by

$$E_{31} = \begin{bmatrix} 1 & 0 & 0 \\ 0 & 1 & 0 \\ -1 & 0 & 1 \end{bmatrix} \tag{8.1.13}$$

to make the $a_{31} = 1$ entry 0, and produce

$$E_{31}E_{21}A = \begin{bmatrix} 1 & 2 & 0 \\ 0 & 0 & -1 \\ 0 & 0 & 1 \end{bmatrix}. \tag{8.1.14}$$

Finally, multiplication by

$$E_{32} = \begin{bmatrix} 1 & 0 & 0 \\ 0 & 1 & 0 \\ 0 & 1 & 1 \end{bmatrix} \tag{8.1.15}$$

[2]We denote this matrix by E_{21} to indicate the location of its sole nonzero offdiagonal entry.

gives the echelon matrix

$$U = E_{32}E_{31}E_{21}A = \begin{bmatrix} 1 & 2 & 0 \\ 0 & 0 & -1 \\ 0 & 0 & 0 \end{bmatrix}. \tag{8.1.16}$$

Now each of the matrices E_{21}, E_{31}, and E_{32} is invertible, with the inverse obtained simply by changing the sign of the nonzero offdiagonal entry. Thus their product is also invertible, and from the first part of Equation (8.1.16) we obtain $A = LU$ with

$$L = E_{21}^{-1}E_{31}^{-1}E_{32}^{-1} = \begin{bmatrix} 1 & 0 & 0 \\ 3 & 1 & 0 \\ 0 & 0 & 1 \end{bmatrix}\begin{bmatrix} 1 & 0 & 0 \\ 0 & 1 & 0 \\ 1 & 0 & 1 \end{bmatrix}\begin{bmatrix} 1 & 0 & 0 \\ 0 & 1 & 0 \\ 0 & -1 & 1 \end{bmatrix}$$

$$= \begin{bmatrix} 1 & 0 & 0 \\ 3 & 1 & 0 \\ 1 & 0 & 1 \end{bmatrix}\begin{bmatrix} 1 & 0 & 0 \\ 0 & 1 & 0 \\ 0 & -1 & 1 \end{bmatrix} = \begin{bmatrix} 1 & 0 & 0 \\ 3 & 1 & 0 \\ 1 & -1 & 1 \end{bmatrix}. \tag{8.1.17}$$

Notice that very luckily this product too is lower diagonal. Also, the entries l_{ij} below the diagonal are exactly the multipliers of the rows occurring in forward elimination; that is, it is $l_{ij}\boldsymbol{a}^j$ that we would subtract from $\boldsymbol{a}^i$ in forward elimination. *This is always the case, and so we never need to compute L separately; we can just assemble it from the coefficients that occur in forward elimination.* (This is generally not the case for the product of the E matrices as in Equation (8.1.16), but fortunately we do not need that product anyway. See also Exercise 8.1.1.)

Now let us use the matrix L we have found to obtain $\boldsymbol{c}$: The equation $L\boldsymbol{c} = \boldsymbol{b}$ becomes

$$\begin{bmatrix} 1 & 0 & 0 \\ 3 & 1 & 0 \\ 1 & -1 & 1 \end{bmatrix}\begin{bmatrix} c_1 \\ c_2 \\ c_3 \end{bmatrix} = \begin{bmatrix} 2 \\ 8 \\ 0 \end{bmatrix}. \tag{8.1.18}$$

Hence $c_1 = 2$, $3 \cdot 2 + c_2 = 8$, $c_2 = 2$, $1 \cdot 2 - 1 \cdot 2 + c_3 = 0$, and $c_3 = 0$.

Thus the equation $U\mathbf{x} = \boldsymbol{c}$ becomes

$$\begin{bmatrix} 1 & 2 & 0 \\ 0 & 0 & -1 \\ 0 & 0 & 0 \end{bmatrix}\begin{bmatrix} x_1 \\ x_2 \\ x_3 \end{bmatrix} = \begin{bmatrix} 2 \\ 2 \\ 0 \end{bmatrix}. \tag{8.1.19}$$

This is the equation corresponding to the augmented matrix in Equation (2.1.19) on page 41 and has, of course, the same solution, as given in Equation (2.1.20) there.

<<>>

We can now summarize the main points of the foregoing discussion in the following theorem:

THEOREM 8.1.1. If in the forward phase of the Gaussian elimination algorithm for an $m \times n$ matrix A no row exchanges are used, then A can be written as a product LU, where L is an $m \times m$ lower triangular matrix with 1's along its main diagonal, and U the $m \times n$ echelon matrix obtained by the algorithm.

Furthermore, each entry l_{ij} of L below the main diagonal is the coefficient of $\boldsymbol{a}^j$ in the product $l_{ij}\boldsymbol{a}^j$ that is subtracted from $\boldsymbol{a}^i$ in forward elimination.

Also, once L and U are known, the system $A\mathbf{x} = \boldsymbol{b}$, for any $\boldsymbol{b}$, can be replaced by the two special systems $L\boldsymbol{c} = \boldsymbol{b}$ and $U\mathbf{x} = \boldsymbol{c}$.

Proof We prove only the case of A being $3 \times n$ and each E_{ij} and the matrix L being 3×3. For other dimensions the argument would be similar.

The elementary matrix representing the first step of the elimination algorithm is

$$E_{21} = \begin{bmatrix} 1 & 0 & 0 \\ -l_{21} & 1 & 0 \\ 0 & 0 & 1 \end{bmatrix}, \tag{8.1.20}$$

since

$$E_{21}A = \begin{bmatrix} 1 & 0 & 0 \\ -l_{21} & 1 & 0 \\ 0 & 0 & 1 \end{bmatrix} \begin{bmatrix} \boldsymbol{a}^1 \\ \boldsymbol{a}^2 \\ \boldsymbol{a}^3 \end{bmatrix} = \begin{bmatrix} \boldsymbol{a}^1 \\ \boldsymbol{a}^2 - l_{21}\boldsymbol{a}^1 \\ \boldsymbol{a}^3 \end{bmatrix}. \tag{8.1.21}$$

Thus, on the one hand, the l_{21} in the matrix E_{21} is the coefficient of $\boldsymbol{a}^1$ in the product that is subtracted from $\boldsymbol{a}^2$ in forward elimination. The analogous statement for any other l_{ij} would follow similarly.

On the other hand, to construct L and to show that each l_{ij} is also the appropriate entry of L, we can proceed as follows: From Equation (8.1.20) we get

$$E_{21}^{-1} = \begin{bmatrix} 1 & 0 & 0 \\ l_{21} & 1 & 0 \\ 0 & 0 & 1 \end{bmatrix}, \tag{8.1.22}$$

and similarly we have

$$E_{31}^{-1} = \begin{bmatrix} 1 & 0 & 0 \\ 0 & 1 & 0 \\ l_{31} & 0 & 1 \end{bmatrix}, \tag{8.1.23}$$

and

$$E_{32}^{-1} = \begin{bmatrix} 1 & 0 & 0 \\ 0 & 1 & 0 \\ 0 & l_{32} & 1 \end{bmatrix}. \tag{8.1.24}$$

Since U is defined as the echelon matrix obtained by the forward-elimination algorithm embodied in the E_{ij} matrices, we must have $U = E_{32}E_{31}E_{21}A$. From this we obtain $A = E_{21}^{-1}E_{31}^{-1}E_{32}^{-1}U$ and, since L is defined by $A = LU$, we find that $L = E_{21}^{-1}E_{31}^{-1}E_{32}^{-1}$ must hold. Thus we compute LU as follows: First we write

$$E_{32}^{-1}U = \begin{bmatrix} 1 & 0 & 0 \\ 0 & 1 & 0 \\ 0 & l_{32} & 1 \end{bmatrix} \begin{bmatrix} \boldsymbol{u}^1 \\ \boldsymbol{u}^2 \\ \boldsymbol{u}^3 \end{bmatrix} = \begin{bmatrix} \boldsymbol{u}^1 \\ \boldsymbol{u}^2 \\ l_{32}\boldsymbol{u}^2 + \boldsymbol{u}^3 \end{bmatrix}. \tag{8.1.25}$$

Next

$$E_{31}^{-1}E_{32}^{-1}U = \begin{bmatrix} 1 & 0 & 0 \\ 0 & 1 & 0 \\ l_{31} & 0 & 1 \end{bmatrix} \begin{bmatrix} \boldsymbol{u}^1 \\ \boldsymbol{u}^2 \\ l_{32}\boldsymbol{u}^2 + \boldsymbol{u}^3 \end{bmatrix} = \begin{bmatrix} \boldsymbol{u}^1 \\ \boldsymbol{u}^2 \\ l_{31}\boldsymbol{u}^1 + l_{32}\boldsymbol{u}^2 + \boldsymbol{u}^3 \end{bmatrix} \tag{8.1.26}$$

and

$$LU = E_{21}^{-1}E_{31}^{-1}E_{32}^{-1}U = \begin{bmatrix} 1 & 0 & 0 \\ l_{21} & 1 & 0 \\ 0 & 0 & 1 \end{bmatrix} \begin{bmatrix} \boldsymbol{u}^1 \\ \boldsymbol{u}^2 \\ l_{31}\boldsymbol{u}^1 + l_{32}\boldsymbol{u}^2 + \boldsymbol{u}^3 \end{bmatrix}$$

$$= \begin{bmatrix} \boldsymbol{u}^1 \\ l_{21}\boldsymbol{u}^1 + \boldsymbol{u}^2 \\ l_{31}\boldsymbol{u}^1 + l_{32}\boldsymbol{u}^2 + \boldsymbol{u}^3 \end{bmatrix} = \begin{bmatrix} 1 & 0 & 0 \\ l_{21} & 1 & 0 \\ l_{31} & l_{32} & 1 \end{bmatrix} \begin{bmatrix} \boldsymbol{u}^1 \\ \boldsymbol{u}^2 \\ \boldsymbol{u}^3 \end{bmatrix}. \tag{8.1.27}$$

Thus indeed,

$$L = \begin{bmatrix} 1 & 0 & 0 \\ l_{21} & 1 & 0 \\ l_{31} & l_{32} & 1 \end{bmatrix}. \tag{8.1.28}$$

The calculation above shows why the l_{ij} coefficients from the forward-elimination process show up intact in L: In the multiplications above, we first added $l_{32}\boldsymbol{u}^2$ to $\boldsymbol{u}^3$, then $l_{31}\boldsymbol{u}^1$ to the sum, without disturbing anything else. Then we added $l_{21}\boldsymbol{u}^1$ to $\boldsymbol{u}^2$, again without disturbing anything else.

This step finishes the proof of the first two statements of Theorem 8.1.1. The last statement has been proved in the second paragraph of this section, on page 255.

<<>>

So far in this section we have used only one kind of elementary row operation on matrices: subtracting a multiple of one row from another. We now discuss briefly how the other two kinds are sometimes incorporated into the LU factorization.

The elementary row operation of multiplying a row by some number (without subtraction from another row) is necessary in Gauss-Jordan elimination only to obtain 1's as pivots. Corresponding to this observation, we can write the matrix U of the LU factorization as $U = DU'$, where D is a diagonal matrix with the pivots of U as its diagonal elements, and U' is an echelon matrix with 1's as pivots. Since the effect of multiplication of U' by D is multiplication of each row of U' by the corresponding diagonal element of D, the rows of U' are obtained from those of U by factoring out the pivots. The entries of U' are the coefficients that show up in Gauss-Jordan elimination when the entries of A above the pivots are annihilated, as the l_{ij} coefficients show up in the proof above.

It is customary to omit the prime from U' and to speak of the LDU factorization of A. For the matrix of Example 8.1.1 we can compute this factorization as follows:

EXAMPLE 8.1.3. From Example 8.1.1 we have

$$\begin{bmatrix} 1 & 3 \\ 2 & 4 \end{bmatrix} = \begin{bmatrix} 1 & 0 \\ 2 & 1 \end{bmatrix} \begin{bmatrix} 1 & 3 \\ 0 & -2 \end{bmatrix}. \tag{8.1.29}$$

Now we can factor out the -2 from the U, to get

$$A = \begin{bmatrix} 1 & 3 \\ 2 & 4 \end{bmatrix} = \begin{bmatrix} 1 & 0 \\ 2 & 1 \end{bmatrix} \begin{bmatrix} 1 & 0 \\ 0 & -2 \end{bmatrix} \begin{bmatrix} 1 & 3 \\ 0 & 1 \end{bmatrix} = LDU. \tag{8.1.30}$$

<<>>

We summarize the LDU factorization as

COROLLARY 8.1.1. If in the forward phase of the Gaussian elimination algorithm for an $m \times n$ matrix A no row exchanges are used, then A can be written as a product LDU, where L is an $m \times m$ lower triangular matrix with 1's along its main diagonal, D is an $m \times m$ diagonal matrix, and U is an $m \times n$ echelon matrix with 1's as pivots. The D and U matrices here can be obtained by appropriately factoring the U of the LU factorization of A.

The third elementary row operation, the exchange of rows, is necessary if we encounter a zero when looking for a pivot. In this case we can imagine all the necessary row exchanges to be done first. If P is the permutation matrix that represents these row exchanges (see Exercise 2.3.17 on page 79), then we can apply the LU or LDU factorization to PA, instead of to A, since for PA no more row exchanges are needed.

To conclude this section we present a brief quantitative discussion comparing the efficiency of LU factorization with that of straightforward Gaussian elimination for an $n \times n$ matrix A.

When n is large, even computers may need considerable time to perform the necessary calculations, and so it is of great practical importance to know the length of time needed for any algorithm. Present-day computers take about the same time for every multiplication, division, and multiplication-addition combination. We call these *long* operations as opposed to the *short* operations of addition, subtraction, and comparison. For all practical purposes the length of time needed for our algorithms is proportional to the number of long operations, and so we want to count these.

In the forward phase of Gaussian elimination, assuming no row exchanges are needed, to get a 0 in place of a_{21}, we compute $l_{21} = a_{21}/a_{11}$ and subtract $l_{21}a_{1j}$ from each element a_{2j} in the second row, for $j = 2, 3, \ldots, n$. (The 0 we do not need to compute.) This procedure uses n long operations on the left side of $A\mathbf{x} = \boldsymbol{b}$.

Next, we do the same for each of the other rows below the first row. Thus to get all the $n-1$ zeroes in the first column requires $n(n-1) = n^2 - n$ long operations.

Now we do the same for the $(n-1) \times (n-1)$ submatrix below and to the right of a_{11}. For this we need $(n-1)^2 - (n-1)$ long operations.

Continuing in this manner, we find that the number of long operations needed for forward elimination on the left side of $A\mathbf{x} = \boldsymbol{b}$ is

$$\sum_{k=1}^{n}(k^2 - k) = \frac{n(n+1)(2n+1)}{6} - \frac{n(n+1)}{2} = \frac{n^3 - n}{3} \approx \frac{n^3}{3}. \qquad (8.1.31)$$

Because the same calculations produce L and U as well, this is also the number of long operations needed for the LU factorization of A.

To reduce the right-hand side of $A\mathbf{x} = \boldsymbol{b}$ along with A, we do $n-1$ multiplications $l_{k1}b_1$ and subtractions from b_k when we produce the zeroes in the first column of A. Then $n-2$ such operations, when we produce the zeroes in the second column, and so on. Thus altogether the right-hand side requires

$$\sum_{k=1}^{n-1} k = \frac{n(n-1)}{2} \approx \frac{n^2}{2} \qquad (8.1.32)$$

long operations. If n is large, this number is negligible next to $n^3/3$, and so we usually consider the latter as the approximate number of long operations needed for the whole of Gaussian elimination.

The number of operations is the same $n^2/2$ whether we reduce the right-hand side of $A\mathbf{x} = \boldsymbol{b}$ along with A or we solve $L\boldsymbol{c} = \boldsymbol{b}$ only afterward. Clearly, the number of long operations needed to solve $U\mathbf{x} = \boldsymbol{c}$ is also $n^2/2$, and so, once L and U are known, we can obtain $\mathbf{x}$ for a new $\boldsymbol{b}$ in just n^2 long operations. This is the main advantage of LU factorization.

Exercises 8.1

8.1.1. Compute the product of the E matrices in Equation (8.1.16) and compare it to L.

8.1.2. Compute the LDU factorization of

$$A = \begin{bmatrix} 1 & 2 \\ 2 & 4 \end{bmatrix}.$$

8.1.3. Show that for a symmetric matrix A the matrix U in the LDU factorization satisfies $U = L^T$.

8.1.4. Compute the time needed for a computer to solve an $n \times n$ system by Gaussian elimination for $n = 1000$ if it can do 10^5 long operations per second.

8.1.5. Compute the number of long operations needed for the LU factorization of an $m \times n$ matrix A.

8.1.6. Compute the number of long operations needed to solve $L\boldsymbol{c} = \boldsymbol{b}$ and $U\mathbf{x} = \boldsymbol{c}$, once L and U are known and A is $m \times n$.

8.1.7. Show that the number of long operations in solving an $n \times n$ system $A\mathbf{x} = \boldsymbol{b}$ by Gauss-Jordan elimination is approximately $n^3/3$ when n is large, provided A is first changed to echelon form and then to reduced echelon form from the bottom up.

8.1.8. Show that the number of long operations in inverting an $n \times n$ matrix A by the Gauss-Jordan elimination algorithm in Section 2.3 is approximately n^3 when n is large.

8.1.9. Show that the number of long operations in computing the determinant of an $n \times n$ matrix A by reducing it to upper triangular form and multiplying the diagonal entries is approximately $n^3/3$ when n is large.

MATLAB Exercises 8.1

In MATLAB, the LU factorization is provided by the command $[L, U] =$ **lu**(A). However, the matrix L in this command usually is only a product of a permutation matrix and a lower triangular matrix because of row exchanges. The latter are introduced to minimize roundoff errors, as explained in the next section. Thus, if a genuine lower triangular matrix is required, then it is better to use the command $[L, U, P] =$ **lu**(A). This command produces a lower triangular matrix L, an upper triangular matrix U and a permutation matrix P such that $LU = PA$.

ML8.1.1. For five instances of $A =$ **round**$(10 *$ **rand**$(4, 5))$, find an LU factorization of A, using $[L, U, P] =$ **lu**(A), and change it using the **diag** command to the corresponding LDU factorization. Check that $LDU = PA$ holds.

8.2. SCALED PARTIAL PIVOTING

As we have seen, in Gaussian elimination we need a row exchange whenever a candidate for a pivot is zero. In machines, because of roundoff errors, we need an exchange also when such an entry is *near* zero, not just when it is exactly zero. The objective of this section is to present the standard procedure for dealing with this problem, but first we give an example of the kind of trouble we may encounter when an entry is near zero.

EXAMPLE 8.2.1. Let us imagine that we have a machine that rounds every number to two significant decimal digits; that is, to a number of the form $\pm 0.a_1a_2 \times 10n$, where a_1 and a_2 are single digits with $a_1 \neq 0$, and n is an arbitrary integer. (Although actual machines compute with much greater accuracy and round to a fixed number of binary rather than decimal digits, this setting illustrates the phenomenon quite well and avoids technical complications.) Let us see how our machine would solve the system $A\boldsymbol{x} = \boldsymbol{b}$, with

$$[A|\boldsymbol{b}] = \left[\begin{array}{cc|c} 0.001 & 1 & 1 \\ 1 & 1 & 2 \end{array}\right]. \tag{8.2.1}$$

The first step in Gaussian Elimination would produce

$$\left[\begin{array}{cc|c} 0.001 & 1 & 1 \\ 0 & -999 & -998 \end{array}\right] \tag{8.2.2}$$

and our machine would round both 999 and 998 to 1000, giving

$$\left[\begin{array}{cc|c} 0.001 & 1 & 1 \\ 1 & -1000 & -1000 \end{array}\right]. \tag{8.2.3}$$

We can now easily solve this system by back substitution and obtain $x_2 = 1$ and $x_1 = 0$. But this solution is wrong. The correct solution, from the matrix in (8.2.2), is

$$x_2 = \frac{998}{999} = 0.9989\ldots \text{ and } x_1 = \frac{1}{0.999} = 1.001\ldots.$$

Thus, although the machine's answer for x_2 is close enough, for x_1 it is way off.

So what has happened? It is this: In the first step of the back substitution the machine rounded $x_2 = 0.9989\ldots$ to 1. This step, in itself, is certainly all right, but in the next step we had to divide x_2 by 0.001 in solving for x_1. Here the small roundoff error, hidden in taking x_2 as 1, got magnified a thousandfold. Thus, somehow, we must avoid dividing a rounded number by a very small quantity, or multiplying it by a large quantity. In the present example we can achieve this goal by switching the two rows: If we reduce the matrix

$$[A|\boldsymbol{b}]' = \left[\begin{array}{cc|c} 1 & 1 & 2 \\ 0.001 & 1 & 1 \end{array}\right], \tag{8.2.4}$$

we get

$$\left[\begin{array}{cc|c} 1 & 1 & 2 \\ 0 & 0.999 & 0.998 \end{array}\right]. \tag{8.2.5}$$

This matrix is rounded by the machine to

$$\left[\begin{array}{cc|c} 1 & 1 & 2 \\ 0 & 1 & 1 \end{array}\right], \tag{8.2.6}$$

which leads to the correct approximate solution $x_2 = 1$ and $x_1 = 1$. This time the pivot in the first row was large and did not magnify the roundoff error in x_2 when we solved for x_1.

Now one may think that all we need is a large pivot in the first row, and that can be achieved more simply by multiplying the first row of the matrix in (8.2.1) by 1000. That would result in

$$\left[\begin{array}{cc|c} 1 & 1000 & 1000 \\ 1 & 1 & 2 \end{array}\right], \tag{8.2.7}$$

which would then be reduced to

$$\left[\begin{array}{cc|c} 1 & 1000 & 1000 \\ 0 & -999 & -998 \end{array}\right]. \tag{8.2.8}$$

The machine would round this result to

$$\left[\begin{array}{cc|c} 1 & 1000 & 1000 \\ 0 & -1000 & -1000 \end{array}\right], \tag{8.2.9}$$

from which we get $x_2 = 1$ and $x_1 + 1000x_2 = 1000$. Thus, in solving this equation for x_1, the small roundoff error in $x_2 = 1$ is again magnified by a factor of 1000 and results in the same wrong answer, $x_1 = 0$, as before. This example shows that in general it is the small value of a_{11}/a_{12} that magnifies the roundoff error, not just the small value of a_{11} alone. Since in the second row the corresponding ratio a_{21}/a_{22} is big, we can avoid the problem, as we have seen, by putting that row on top.

<<>>

Considerations like those in Example 8.2.1 have led to the following strategy to minimize the magnification of roundoff errors in Gaussian elimination:

For any $m \times n$ matrix A,

1. Compute a scale factor for each row as the largest absolute value of the entries in that row. In other words, compute

$$s_i = \max_{1 \le j \le n} |a_{ij}|$$

for each i.
2. Compute the ratio of the absolute value of the first entry in each row to its scale factor; that is, compute $r_i = |a_{i1}|/s_i$ for each i. (Ignore rows with $s_i = 0$.)

3. Find a row for which r_i is maximal and put it on top. Use the first entry in this row as the pivot to produce zeroes below it as usual. (If all r_i are 0, then go to the next column, and so on.) Note that in actual machine programs we do not really move the rows, but just keep track of which one is to be used as the pivot row.
4. Repeat the steps above on the submatrix obtained by deleting the first row and the first column, until we run out of rows or columns.

The procedure above is called *scaled partial pivoting*. The word "scaled" refers to the scaling used in Step 2 above, "pivoting" refers to the whole pivot-selection procedure by reordering the rows, and the adjective "partial" indicates that we do not consider a reordering of the columns as well. (The latter has been tried, but did not result in significant improvements.)

EXAMPLE 8.2.2. To see how this procedure works, consider the system given by

$$[A\,|\,\boldsymbol{b}] = \left[\begin{array}{rrrr|r} 15 & 13 & -22 & 1 & 2 \\ -7 & -11 & 53 & 32 & 12 \\ 12 & 7 & 4 & 8 & 44 \\ 0 & 12 & -7 & 1 & 11 \end{array}\right]. \tag{8.2.10}$$

Then the scale factors are $s_1 = 22$, $s_2 = 53$, $s_3 = 12$, and $s_4 = 12$. The corresponding ratios in the first column are $r_1 = 15/22$, $r_2 = 7/53$, $r_3 = 12/12 = 1$, and $r_4 = 0$. Since r_3 is the biggest of these, we put the third row on top, and then proceed with reduction of the first column as usual:

$$[A\,|\,\boldsymbol{b}]' = \left[\begin{array}{rrrr|r} 12 & 7 & 4 & 8 & 44 \\ 15 & 13 & -22 & 1 & 2 \\ -7 & -11 & 53 & 32 & 12 \\ 0 & 12 & -7 & 1 & 11 \end{array}\right]. \tag{8.2.11}$$

Subtracting appropriate multiples of the first row from the others, we first reduce this matrix to

$$\left[\begin{array}{rrrr|r} 12 & 7 & 4 & 8 & 44 \\ 0 & 17/4 & -27 & -9 & -53 \\ 0 & -83/12 & 166/3 & 110/3 & 113/3 \\ 0 & 12 & -7 & 1 & 11 \end{array}\right]. \tag{8.2.12}$$

Next, we should rescale the last three rows. But in practice this rescaling is usually not done because people have observed that it is not worth the effort. In this example, as the reader could easily check, rescaling would lead to the same result, namely that the fourth row should be the next pivot row. Thus we put the fourth row in second place, and proceed as follows:

$$\left[\begin{array}{cccc|c} 12 & 7 & 4 & 8 & 44 \\ 0 & 12 & -7 & 1 & 11 \\ 0 & 17/4 & -27 & -9 & -53 \\ 0 & -83/12 & 166/3 & 110/3 & 113/3 \end{array}\right] \tag{8.2.13}$$

$$\to \left[\begin{array}{cccc|c} 12 & 7 & 4 & 8 & 44 \\ 0 & 12 & -7 & 1 & 11 \\ 0 & 0 & -1177/48 & -449/48 & -2731/48 \\ 0 & 0 & 7387/144 & 5363/144 & 6337/144 \end{array}\right]. \tag{8.2.14}$$

Now 7387/144 > 1177/48, and so we swap the last two rows to get

$$\to \left[\begin{array}{cccc|c} 12 & 7 & 4 & 8 & 44 \\ 0 & 12 & -7 & 1 & 11 \\ 0 & 0 & 7387/144 & 5363/144 & 6337/144 \\ 0 & 0 & -1177/48 & -449/48 & -2731/48 \end{array}\right] \tag{8.2.15}$$

and

$$\to \left[\begin{array}{cccc|c} 12 & 7 & 4 & 8 & 44 \\ 0 & 12 & -7 & 1 & 11 \\ 0 & 0 & 7387/144 & 5363/144 & 6337/144 \\ 0 & 0 & 0 & 1546/183 & -1542/43 \end{array}\right]. \tag{8.2.16}$$

From here we proceed with regular back substitution to obtain $x_4 = -2653/625$, $x_3 = 3392/861$, $x_2 = 2266/635$, and $x_1 = 5276/1701$.

<<>>

It should be obvious that no algorithm can completely prevent the magnification of roundoff errors in the solution of linear systems. J. M. Wilkinson proved in the 1960s, however, that Gaussian Elimination with partial pivoting is as good as we can get, that is, in this procedure the magnification depends only on the matrix A, and can be characterized by the so-called *condition number of A*. If A is a symmetric nonsingular matrix, then the condition number is given by the simple formula $c = |\lambda_n/\lambda_1|$, where λ_n is the eigenvalue of largest absolute value and λ_1 of the smallest. For other types of matrices the condition number is more difficult to compute; we do not go into this problem. In general, c is the factor by which a relative error $|\Delta \boldsymbol{b}|/|\boldsymbol{b}|$ in the right-hand side of $A\mathbf{x} = \boldsymbol{b}$ is magnified to produce the corresponding relative error $|\Delta\mathbf{x}|/|\mathbf{x}|$ in the solution $\boldsymbol{x}$.

Exercises 8.2

8.2.1. Show how the machine of Example 8.2.1 would solve the system $A\mathbf{x} = \boldsymbol{b}$, with

$$[A|\boldsymbol{b}] = \left[\begin{array}{cc|c} 0.002 & 1 & 4 \\ 6 & -1 & 2 \end{array}\right],$$

compare the result to the correct solution, and explain the discrepancy.

8.2.2. Show how the machine of Example 8.2.1 would solve the system $A\mathbf{x} = \boldsymbol{b}$, with

$$[A|\boldsymbol{b}] = \left[\begin{array}{cc|c} 2 & 1000 & 4000 \\ 6 & -1 & 2 \end{array}\right].$$

compare the result to the correct solution, and explain the discrepancy.

8.2.3. Solve the system of Exercise 8.2.1 by the method of partial pivoting. Show all intermediate results, including the scale factors s_i and the ratios r_i. Compare the result to the correct solution, and explain why it is a good approximation.

8.2.4. Solve the system

$$\begin{bmatrix} 2 & 4 & -2 \\ 1 & 3 & 4 \\ 5 & 2 & 1 \end{bmatrix}\begin{bmatrix} x_1 \\ x_2 \\ x_3 \end{bmatrix} = \begin{bmatrix} 6 \\ -1 \\ 2 \end{bmatrix}.$$

by the method of partial pivoting. Show all intermediate results, including the scale factors s_i and the ratios r_i.

8.2.5. Solve the system

$$\begin{bmatrix} 1 & 2 & 5 \\ 4 & -7 & 11 \\ 5 & 8 & 9 \end{bmatrix}\begin{bmatrix} x_1 \\ x_2 \\ x_3 \end{bmatrix} = \begin{bmatrix} 1 \\ 0 \\ 1 \end{bmatrix}.$$

by the method of partial pivoting. Show all intermediate results, including the scale factors s_i and the ratios r_i.

8.2.6. **a.** Show that the condition number c of a symmetric matrix satisfies $c \geq 1$.

b. Find all symmetric 2×2 matrices with $c = 1$.

c. Find the condition number of the matrix A in Equation (8.2.1).

d. Find the condition number of the matrix A in Equation (8.2.4).

e. What conclusions can you draw from the answers to Parts (a)–(d)?

MATLAB Exercises 8.2

ML8.2.1. Enter

$$A = \begin{bmatrix} 1 & 999999 & 999999 \\ 1 & 1 & 2 \end{bmatrix} \tag{8.2.17}$$

and

$$B = \begin{bmatrix} 1 & 1 & 2 \\ 1 & 999999 & 999999 \end{bmatrix}. \tag{8.2.18}$$

Run the commands **rrefmovie**(A) and **rrefmovie**(B). Which one gives the better solution to the system represented by these as augmented matrices? Why?

ML8.2.2. Enter the augmented matrix of Equation (8.2.11) as A in MATLAB and run the command **rrefmovie**(A).

a. Compare the observed sequence of operations to achieve Gauss-Jordan reduction to the one suggested in Exercise 8.1.7 on page 263. Which one is more efficient? How many long operations are needed in this method? (You can actually count *all* operations by using the MATLAB command **flops**. See **help flops**.)

b. Are the rows used in the same order as in Example 8.2.2? What is the difference? Which method is preferable, in general?

ML8.2.3. Enter the MATLAB commands A = **hilb**(12), $\mathbf{c}$ = **ones**(12, 1) and $\mathbf{b} = A * \mathbf{c}$. The matrix A is the so-called *Hilbert matrix* of order 12, defined by $a_{ij} = (i + j - 1)^{-1}$ for $i, j = 1, \ldots, 12$. It is extremely ill-conditioned; that is, it has a very high condition number. You can check this fact by entering **cond**(A). The equation $A\mathbf{x} = \mathbf{b}$ obviously should have the solution $\mathbf{x} = \mathbf{c}$. See what the MATLAB commands $\mathbf{x} = A\backslash\mathbf{b}$ and $\mathbf{x}$ = **rref**[A $\mathbf{b}$] produce.

8.3. COMPUTING EIGENVALUES AND EIGENVECTORS

In Chapter 7 the eigenvalues of a matrix A were always computed from the characteristic equation of A. Though indispensable for the theory, this is a very inefficient procedure for almost all but the very smallest matrices, for several reasons. First, the expansion of an $n \times n$ determinant has $n!$ terms, which is already enormous for moderately large values of n. Second, the characteristic equation is an algebraic equation of degree n, and nth-degree equations can be solved only by approximate methods anyway if n is 5 or more, and so it might be better to use approximate methods designed directly for computing eigenvalues. Third, the solutions of high-degree equations usually are very dependent on roundoff errors in the coefficients.

There exist several numerical procedures for computing eigenvalues and eigenvectors. We shall consider only the so-called *power method* and some of its variants. In this method we reverse the procedure of using diagonalization to compute powers of a matrix, and use the powers to obtain the eigenvalues. For $n \leq 100$ or so, this technique is quite feasible with modern computers, which can compute such powers directly with great speed. To begin our discussion, let us take another look at Example 7.1.1on page 218.

EXAMPLE 8.3.1. The matrix

$$A = \begin{bmatrix} 1/2 & 3/2 \\ 3/2 & 1/2 \end{bmatrix} \tag{8.3.1}$$

has unit eigenvectors $\mathbf{s}_1 = \frac{1}{\sqrt{2}}(1, 1)^T$ and $\mathbf{s}_2 = \frac{1}{\sqrt{2}}(-1, 1)^T$ with corresponding eigenvalues $\lambda_1 = 2$ and $\lambda_2 = -1$. Thus, if we write $\mathbf{x}$ in terms of the basis $\{\mathbf{s}_1, \mathbf{s}_2\}$ as

$$\mathbf{x} = x_{S1}\mathbf{s}_1 + x_{S2}\mathbf{s}_2, \tag{8.3.2}$$

then we get

$$A^n\mathbf{x} = x_{S1}A^n\mathbf{s}_1 + x_{S2}A^n\mathbf{s}_2 = 2^n x_{S1}\mathbf{s}_1 + (-1)^n x_{S2}\mathbf{s}_2. \tag{8.3.3}$$

Thus, for large values of n, the first term dominates (we say $\lambda_1 = 2$ is a dominant eigenvalue), and $A^n\mathbf{x}$ will point approximately in the direction of $\mathbf{s}_1$ and will have length $2^n x_{S1}$ approximately. This result can also be seen from Figure 8.1, by observing that the direction of the vectors $\mathbf{x}$, $A\mathbf{x}$, $A^2\mathbf{x}$, ... approaches that of $\mathbf{s}_1$, and their length nearly doubles with each step.

FIGURE 8.1

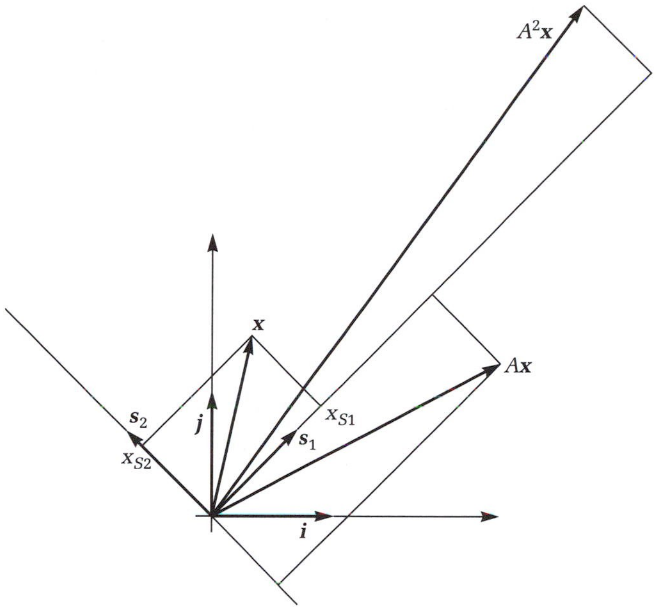

If the eigenvalue λ_1 were not known, we could use Equation (8.3.3) in various ways to give us an approximation to it. For instance, if we consider the ratio of the first components of the vectors $A^{n+1}\mathbf{x}$ and $A^n\mathbf{x}$, then we find that

$$\lim_{n\to\infty} \frac{(A^{n+1}\mathbf{x})_1}{(A^n\mathbf{x})_1} = \lim_{n\to\infty} \frac{2^{n+1}x_{S1}/\sqrt{2} + (-1)^{n+2}x_{S2}/\sqrt{2}}{2^n x_{S1}/\sqrt{2} + (-1)^n x_{S2}/\sqrt{2}}$$

$$= \lim_{n\to\infty} \frac{2x_{S1} + (-1/2)^n x_{S2}}{x_{S1} + (-1/2)^n x_{S2}} = 2 \text{ if } x_{S1} \neq 0. \tag{8.3.4}$$

Thus the ratio above provides an approximation to $\lambda_1 = 2$ when n is large.

<<>>

In this example above we could have used the second components or the lengths of the same vectors. However, because all these can become very large or very small, the preferred procedure is to scale the vectors so that, for each value of n, we divide through by a selected component. It is this method that we summarize in the next theorem:

THEOREM 8.3.1. Let A be a diagonalizable matrix with a dominant eigenvalue λ_1; that is, an eigenvalue such that $|\lambda_1| > |\lambda_j|$ for all $j \neq 1$. Assume also that λ_1 has multiplicity 1, and that $\mathbf{s}_1$ is a corresponding eigenvector with a nonzero kth component, for some fixed k. Choose an arbitrary vector $\mathbf{x}_0$ with a nonzero kth component, and set successively $\mathbf{x}_i' = \mathbf{x}_i/(\mathbf{x}_i)_k$ and $\mathbf{x}_{i+1} = A\mathbf{x}_i'$ for $i = 0, 1, 2, \ldots,$ assuming also that $(\mathbf{x}_i)_k \neq 0$ for any i. This choice makes $(\mathbf{x}_i')_k = 1$ in each step, and $(\mathbf{x}_i)_k$ will approach the dominant eigenvalue λ_1, while the vectors $\mathbf{x}_i'$ and $\mathbf{x}_i$ will approach multiples of $\mathbf{s}_1$, unless the decomposition of the initial vector $\mathbf{x}_0$ relative to a basis of eigenvectors had no component in the direction of $\mathbf{s}_1$.

Proof Let A be $n \times n$ and $\mathbf{s}_1, \mathbf{s}_2, \ldots, \mathbf{s}_n$ a complete set of eigenvectors. Then any $\mathbf{x}_0$ can be written as

$$\mathbf{x}_0 = x_{0S1}\mathbf{s}_1 + x_{0S2}\mathbf{s}_2 + \cdots + x_{0Sn}\mathbf{s}_n, \tag{8.3.5}$$

and

$$\mathbf{x}_0' = \frac{1}{x_{0k}}(x_{0S1}\mathbf{s}_1 + x_{0S2}\mathbf{s}_2 + \cdots + x_{0Sn}\mathbf{s}_n), \tag{8.3.6}$$

where x_{0k} stands for the kth component of $\mathbf{x}_0$ relative to the standard basis.

Then

$$\mathbf{x}_1 = A\mathbf{x}_0' = \frac{1}{x_{0k}}(\lambda_1 x_{0S1}\mathbf{s}_1 + \lambda_2 x_{0S2}\mathbf{s}_2 + \cdots + \lambda_n x_{0Sn}\mathbf{s}_n), \tag{8.3.7}$$

and similarly

$$\begin{aligned}\mathbf{x}_{i+1} = A\mathbf{x}_i' &= \frac{1}{x_{0k}x_{1k}\cdots x_{ik}}(\lambda_1^i x_{0S1}\mathbf{s}_1 + \lambda_2^i x_{0S2}\mathbf{s}_2 + \cdots + \lambda_n^i x_{0Sn}\mathbf{s}_n) \\ &= \frac{\lambda_1^i}{x_{0k}x_{1k}\cdots x_{ik}}\left(x_{0S1}\mathbf{s}_1 + \frac{\lambda_2^i}{\lambda_1^i}x_{0S2}\mathbf{s}_2 + \cdots + \frac{\lambda_n^i}{\lambda_1^i}x_{0Sn}\mathbf{s}_n\right) \\ &\approx \frac{\lambda_1^i x_{0S1}}{x_{0k}x_{1k}\cdots x_{ik}}\mathbf{s}_1 \end{aligned} \tag{8.3.8}$$

for large i, because of the dominance of λ_1. (We can see from this result that the speed of convergence is determined by the magnitude of $|\lambda_2/\lambda_1|$, where λ_2 is the second largest eigenvalue in absolute value.)

Denoting the first component of $\mathbf{s}_1$ by s_{11}, we get

$$x_{i+1,1} \approx \frac{\lambda_1^i x_{0S1}s_{11}}{x_{0k}x_{1k}\cdots x_{ik}} \tag{8.3.9}$$

and

$$\mathbf{x}_{i+1}' = \frac{\mathbf{x}_{i+1}}{x_{i+1,1}} \approx \frac{\mathbf{s}_1}{s_{11}}. \tag{8.3.10}$$

Hence

$$\mathbf{x}_{i+2} = A\mathbf{x}'_{i+1} \approx \frac{\lambda_1 \mathbf{s}_1}{s_{11}} \tag{8.3.11}$$

and

$$x_{i+2,1} \approx \lambda_1. \tag{8.3.12}$$

<<>>

As we have mentioned, the method seems to fail if by bad luck the decomposition of the initial vector $\mathbf{x}_0$ relative to a basis of eigenvectors has no component in the direction of $\mathbf{s}_1$. Although this failure is certainly true in theory, in practice after a few steps roundoff errors will usually introduce a sufficiently large component in the required direction, which will eventually swamp the other components.

We illustrate the method by an example, in which we used MATLAB for the computations.

EXAMPLE 8.3.2. Let $k = 1$ and

$$A = \begin{bmatrix} 4 & 2 \\ 3 & -1 \end{bmatrix} \text{ and } \mathbf{x}_0 = \mathbf{x}'_0 = \begin{bmatrix} 1 \\ 1 \end{bmatrix}. \tag{8.3.13}$$

Then

$$\mathbf{x}_1 = A\mathbf{x}'_0 = \begin{bmatrix} 6.0000 \\ 2.0000 \end{bmatrix} \text{ and } \mathbf{x}'_1 = \begin{bmatrix} 1 \\ 0.3333 \end{bmatrix},$$

$$\mathbf{x}_2 = A\mathbf{x}'_1 = \begin{bmatrix} 4.6667 \\ 2.6667 \end{bmatrix} \text{ and } \mathbf{x}'_2 = \begin{bmatrix} 1 \\ 0.5714 \end{bmatrix},$$

$$\mathbf{x}_3 = A\mathbf{x}'_2 = \begin{bmatrix} 5.1429 \\ 2.4286 \end{bmatrix} \text{ and } \mathbf{x}'_3 = \begin{bmatrix} 1 \\ 0.4722 \end{bmatrix},$$

$$\mathbf{x}_4 = A\mathbf{x}'_3 = \begin{bmatrix} 4.9444 \\ 2.5278 \end{bmatrix} \text{ and } \mathbf{x}'_4 = \begin{bmatrix} 1 \\ 0.5112 \end{bmatrix},$$

$$\mathbf{x}_5 = A\mathbf{x}'_4 = \begin{bmatrix} 5.0225 \\ 2.4888 \end{bmatrix} \text{ and } \mathbf{x}'_5 = \begin{bmatrix} 1 \\ 0.4955 \end{bmatrix}. \tag{8.3.14}$$

Thus we see that $(\mathbf{x}_i)_1 \to 5$ as $i \to \infty$ and so $\lambda_1 = 5$. Similarly, $\mathbf{s}_1 \sim (1, 0.5)^T$.
<<>>

The direct power method has the obvious drawback that it computes only dominant eigenvalues and corresponding eigenvectors. This problem can be alleviated by observing that, for each eigenvalue λ of A, the matrix $B = A - cI$ has an eigenvalue $\lambda - c$ with the same eigenvectors as those belonging to λ. (Clearly, if $A\mathbf{s} = \lambda\mathbf{s}$, then $(A - cI)\mathbf{s} = (\lambda - c)\mathbf{s}$ and vice versa.) Thus we can undo the dominance of any eigenvalue λ_1 by changing over to the matrix $B = A - \lambda_1 I$. It is possible, however, that the matrix B will not have a dominant eigenvalue, just as A itself did not have to have one. (This situation occurs if there are several eigenvalues with the same maximal absolute value.) Also, we may not be able to make every eigenvalue dominant by this method (see Exercise 8.3.2), and so we cannot compute such an eigenvalue in this way.

We can, however, modify the power method to yield the eigenvalue *nearest to* 0, if it is unique. This *inverse power method* consists of defining the recursion by

$$A\mathbf{x}_{i+1} = \mathbf{x}_i', \tag{8.3.15}$$

with the A on the left rather than on the right. Indeed, for nonsingular A, Equation (8.3.15) is equivalent to

$$\mathbf{x}_{i+1} = A^{-1}\mathbf{x}_i' \tag{8.3.16}$$

and we know that the eigenvalues of A^{-1} are the reciprocals of the eigenvalues of A with the same eigenvectors. (See Exercise 7.1.13 on page 226.) Thus, if μ_1 is a dominant eigenvalue of A^{-1}, then we can compute it with the direct power method of Theorem 8.3.1. Then A will have a unique eigenvalue nearest to 0, given by $\lambda_1 = 1/\mu_1$. In practice we use Equation (8.3.15), rather than (8.3.16), since A^{-1} is difficult to compute and it is more efficient to solve Equation (8.3.15) for each i using LU factorization.

Coupled with shifting by an appropriate c, the inverse power method enables us to compute every eigenvalue of a diagonalizable matrix; because if c is nearest to a given eigenvalue λ of A, but $c \neq \lambda$, then $B = A - cI$ will be nonsingular (see Exercise 8.3.3) and $\lambda - c$ will be the unique eigenvalue of B nearest to 0. The only problem is how to find good values for c. There are some prescriptions for this choice, but, except for one rather special suggestion in Example 8.3.4 below, they are left to more advanced texts. So too is the most popular iterative method, the so-called QR algorithm, in which all the eigenvalues are calculated simultaneously and no such choices are needed.

EXAMPLE 8.3.3. As in Example 8.3.2, let $k = 1$ and

$$A = \begin{bmatrix} 4 & 2 \\ 3 & -1 \end{bmatrix} \text{ and } \mathbf{x}_0 = \mathbf{x}_0' = \begin{bmatrix} 1 \\ 1 \end{bmatrix}, \tag{8.3.17}$$

and use Equation (8.3.15) to find the smaller eigenvalue λ_2 of A and a corresponding eigenvector $\mathbf{s}_2$. Then

$$A\mathbf{x}_1 = \mathbf{x}_0', \ \mathbf{x}_1 = \begin{bmatrix} 0.3000 \\ -0.1000 \end{bmatrix} \text{ and } \mathbf{x}_1' = \begin{bmatrix} 1 \\ -0.3333 \end{bmatrix},$$

$$A\mathbf{x}_2 = \mathbf{x}_1', \ \mathbf{x}_2 = \begin{bmatrix} 0.0333 \\ 0.4333 \end{bmatrix} \text{ and } \mathbf{x}_2' = \begin{bmatrix} 1 \\ 13 \end{bmatrix},$$

$$A\mathbf{x}_3 = \mathbf{x}_2', \ \mathbf{x}_3 = \begin{bmatrix} 2.7000 \\ -4.9000 \end{bmatrix} \text{ and } \mathbf{x}_3' = \begin{bmatrix} 1 \\ -1.8148 \end{bmatrix},$$

$$A\mathbf{x}_4 = \mathbf{x}_3', \ \mathbf{x}_4 = \begin{bmatrix} -0.2630 \\ 1.0259 \end{bmatrix} \text{ and } \mathbf{x}_4' = \begin{bmatrix} 1 \\ -3.9014 \end{bmatrix},$$

$$A\mathbf{x}_5 = \mathbf{x}_4', \ \mathbf{x}_5 = \begin{bmatrix} -0.6803 \\ 1.8606 \end{bmatrix} \text{ and } \mathbf{x}_5' = \begin{bmatrix} 1 \\ -2.7350 \end{bmatrix},$$

$$A\mathbf{x}_6 = \mathbf{x}_5', \ \mathbf{x}_6 = \begin{bmatrix} -0.4470 \\ 1.3940 \end{bmatrix} \text{ and } \mathbf{x}_6' = \begin{bmatrix} 1 \\ -3.1186 \end{bmatrix},$$

$$A\mathbf{x}_7 = \mathbf{x}_6', \ \mathbf{x}_7 = \begin{bmatrix} -0.5237 \\ 1.5474 \end{bmatrix} \text{ and } \mathbf{x}_7' = \begin{bmatrix} 1 \\ -2.9547 \end{bmatrix}. \quad (8.3.18)$$

Thus we see that $(\mathbf{x}_i)_1 \to -0.5$ as $i \to \infty$ and so $\mu_1 = -0.5$ is the dominant eigenvalue of A^{-1} and $\lambda_1 = -2$ is the corresponding eigenvalue of A. Clearly, $\mathbf{s}_2 \sim (1, -3)^T$.

<<>>

EXAMPLE 8.3.4. Using the power method, find the eigenvalues and eigenvectors of the matrix

$$A = \begin{bmatrix} 1 & 2 & 3 \\ 2 & 1 & 2 \\ 3 & 2 & 1 \end{bmatrix}. \quad (8.3.19)$$

First we use the direct method to obtain the largest eigenvalue, as in Example 8.3.2 above: Letting

$$\mathbf{x}_0 = \mathbf{x}_0' = \begin{bmatrix} 1 \\ 1 \\ 1 \end{bmatrix}, \quad (8.3.20)$$

we find

$$\mathbf{x}_1 = A\mathbf{x}_0' = \begin{bmatrix} 6 \\ 5 \\ 6 \end{bmatrix} \text{ and } \mathbf{x}_1' = \begin{bmatrix} 1 \\ 0.8333 \\ 1 \end{bmatrix},$$

$$\mathbf{x}_2 = A\mathbf{x}_1' = \begin{bmatrix} 5.6667 \\ 4.8333 \\ 5.6667 \end{bmatrix} \quad \text{and} \quad \mathbf{x}_2' = \begin{bmatrix} 1 \\ 0.8529 \\ 1 \end{bmatrix},$$

$$\mathbf{x}_3 = A\mathbf{x}_2' = \begin{bmatrix} 5.7059 \\ 4.8529 \\ 5.7059 \end{bmatrix} \quad \text{and} \quad \mathbf{x}_3' = \begin{bmatrix} 1 \\ 0.8505 \\ 1 \end{bmatrix},$$

$$\mathbf{x}_4 = A\mathbf{x}_3' = \begin{bmatrix} 5.7010 \\ 4.8505 \\ 5.7010 \end{bmatrix} \quad \text{and} \quad \mathbf{x}_4' = \begin{bmatrix} 1 \\ 0.8508 \\ 1 \end{bmatrix},$$

$$\mathbf{x}_5 = A\mathbf{x}_4' = \begin{bmatrix} 5.7016 \\ 4.8508 \\ 5.7016 \end{bmatrix} \quad \text{and} \quad \mathbf{x}_5' = \begin{bmatrix} 1 \\ 0.8508 \\ 1 \end{bmatrix}. \tag{8.3.21}$$

Thus $\lambda_1 \approx 5.7016$ and $\mathbf{s}_1 \approx (1, 0.8508, 1)^T$.

Next we compute the eigenvalue of smallest absolute value by the inverse power method: Using the same $\mathbf{x}_0$ as above, we get

$$A\mathbf{x}_1 = \mathbf{x}_0', \;\; \mathbf{x}_1 = \begin{bmatrix} 0.2500 \\ 0 \\ 0.2500 \end{bmatrix} \quad \text{and} \quad \mathbf{x}_1' = \begin{bmatrix} 1 \\ 0 \\ 1 \end{bmatrix},$$

$$A\mathbf{x}_2 = \mathbf{x}_1', \;\; \mathbf{x}_2 = \begin{bmatrix} -0.2500 \\ 1 \\ -0.2500 \end{bmatrix} \quad \text{and} \quad \mathbf{x}_2' = \begin{bmatrix} -1 \\ 4 \\ -1 \end{bmatrix},$$

$$A\mathbf{x}_3 = \mathbf{x}_2', \;\; \mathbf{x}_3 = \begin{bmatrix} 2.2500 \\ -5 \\ 2.2500 \end{bmatrix} \quad \text{and} \quad \mathbf{x}_3' = \begin{bmatrix} 1 \\ -2.2222 \\ 1 \end{bmatrix},$$

$$A\mathbf{x}_4 = \mathbf{x}_3', \;\; \mathbf{x}_4 = \begin{bmatrix} -1.3611 \\ 3.2222 \\ -1.3611 \end{bmatrix} \quad \text{and} \quad \mathbf{x}_4' = \begin{bmatrix} 1 \\ -2.3673 \\ 1 \end{bmatrix},$$

$$A\mathbf{x}_5 = \mathbf{x}_4', \;\; \mathbf{x}_5 = \begin{bmatrix} -1.4337 \\ 3.3673 \\ -1.4337 \end{bmatrix} \quad \text{and} \quad \mathbf{x}_5' = \begin{bmatrix} 1 \\ -2.3488 \\ 1 \end{bmatrix},$$

$$A\mathbf{x}_6 = \mathbf{x}_5', \;\; \mathbf{x}_6 = \begin{bmatrix} -1.4244 \\ 3.3673 \\ -1.4244 \end{bmatrix} \quad \text{and} \quad \mathbf{x}_6' = \begin{bmatrix} 1 \\ -2.3510 \\ 1 \end{bmatrix},$$

$$A\mathbf{x}_7 = \mathbf{x}_6', \;\; \mathbf{x}_7 = \begin{bmatrix} -1.4255 \\ 3.3510 \\ -1.4255 \end{bmatrix} \quad \text{and} \quad \mathbf{x}_7' = \begin{bmatrix} 1 \\ -2.3508 \\ 1 \end{bmatrix}, \tag{8.3.22}$$

Thus $\lambda_3 \approx -1/1.4255 \approx -0.7016$ and $\boldsymbol{s}_3 \approx (1, -2.3508, 1)^T$.

Experimentation with different initial vectors would show that the eigenvalues above are simple. (See Exercise 8.3.1.) Thus we still have to find a third eigenvalue whose absolute value must fall between 0.7016 and 5.7016. Since A is symmetric, we know that its eigenvalues are real, and so we can look for one near 3 or –3. Using $c = 3$ for shifting would yield λ_1 again, and so we use $c = -3$. Another argument that suggests a negative λ_2 is the following: We know that the eigenvectors of a symmetric matrix belonging to different eigenvectors are orthogonal to each other, and the vector $\mathbf{x}_0 = (1, 0.5, -1.5)^T$ is easily seen to be approximately orthogonal to the $\boldsymbol{s}_1$ and $\boldsymbol{s}_3$ found above. (We could, of course, find a vector exactly orthogonal to $\boldsymbol{s}_1$ and $\boldsymbol{s}_3$, and that would be an appropriate $\boldsymbol{s}_2$, but we want to illustrate the shifted power method.) Now $A\mathbf{x}_0 \sim -\mathbf{x}_0$, and so λ_2 should be negative. Thus we apply the inverse power method to the matrix

$$B = A + 3I = \begin{bmatrix} 4 & 2 & 3 \\ 2 & 4 & 4 \\ 3 & 2 & 1 \end{bmatrix}. \tag{8.3.23}$$

Setting

$$\mathbf{x}_0 = \mathbf{x}'_0 = \begin{bmatrix} 1 \\ 0.5 \\ -1.5 \end{bmatrix}, \tag{8.3.24}$$

we get

$$B\mathbf{x}_1 = \mathbf{x}'_0, \quad \mathbf{x}_1 = \begin{bmatrix} 1.1500 \\ 0.2250 \\ -1.3500 \end{bmatrix} \text{ and } \mathbf{x}'_1 = \begin{bmatrix} 1 \\ 0.1957 \\ -1.1739 \end{bmatrix},$$

$$B\mathbf{x}_2 = \mathbf{x}'_1, \quad \mathbf{x}_2 = \begin{bmatrix} 1.0500 \\ 0.0859 \\ -1.1239 \end{bmatrix} \text{ and } \mathbf{x}'_2 = \begin{bmatrix} 1 \\ 0.0818 \\ -1.0704 \end{bmatrix},$$

$$B\mathbf{x}_3 = \mathbf{x}'_2, \quad \mathbf{x}_3 = \begin{bmatrix} 1.0200 \\ 0.0357 \\ -1.0504 \end{bmatrix} \text{ and } \mathbf{x}'_3 = \begin{bmatrix} 1 \\ 0.0350 \\ -1.0298 \end{bmatrix},$$

$$B\mathbf{x}_4 = \mathbf{x}'_3, \quad \mathbf{x}_4 = \begin{bmatrix} 1.0084 \\ 0.0152 \\ -1.0214 \end{bmatrix} \text{ and } \mathbf{x}'_4 = \begin{bmatrix} 1 \\ 0.0151 \\ -1.0129 \end{bmatrix},$$

$$B\mathbf{x}_5 = \mathbf{x}'_4, \quad \mathbf{x}_5 = \begin{bmatrix} 1.0036 \\ 0.0066 \\ -1.0092 \end{bmatrix} \text{ and } \mathbf{x}'_5 = \begin{bmatrix} 1 \\ 0.0065 \\ -1.0056 \end{bmatrix},$$

$$B\mathbf{x}_6 = \mathbf{x}'_5, \quad \mathbf{x}_6 = \begin{bmatrix} 1.0016 \\ 0.0028 \\ -1.0400 \end{bmatrix} \text{ and } \mathbf{x}'_6 = \begin{bmatrix} 1 \\ 0.0028 \\ -1.0240 \end{bmatrix}. \tag{8.3.25}$$

Hence we find $\lambda_2(B) = 1$, and so $\lambda_2(A) = -2$ and $\mathbf{s}_2 = (1, 0, -1)^T$.

<<>>

Exercises 8.3

8.3.1. **a.** Apply the method of Theorem 8.3.1 to the matrix

$$A = \begin{bmatrix} 3 & 0 & 1 \\ 0 & 3 & 0 \\ 0 & 0 & 1 \end{bmatrix}$$

of Example 7.1.3 of page 222. Use $k = 1$, first with the initial vector $\mathbf{x}_0 = (1, 1, 1)^T$ and second, with the initial vector $\mathbf{x}_0 = (1, -1, 1)^T$.

b. How would you modify Theorem 8.3.1 in case there is a dominant eigenvalue of multiplicity 2 or more?

8.3.2. Show that if the eigenvalues of A are 1, 2, and 3, then there is no c, whether real or complex, which would make $2 - c$ a dominant eigenvalue of $B = A - cI$.

8.3.3. Show that if c is not an eigenvalue of A, then $B = A - cI$ is nonsingular.

MATLAB Exercises 8.3

ML8.3.1. Let A = **hilb**(3) and compute A^8./A^7. Explain what you get.

ML8.3.2. Use the methods of this section to compute the eigenvalues and eigenvectors of

a. A = **hilb**(3),

b. A = **ones**(3),

c. A = **ones**(4),

d. A = **hadamard**(4).

APPENDIX <<1>>

IMPLICATION AND EQUIVALENCE

In this section we discuss in a very informal manner two basic relations of logic and the ways in which they are used in mathematics. These relations apply to statements that are either true or false, such as "$2 + 2 = 4$," "the sun is shining," "$x + 1 = 4$." Here the first statement is always true, the second one is occasionally true, and "$x + 1 = 4$" is true if x is 3 and is false for all other values of x. However, in mathematics, as in common speech, the statements we make are generally considered to be true, in contrast to some discussions in formal logic where they may be either true or false. Thus the equation above is usually considered to mean that x is 3.

Statements can be connected by various logical operations to form new statements. The connectives "and," "or," "not," "if ... then" indicate the simplest of these. The first three are fairly straightforward and well known, but the last one is frequently misused and misunderstood. Also, because it is used in the theorems and proofs throughout the book, we now examine it in detail.

Let p and q denote arbitrary statements. We call the compound statement "if p then q" an *implication*. Note that when we make such a statement we mean the implication to be true, but that does not say anything about the truth of p and q themselves. For instance, "if it rains, I will take my umbrella" does not say anything about rain or shine on any particular day, nor whether I will take my umbrella tomorrow.

It is sometimes helpful to illustrate implications by a so-called Euler diagram,[1] as shown in Figure A.1.

FIGURE A.1

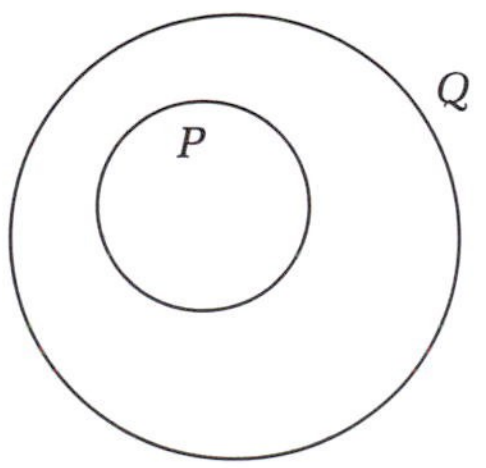

[1]Euler first used these diagrams around 1770, more than a century before Venn's more familiar diagrams were published.

This diagram is meant to be interpreted as saying that for the points of the set P the statement p is true, and for those of Q the statement q is true. Clearly, for this configuration, if a point is in P, then it is in Q. The diagram also shows that for the points outside the set P the statement "not p" is true, and for the points outside the set Q the statement "not q" is true.

There are various equivalent ways of expressing an implication. Here is a list:

1. If p then q
2. q if p
3. $p \Rightarrow q$
4. p implies q
5. p only if q (look at the diagram)
6. q is a necessary condition for p
7. p is a sufficient condition for q
8. q follows from p.

An important fact is that implication is *transitive*; that is, if p implies q, and q implies r, then p implies r.

Although this list merely shows different possible language constructions for the same relation, we can also make a logical change to produce a new relation that will be true exactly when the implication above is true. We obtain this by interchanging p and q with (not q) and (not p), respectively. Thus $p \Rightarrow q$ becomes (not q) $\Rightarrow$ (not p). The latter statement is called the *contrapositive* of the former and is logically equivalent to it, as can perhaps best be seen by looking at the Euler diagram. Of course, the contrapositive can also be expressed in all the various equivalent ways that we had for the original implication.

EXAMPLE A1.1. In calculus we have encountered the simple theorem:

$$\text{If } \sum_{n=1}^{\infty} a_n \text{ converges, then } \lim_{n \to \infty} a_n \text{ exists and equals 0.}$$

The contrapositive of this implication is the equivalent statement:

$$\text{If it is not true that } \lim_{n \to \infty} a_n \text{ exists and equals 0, then } \sum_{n=1}^{\infty} a_n \text{ diverges.}$$

<<>>

Note that, because of the logical equivalence of contrapositives, it is sufficient to prove only the first statement to establish the truth of the second one as well. The first statement is easy to prove; while the second would be hard to prove directly, but is the one we need in most applications.

Implication is not symmetric. That is, $p \Rightarrow q$ is not equivalent to $q \Rightarrow p$; they say different things. The latter is called the *converse* of the former and vice versa. Depending on what p and q stand for, it can happen that one of the two implications is true and the other is not, that both are true, or that both are false. For instance the statement "If an animal is a dog, then it is a canine" is true, but its converse, "If an animal is a canine, then it is a dog" is false, because it could be a fox. Notice that the converse too can be expressed in all the equivalent ways that were possible for the original implication.

EXAMPLE A1.2. The statement in Example A1.1 also illustrates the case in which the original implication is true but its converse is not: The original statement is:

$$\text{If } \sum_{n=1}^{\infty} a_n \text{ converges, then } \lim_{n\to\infty} a_n \text{ exists and equals } 0.$$

The converse is:

$$\text{If } \lim_{n\to\infty} a_n \text{ exists and equals } 0\text{, then } \sum_{n=1}^{\infty} a_n \text{ converges.}$$

As we know, this statement is false, because the harmonic series $\sum_{n=1}^{\infty} \frac{1}{n}$ provides a counterexample.

<<>>

EXAMPLE A1.3. The Theorem of Pythagoras illustrates the case in which both statements are true: Letting a, b, c stand for the lengths of the sides and the hypotenuse of a triangle, "If the triangle is a right triangle, then $a^2 + b^2 = c^2$" and its converse "If $a^2 + b^2 = c^2$, then the triangle is a right triangle" are both true.

<<>>

The case of both a statement and its converse being true occurs so often that it has a new name and new language associated with it: The new relation between p and q is called *equivalence,* and can be expressed by combining any of the forms in the list of expressions for implication for the two statements $p \Rightarrow q$ and $q \Rightarrow p$ and for their contrapositives. The most common ways are:[2]

1. p is equivalent to q
2. $p \Leftrightarrow q$
3. p if and only if q
4. $(p \Rightarrow q)$ and $(q \Rightarrow p)$
5. p iff q
6. p is a necessary and sufficient condition for q
7. $(p \Rightarrow q)$ and $((\text{not } p) \Rightarrow (\text{not } q))$.

[2]The numbers in this list have only a vague connection to the numbers in the preceding list.

Observe that "p if and only if q" is an abbreviation for "(p if q) and (p only if q)," which is equivalent to "(p if q) and (q if p)." This statement can be written symbolically as "($q \Rightarrow p$) and ($p \Rightarrow q$)," which means the same as Statement 4 above, since the "and" operation is commutative. Thus, when we want to prove an equivalence, we must prove two implications. We usually do this in either the form 4 or in the form 7, which is obtained from 4 by replacing the second part by its contrapositive.

APPENDIX

<<2>>

COMPLEX NUMBERS

In this appendix we briefly review complex numbers and exponential functions.

In the sixteenth century an Italian mathematician named Tartaglia discovered a formula for solving cubic equations (usually called Cardano's Formula, for the mathematician in whose book it first appeared, with full credit to Tartaglia, though). This formula had the interesting property that, even when the equation had three real roots, it would give those only if the square roots of negative numbers were used in the computations. This discovery started the exploration of such numbers, which were named *imaginary,* as opposed to the familiar *real* numbers, and of sums of the two kinds named *complex.* These unfortunate names have stuck, although we now regard imaginary numbers as no more imaginary than reals or even than natural numbers. The theory of complex numbers was fully developed only at the beginning of the nineteenth century and that of complex matrices at the end of it.

DEFINITION A2.1.

The set $\mathbb{C}$ of *complex numbers* is a two-dimensional vector space[1] with a multiplication defined as follows: Two basis vectors of $\mathbb{C}$ are denoted by 1 and i, and each element of $\mathbb{C}$ is usually written as $z = a + bi$, where a and b are real numbers and a is an abbreviation for $a \cdot 1$. The vector 1 is identified with the real number 1, and its real multiples with the real numbers. The complex numbers of the form bi, for any real b, are called *imaginary numbers.* The product of two complex numbers is defined to be commutative, associative and distributive, and for the basis vectors as $1^2 = 1$, $1 \cdot i = i$ and $i^2 = -1$.

This definition has several simple consequences.

First, the original problem of the square roots of negative numbers is now solved: We define $\sqrt{c}$, for any complex number c, as any complex number z whose square is c; that is, $z = \sqrt{c}$ if $c = z^2$. Then $\sqrt{-1} = \pm i$, for

[1]To be more precise, it is a two-dimensional vector space *over the reals*, meaning that it is two-dimensional with real numbers as the scalars in the definition of a vector space.

$(\pm i)^2 = (\pm 1)^2 \cdot i^2 = 1 \cdot (-1) = -1$. The square root of any other negative number can now be computed as follows: If a is any positive number, then $\sqrt{-a} = \pm i\sqrt{a}$, for then $(\pm i\sqrt{a})^2 = (\pm i)^2(\sqrt{a})^2 = (-1)a = -a$. We shall see shortly that every complex number except zero has exactly two square roots, and exactly n nth roots. (Every root of 0 is 0.) Notice that, when dealing with complex numbers, the symbol $\sqrt{c}$ is used ambiguously for either of the two roots, unlike the case of real positive c, where it stands for the positive root. A similar convention holds for nth roots. The correct meaning should always be clear from the context.

Division of complex numbers is also possible, with the not unexpected exception of division by zero. For any complex numbers a, b, with $b \neq 0$, we say $\frac{a}{b} = z$ if $a = bz$. To find z let us write $a = a_1 + a_2 i$, $b = b_1 + b_2 i$, and $z = x + yi$, where a_1, a_2, b_1, b_2, x, y are real. Then $a = bz$ becomes

$$a_1 + a_2 i = b_1 x - b_2 y + (b_1 y + b_2 x)i$$

This formula is equivalent to the two real equations

$$b_1 x - b_2 y = a_1 \tag{A2.1}$$

and

$$b_2 x + b_1 y = a_2. \tag{A2.2}$$

Their solution is

$$x = \frac{a_1 b_1 + a_2 b_2}{b_1^2 + b_2^2} \quad \text{and} \quad y = \frac{a_2 b_1 + a_1 b_2}{b_1^2 + b_2^2}. \tag{A2.3}$$

These always exist, because $b \neq 0$ implies $b_1^2 + b_2^2 \neq 0$.

Now that we know that $\frac{a}{b} = z$ is well defined, we can obtain it more simply by multiplying both numerator and denominator of

$$\frac{a}{b} = \frac{a_1 + a_2 i}{b_1 + b_2 i} \tag{A2.4}$$

by $\overline{b} = b_1 - b_2 i$, as in the following example:

EXAMPLE A2.1.

$$\frac{1 + 2i}{3 + 4i} = \frac{(1 + 2i)(3 - 4i)}{(3 + 4i)(3 - 4i)} = \frac{1 \cdot 3 + 2 \cdot 4 + (2 \cdot 3 - 1 \cdot 4)i}{3^2 + 4^2} = \frac{11}{25} + \frac{2}{25}i. \tag{A2.5}$$

<<>>

The multiplier we used in the denominator occurs in many other situations as well, and so it has a name: For any complex number $z = x + yi$ we call the complex number $\overline{z} = x - yi$ the *complex conjugate of z*.

The main properties of the complex conjugate are:

$$z\bar{z} = x^2 + y^2, \tag{A2.6}$$

$$z + \bar{z} = 2x, \tag{A2.7}$$

$$z - \bar{z} = 2yi, \tag{A2.8}$$

$$\bar{\bar{z}} = z. \tag{A2.9}$$

$$\overline{z_1 + z_2} = \bar{z}_1 + \bar{z}_2, \tag{A2.10}$$

and

$$\overline{z_1 z_2} = \bar{z}_1 \bar{z}_2. \tag{A2.11}$$

Because $\mathbb{C}$ is a two-dimensional vector space, it can be represented by the points or vectors of a plane. Length, addition, subtraction, and multiplication by reals have the same geometrical meaning as in any other vector space. We have, however, new constructions for products, quotients, powers, and roots of complex numbers, as will now be described.

For the complex number $z = x + yi$ we write $x = \Re z$ and $y = \Im z$ for the real and imaginary parts of z, respectively. The *absolute value* or *modulus of* z is defined as

$$|z| = \sqrt{x^2 + y^2}, \tag{A2.12}$$

and any one of the angles from the positive real axis to the vector z is called the *argument of* z and is denoted by arg z or arc z. Thus in polar coordinates any $z \neq 0$ is represented by

$$r = |z| \text{ and } \phi = \arg z \tag{A2.13}$$

and

$$z = r(\cos\phi + i \sin\phi). \tag{A2.14}$$

The polar form of z leads easily to the geometric meaning of the product: Let

$$z_1 = r_1(\cos\phi_1 + i \sin\phi_1), \tag{A2.15}$$

and

$$z_2 = r_2(\cos\phi_2 + i \sin\phi_2). \tag{A2.16}$$

Then

$$z_1z_2 = r_1r_2[(\cos\phi_1\cos\phi_2 - \sin\phi_1\sin\phi_2) + i(\sin\phi_1\cos\phi_2 + \cos\phi_1\sin\phi_2)] \quad \text{(A2.17)}$$

and so

$$z_1z_2 = r_1r_2[\cos(\phi_1+\phi_2) + i\sin(\phi_1+\phi_2)]. \quad \text{(A2.18)}$$

Thus in multiplying complex numbers the absolute values are multiplied and the arguments are added.

The following properties of the absolute value can easily be deduced:

$$|z|^2 = z\bar{z}, \quad \text{(A2.19)}$$

$$|z_1z_2| = r_1r_2 = |z_1||z_2|, \quad \text{(A2.20)}$$

and

$$||z_1| - |z_2|| \le |z_1 + z_2| \le |z_1| + |z_2|. \quad \text{(A2.21)}$$

For a sequence of complex numbers z_n we say that

$$z_n \to z \text{ as } n \to \infty \text{ if } |z_n - z| \to 0 \text{ as } n \to \infty; \quad \text{(A2.22)}$$

that is, if the real valued sequence $|z_n - z|$ converges to zero.

As for reals, a series $\sum_{n=0}^{\infty} z_n$ is said to be convergent if the sequence of its partial sums $s_k = \sum_{n=0}^{k} z_n$ converges, and absolutely convergent if $\sum_{n=0}^{\infty}|z_n|$ is convergent. It is easy to prove that absolute convergence of a series implies its convergence, just as for reals. (See Exercises A2.4 and A2.5.)

The series $\sum_{n=0}^{\infty}\frac{z^n}{n!}$ is absolutely convergent for every value of z, because we know that the real series $\sum_{n=0}^{\infty}\frac{|z|^n}{n!}$ is convergent for every value of $|z|$. The sum of this series is e^z when z is real, and so we use it to define e^z for complex z:

$$e^z = \sum_{n=0}^{\infty}\frac{z^n}{n!} \quad \text{(A2.23)}$$

for every $z \in \mathbb{C}$. This definition preserves the multiplication property $e^{z_1}e^{z_2} = e^{z_1+z_2}$ for complex exponents too:[2]

[2]In this derivation we make use of the fact (without proof) that absolutely convergent series can be multiplied term by term and the terms may be rearranged arbitrarily.

$$e^{z_1}e^{z_2} = \left(1 + z_1 + \frac{z_1^2}{2!} + \frac{z_1^3}{3!} + \cdots\right)\left(1 + z_2 + \frac{z_2^2}{2!} + \frac{z_2^3}{3!} + \cdots\right)$$

$$= 1 + (z_1 + z_2) + \frac{1}{2!}(z_1^2 + 2z_1z_2 + z_2^2) + \frac{1}{3!}(z_1^3 + 3z_1^2z_2 + 3z_1z_2^2 + z_2^3) + \cdots$$

$$= 1 + (z_1 + z_2) + \frac{1}{2!}(z_1 + z_2)^2 + \frac{1}{3!}(z_1 + z_2)^3 + \cdots = e^{z_1+z_2}. \qquad \text{(A2.24)}$$

If we use the definition above of the exponential function with $z = i\phi$, where ϕ is real, then we get Euler's Formula:

$$e^{i\phi} = 1 + i\phi + \frac{1}{2!}(i\phi)^2 + \frac{1}{3!}(i\phi)^3 + \cdots = 1 + i\phi - \frac{\phi^2}{2!} - i\frac{\phi^3}{3!} + \frac{\phi^4}{4!} + \cdots$$

$$= \left(1 - \frac{\phi^2}{2!} + \frac{\phi^4}{4!} + \cdots\right) + i\left(\phi - \frac{\phi^3}{3!} + \frac{\phi^5}{5!} + \cdots\right); \qquad \text{(A2.25)}$$

that is,

$$e^{i\phi} = \cos\phi + i\sin\phi. \qquad \text{(A2.26)}$$

From this equation we get

$$\cos\phi = \frac{e^{i\phi} + e^{-i\phi}}{2} \text{ and } \sin\phi = \frac{e^{i\phi} - e^{-i\phi}}{2i}, \qquad \text{(A2.27)}$$

and the polar form of any z as

$$z = re^{i\phi}. \qquad \text{(A2.28)}$$

The polar form of $\overline{z}$ can be written similarly as

$$\overline{z} = re^{-i\phi}. \qquad \text{(A2.29)}$$

From Equation (A2.28) and the multiplication property for complex exponentials we obtain the following important rule for multiplying complex numbers: If $z_1 = r_1e^{i\phi_1}$ and $z_2 = r_2e^{i\phi_2}$, then

$$z_1z_2 = r_1r_2e^{i(\phi_1+\phi_2)}. \qquad \text{(A2.30)}$$

In other words, to multiply two complex numbers, we multiply their absolute values and add their arguments. Using Euler's Formula for the exponential here, we can reproduce Equation (A2.18). Thus the derivation above of Equation (A2.30) provides a new proof of the trigonometric formulas for the sine and cosine of the sum of two angles, based on the Taylor series. Alternatively, Equation (A2.30) could be obtained from Equation (A2.18) and Euler's Formula.

Repeated application of Equation (A2.30) to the same $z = re^{i\phi}$ leads to the power rule

$$z^n = r^n e^{in\phi} \tag{A2.31}$$

for any positive integer n. In Exercise A2.6 this formula will be extended to roots as well.

Exercises A2

A2.1. Prove Equation (A2.20).

A2.2. Prove Inequality (A2.21).

A2.3. Let $P(z) = \sum_{k=0}^{n} a_k z^k$ be a polynomial with real coefficients a_k.

Show that if z_0 is a zero of P; that is if $P(z_0) = 0$, then so too is $\bar{z}_0$. In other words, complex roots of algebraic equations with real coefficients always come in complex conjugate pairs.

A2.4. Prove that a complex series $\sum_{n=0}^{\infty} z_n$, with $z_n = x_n + iy_n$, converges if and only if the series of the real and imaginary parts $\sum_{n=0}^{\infty} x_n$ and $\sum_{n=0}^{\infty} y_n$ both converge.

A2.5. Prove that absolute convergence of a complex series $\sum_{n=0}^{\infty} z_n$ implies its convergence. (Hint: Use the result of Exercise A2.4, the facts that $|z_n| \geq |x_n|$ and $|z_n| \geq |y_n|$, the Comparison Test, and the corresponding theorem for the real series $\sum_{n=0}^{\infty} x_n$ and $\sum_{n=0}^{\infty} y_n$.)

A2.6. Invert Equation (A2.31) to obtain a formula for nth roots: If $z = w^n$, then we call w an nth root of z and write $w = z^{1/n}$. Letting $z = re^{i(\phi+2k\pi)}$ here, for any integer k, and $w = Re^{i\Phi}$, show that any nth root of z must satisfy

$$z^{1/n} = r^{1/n} e^{i(\phi+2k\pi)/n}, \tag{A2.32}$$

and different values of k result in exactly n distinct nth roots for any $z \neq 0$.

A2.7. Use Formula (A2.32) to find and plot

a. all square roots of i,

b. all square roots of $1 + i$,

c. all cube roots of 1,

d. all cube roots of -1,

e. all fourth roots of i.

SOLUTIONS TO SELECTED EXERCISES

CHAPTER 1 ANALYTIC GEOMETRY OF EUCLIDEAN SPACES

1.1. Vectors 1

1.1.1. $\overrightarrow{PR} = \boldsymbol{q}$, $\overrightarrow{PQ} = \boldsymbol{q} - \boldsymbol{p}$, $\overrightarrow{QC} = \frac{1}{2}(\boldsymbol{p} - \boldsymbol{q})$.

1.1.2. $\boldsymbol{r} = (5, 5)$, $\boldsymbol{p} - \boldsymbol{q} = (-3, 1)$.

1.1.4. Midpoint of the edge from (0, 1, 1) to (1, 1, 1) has position vector $\boldsymbol{c} = (1/2, 1, 1)$. The center of the face through (0, 0, 1), (0, 1, 1), (1, 1, 1), and (1, 0, 1) has position vector $\boldsymbol{f} = (1/2, 1/2, 1)$.

1.1.7. $\boldsymbol{r} = (-0.3, -1.2, 1)$.

1.1.9. The center R of the side BC has position vector $\boldsymbol{r} = \frac{1}{2}(\boldsymbol{b} + \boldsymbol{c})$ and the vector from R to A is $\boldsymbol{a} - \boldsymbol{r}$. Thus the point 1/3 of the way from R to A has position vector

$$\frac{1}{2}(\boldsymbol{b} + \boldsymbol{c}) + \frac{1}{3}(\boldsymbol{a} - \boldsymbol{r}) = \frac{1}{2}\boldsymbol{b} + \frac{1}{2}\boldsymbol{c} + \frac{1}{3}\boldsymbol{a} - \frac{1}{3} \cdot \frac{1}{2}(\boldsymbol{b} + \boldsymbol{c}) = \frac{1}{3}(\boldsymbol{a} + \boldsymbol{b} + \boldsymbol{c}).$$

1.2. Length and Dot Product of Vectors in $\mathbb{R}^n$ 14

1.2.4. Approximately 0.975π.

1.2.5. Approximately 0.49π.

1.2.8. **a.** (−6/29, 15/29),

b. $7/\sqrt{29}$,

c. 7/2.

1.2.17. $\cos\alpha_1 = 3/13$, $\cos\alpha_2 = -4/13$, $\cos\alpha_3 = 12/13$.

1.3. Lines and Planes 25

1.3.1. $x = 1 + 2t$, $y = -2 + 3t$, $z = 4 - 5t$; $\frac{x-1}{2} = \frac{y+2}{3} = \frac{z-4}{-5}$.

1.3.2. $x = 1$, $y = -2 + t$, $z = 4$; the nonparametric equations are: $x = 1$ and $z = 4$ (y arbitrary).

1.3.3. $(x, y, z) = (7, -2, 5) + t(-2, 8, -8)$.

1.3.6. $(x, y, z) = (5, 4, -8) + t(3, -4, 3)$.

1.3.7. $(x, y, z) = (5, 4, -8) + t(4, 3, 0)$.

1.3.12. $3x + 2z = 11$.

1.3.20. $s = -1$, $t = 2$, $\boldsymbol{p} = (7, 0, -5)$.

1.3.21. $s = 3$, $t = -4$, $\boldsymbol{p} = (1, 7, -22)$.

1.3.22. $(-55, 31, 181)$.

1.3.26. $12/\sqrt{17}$.

1.3.27. $1/\sqrt{5}$.

CHAPTER 2 SYSTEMS OF LINEAR EQUATIONS, MATRICES

2.1. Gaussian Elimination 35

2.1.1. $x_3 = -2/7$, $x_2 = 1/2$, $x_1 = -13/14$.

2.1.3. $x_3 = t$, $x_2 = \frac{1}{4} - \frac{7}{8}t$, $x_1 = -\frac{1}{4} + \frac{19}{8}t$.

2.1.4. $x_2 = s$, $x_3 = t$, $x_1 = -s + \frac{3}{2}t$.

2.1.5. $x_1 = x_2 = x_3 = s$.

2.1.13. $\begin{bmatrix} p_1 & * \\ 0 & p_2 \end{bmatrix}, \begin{bmatrix} p_1 & * \\ 0 & 0 \end{bmatrix}, \begin{bmatrix} 0 & p_1 \\ 0 & 0 \end{bmatrix}, \begin{bmatrix} 0 & 0 \\ 0 & 0 \end{bmatrix}$.

2.1.21. $2b_1 - 2b_2 + b_3 = 0$.

2.1.24. $-b_1 + b_2 + b_3 = 0$, $b_4 - 2b_1 = 0$.

2.2. The Algebra of Matrices 54

2.2.3. $AB = 2$, $BA = \begin{bmatrix} 3 & -6 & 9 \\ 2 & -4 & 6 \\ 1 & -2 & 3 \end{bmatrix}$.

2.2.5. $AB = \begin{bmatrix} 17 & -17 \\ 2 & -17 \\ 3 & -26 \end{bmatrix}$, BA does not exist.

2.2.11. $A = \begin{bmatrix} 0 & 0 \\ 0 & 1 \end{bmatrix}$, $B = \begin{bmatrix} 0 & 1 \\ 0 & 0 \end{bmatrix}$.

2.2.15. $A = \begin{bmatrix} 0 & 1 \\ 0 & 0 \end{bmatrix}$ or $A = \begin{bmatrix} 1 & 1 \\ -1 & -1 \end{bmatrix}$.

2.3. The Algebra of Matrices, Continued 69

2.3.1. $A^{-1} = \frac{1}{14}\begin{bmatrix} 1 & 3 \\ 4 & -2 \end{bmatrix}$.

2.3.3. $A^{-1} = \frac{1}{88}\begin{bmatrix} 0 & 16 & 8 \\ 11 & -19 & 18 \\ 11 & 5 & -14 \end{bmatrix}$.

2.3.14. $P = \begin{bmatrix} 0 & 0 & 1 \\ 0 & 1 & 0 \\ 1 & 0 & 0 \end{bmatrix} = P^{-1}$.

2.3.19. In Definition 2.3.1 on page 71, replace A by A^{-1} and X by A.

2.3.20. Write $BC = D$ and apply Theorem 2.3.6 on page 75 to both AD and D.

CHAPTER 3 VECTOR SPACES AND SUBSPACES

3.1. General Vector Spaces 81

3.1.1. No.

3.1.2. Yes.

3.1.3. No.

3.1.4. Yes.

3.1.5. No.

3.1.6. Yes.

3.2. Subspaces, Span, and Independence of Vectors 87

3.2.1. Yes.

3.2.2. No.

3.2.3. No.

3.2.4. Yes.

3.2.5. No.

3.2.6. Yes.

3.2.7. No.

3.2.8. No.

3.2.14. $\boldsymbol{b} = 2\boldsymbol{a}_1 - \boldsymbol{a}_2 + 3\boldsymbol{a}_3$.

3.3. Bases 98

3.3.1. Basis for Row(A): $\{(1, 3, 1)^T, (0, -7, 1)^T\}$, Basis for Col($A$): $\{(1, 3, 2)^T, (3, 2, -1)^T\}$, Basis for Null($A$): $\{(-10, 1, 7)^T\}$.

3.3.5. $A = \begin{bmatrix} 1 & 1 & 1 \\ 1 & 1 & 1 \\ 1 & 0 & 0 \end{bmatrix}$.

3.3.10. Let $B = (\boldsymbol{b}_1, \boldsymbol{b}_2, \ldots, \boldsymbol{b}_n)$ be the list of the given independent vectors and use the standard vectors for the spanning list; that is, let $A = (\boldsymbol{e}_1, \boldsymbol{e}_2, \ldots, \boldsymbol{e}_n)$ in the Exchange Theorem. Then by the theorem we can exchange all the standard vectors for the $\boldsymbol{b}_i$ vectors and they will span $\mathbb{R}^n$.

3.4. Dimension, Orthogonal Complements. 108

3.4.1. dim(Col(A)) = 2, dim(Row(A)) = 2, dim(Null(A)) = 3, dim(Left-null(A)) = 0,

$$\mathbf{x}_0 = \frac{2}{25}\begin{bmatrix}-1\\1\\0\\0\\0\end{bmatrix} - \frac{9}{25}\begin{bmatrix}-4/3\\0\\-2/3\\1\\0\end{bmatrix} + \frac{6}{25}\begin{bmatrix}-4/3\\0\\7/3\\0\\1\end{bmatrix} = \frac{1}{25}\begin{bmatrix}2\\2\\20\\-9\\6\end{bmatrix},$$

$$\mathbf{x}_R = \frac{23}{25}\begin{bmatrix}1\\1\\1\\2\\-1\end{bmatrix} + \frac{6}{25}\begin{bmatrix}0\\0\\-3\\-2\\7\end{bmatrix} = \frac{1}{25}\begin{bmatrix}23\\23\\5\\34\\19\end{bmatrix}.$$

3.4.12. **a.** Col(A) = $\{s\boldsymbol{e}_1 + t\boldsymbol{e}_2\}$, Col($B$) = $\{s\boldsymbol{e}_1 + t\boldsymbol{e}_3\}$, Col($A + B$) = $\{s\boldsymbol{e}_1 + t\boldsymbol{e}_3\}$.

b. Col(A) + Col(B) = $\mathbb{R}^3$, Col(A) $\cap$ Col(B) = $\{s\boldsymbol{e}_1\}$.

c. No.

d. $3 = 2 + 2 - 1$.

3.4.19. $\{\boldsymbol{a}_1, \boldsymbol{a}_2, \boldsymbol{a}_3\}$ is a basis for U, $\{\boldsymbol{a}_3, \boldsymbol{a}_4, \boldsymbol{a}_5\}$ is a basis for V, $\{\boldsymbol{a}_2, \boldsymbol{a}_3\}$ is a basis for $U \cap V$, $\{\boldsymbol{a}_1, \boldsymbol{a}_2, \boldsymbol{a}_3, \boldsymbol{a}_4\}$ is a basis for $U + V = \mathbb{R}^4$, $\{\boldsymbol{e}_4\}$ is a basis for $U^\perp$, and $\{(1, 0, -1, 1)^T\}$ is a basis for $V^\perp$.

3.4.33. From the hint, $U\mathbf{x} = L\boldsymbol{b}$ and $\boldsymbol{0} = M\boldsymbol{b}$. Thus the rows of M must be orthogonal to any vector $\boldsymbol{b}$ in the column space of A, and so their transposes are in the left nullspace of A. Furthermore, the matrix $\begin{bmatrix}L\\M\end{bmatrix}$ has full rank, since it is obtained from I by elementary row operations, which are invertible. Consequently the rows of M are independent. On the other hand, because the dimension of the left nullspace of A is $m - r$ and M has $m - r$ independent rows, their transposes span Left-null(A).

3.4.35. Row-reduce $[A^T \; I]$ to the form $\begin{bmatrix}U & L\\O & M\end{bmatrix}$.

3.4.36. The equation $A\mathbf{x} = \boldsymbol{b}$ can be reformulated as $[\mathbf{x}^T \; 1]\begin{bmatrix}A^T\\-\boldsymbol{b}^T\end{bmatrix} = \boldsymbol{0}$. This equation shows that $\begin{bmatrix}\mathbf{x}\\1\end{bmatrix}$ is in the left nullspace of $\begin{bmatrix}A^T\\-\boldsymbol{b}^T\end{bmatrix}$. Thus the algorithm of Exercise 3.4.33 is applicable. This algorithm would require that we augment $\begin{bmatrix}A^T\\-\boldsymbol{b}^T\end{bmatrix}$ by the unit matrix of order $n + 1$. However, the last column may be omitted, for it does not change in the computation, and this omission just corresponds to the omission of the 1 in $[\mathbf{x}^T \; 1]$. The unit matrix of order $n + 1$ without its last column is the second block column of $\begin{bmatrix}A^T & I\\-\boldsymbol{b}^T & \boldsymbol{0}\end{bmatrix}$. By the algorithm of

Exercise 3.4.33, the reduction of this matrix yields both the particular solution $\mathbf{x}$ from the last row, and the general solution, by adding to $\mathbf{x}$ any linear combination of the transposed rows of L that continue the zero rows of U.

3.4.37.
$$\left[\begin{array}{rrr|rrrr} 1 & 0 & -1 & 1 & 0 & 0 & 0 \\ -2 & 3 & -1 & 0 & 1 & 0 & 0 \\ 3 & -3 & 0 & 0 & 0 & 1 & 0 \\ 2 & 0 & -2 & 0 & 0 & 0 & 1 \\ \hline 5 & -6 & 1 & 0 & 0 & 0 & 0 \end{array}\right] \to \left[\begin{array}{rrr|rrrr} 1 & 0 & -1 & 1 & 0 & 0 & 0 \\ 0 & 3 & -3 & 2 & 1 & 0 & 0 \\ 0 & 0 & 0 & -1 & 1 & 1 & 0 \\ 0 & 0 & 0 & -2 & 0 & 0 & 1 \\ \hline 0 & 0 & 0 & -1 & 2 & 0 & 0 \end{array}\right].$$

Thus $\mathbf{x} = (-1\ \ 2\ \ 0\ \ 0)^T$ is a particular solution, and the general solution is given by

$$\mathbf{x} = \begin{bmatrix} -1 \\ 2 \\ 0 \\ 0 \end{bmatrix} + s\begin{bmatrix} -1 \\ 1 \\ 1 \\ 0 \end{bmatrix} + t\begin{bmatrix} -2 \\ 0 \\ 0 \\ 1 \end{bmatrix}.$$

3.4.38. If we reduce $\begin{bmatrix} A^T & I \\ -\mathbf{e}_i^T & \mathbf{0} \end{bmatrix}$, where $\mathbf{e}_i$ is the ith column of the unit matrix I, we get the transposed solution of $A\mathbf{x} = \mathbf{e}_i$ in the last row of $\begin{bmatrix} U & L \\ \mathbf{0} & \mathbf{x}^T \end{bmatrix}$. The solutions of these equations, for $i = 1, 2, \ldots, n$, are the columns of A^{-1}. We can obtain these simultaneously for all i by reducing $\begin{bmatrix} A^T & I \\ -I & \mathbf{0} \end{bmatrix}$ instead of each $\begin{bmatrix} A^T & I \\ -\mathbf{e}_i^T & \mathbf{0} \end{bmatrix}$ separately. This reduction would, however, result in the transpose of A^{-1} in the lower right corner. Thus to obtain A^{-1} itself there, we just have to use A instead of A^T in the block matrix that we reduce.

3.4.39.
$$\left[\begin{array}{rr|rr} 1 & -2 & 1 & 0 \\ 3 & 4 & 0 & 1 \\ \hline -1 & 0 & 0 & 0 \\ 0 & -1 & 0 & 0 \end{array}\right] \to \left[\begin{array}{rr|rr} 1 & -2 & 1 & 0 \\ 0 & 10 & -3 & 1 \\ \hline 0 & -2 & 1 & 0 \\ 0 & -1 & 0 & 0 \end{array}\right] \to \left[\begin{array}{rr|rr} 1 & -2 & 1 & 0 \\ 0 & 10 & -3 & 1 \\ \hline 0 & 0 & 4/10 & 2/10 \\ 0 & 0 & -3/10 & 1/10 \end{array}\right].$$

Thus $A^{-1} = \frac{1}{10}\begin{bmatrix} 4 & 2 \\ -3 & 1 \end{bmatrix}$.

3.5. Change of Basis 121

3.5.2. $S = A,\ A^{-1} = \frac{1}{4}\begin{bmatrix} -1 & 1 & 2 \\ 2 & -2 & 0 \\ 5 & -1 & -2 \end{bmatrix}, \mathbf{x}_A = \frac{1}{4}\begin{bmatrix} 11 \\ -2 \\ 1 \end{bmatrix}$.

3.5.5. $A = \frac{1}{2}\begin{bmatrix} -1 & -1 & 1 \\ 1 & 1 & 1 \\ -1 & 1 & 1 \end{bmatrix}$.

3.5.6. $M_A = \begin{bmatrix} \cos^2\theta & -\sin\theta\cos\theta \\ -\sin\theta\cos\theta & \sin^2\theta \end{bmatrix}$.

3.5.9. a. $S = \begin{bmatrix} 2 & 3 \\ -1 & -2 \end{bmatrix}$.

3.5.14. For any nonzero t and t' we can choose $S = \begin{bmatrix} t' & 0 \\ 0 & t \end{bmatrix}$, and then $SM(t)S^{-1} = M(t')$. If $t' = 0$, choose $S = \begin{bmatrix} 1 & -t \\ 0 & 1 \end{bmatrix}$.

CHAPTER 4 LINEAR TRANSFORMATIONS

4.1. Representations of Linear Transformations by Matrices 135

4.1.1. b. For any $\mathbf{x}$ we have $\boldsymbol{T}(\boldsymbol{0}) = \boldsymbol{T}(\mathbf{x} - \mathbf{x}) = \boldsymbol{T}(\mathbf{x} + (-1)\mathbf{x}) = \boldsymbol{T}(\mathbf{x}) + \boldsymbol{T}((-1)\mathbf{x}) = \boldsymbol{T}(\mathbf{x}) + (-1)\boldsymbol{T}(\mathbf{x}) = \boldsymbol{T}(\mathbf{x}) - \boldsymbol{T}(\mathbf{x}) = \boldsymbol{0}$.

4.1.5. $[T] = \begin{bmatrix} 1 & -1 \\ 2 & 3 \\ 3 & 2 \end{bmatrix}$.

4.1.8. $\boldsymbol{T}(\mathbf{x}) = \begin{bmatrix} x_1 \\ x_2 \\ x_2 \end{bmatrix}$, $[T] = \begin{bmatrix} 1 & 0 \\ 0 & 1 \\ 0 & 1 \end{bmatrix}$.

4.2. Properties of Linear Transformations 146

4.2.2. a. If $\boldsymbol{a} \neq \boldsymbol{0}$, then Range($\boldsymbol{T}$) = $\mathbb{R}$, Ker($\boldsymbol{T}$) is the $(n-1)$-dimensional subspace $\{\boldsymbol{a}\}^\perp = \{\mathbf{x} : \boldsymbol{a}^T\mathbf{x} = \boldsymbol{0}\}$ of $\mathbb{R}^n$, $\boldsymbol{T}$ is onto but not one-to-one for $n > 1$. If $\boldsymbol{a} = \boldsymbol{0}$, then Range($\boldsymbol{T}$) = $\{0\} \subset \mathbb{R}$, Ker($\boldsymbol{T}$) = $\mathbb{R}^n$, and $\boldsymbol{T}$ is neither onto nor one-to-one.

b. Range($\boldsymbol{T}$) = Col($A - \lambda I$), Ker($\boldsymbol{T}$) = Null($A - \lambda I$). If Rank($A - \lambda I$) = n, then $\boldsymbol{T}$ is onto and one-to-one. If, however, Rank($A - \lambda I$) $< n$, as is the case when λ is an eigenvalue of A, then $\boldsymbol{T}$ is neither onto nor one-to-one.

c. Although this $\boldsymbol{T}$ is not a linear transformation, the concepts in question are still applicable and Range($\boldsymbol{T}$) = $\{\boldsymbol{y} + \boldsymbol{b} : \boldsymbol{y} \in \text{Col}(A)\} \subset \mathbb{R}^m$, and Ker($\boldsymbol{T}$) is the solution set of $A\mathbf{x} = -\boldsymbol{b}$. If $m > n$, then $\boldsymbol{T}$ is not onto and is one-to-one only if Rank(A) = n. If $m = n$, then $\boldsymbol{T}$ is both onto and one-to-one if and only if Rank(A) = n. If $m < n$, then $\boldsymbol{T}$ is not one-to-one and is onto only if Rank(A) = m.

d. Again, this $\boldsymbol{T}$ is not a linear transformation, but Range($\boldsymbol{T}$) = $\{y : y \geq 0\}$ and Ker($\boldsymbol{T}$) = $\{\boldsymbol{0}\}$. $\boldsymbol{T}$ is not onto and is one-to-one only if $n = 0$.

e. Same as Part (a), since $\mathbf{x}^T\boldsymbol{a} = \boldsymbol{a}^T\mathbf{x} = \boldsymbol{a} \cdot \mathbf{x}$.

CHAPTER 5 ORTHOGONAL PROJECTIONS AND BASES

5.1. Orthogonal Projections and Least-Squares Approximations 167

5.1.3. $A = \begin{bmatrix} 1 & 0 \\ 1 & 0 \\ 0 & 1 \end{bmatrix}$.

5.1.4. $\frac{1}{3}(5, -1, 4)^T$.

5.1.6. $P = \begin{bmatrix} 1 & 0 & 0 \\ 0 & 0 & 0 \\ 0 & 0 & 0 \end{bmatrix}$.

5.1.8. $P = \dfrac{1}{a^2 + b^2 + c^2} \begin{bmatrix} a^2 & ab & ac \\ ab & b^2 & bc \\ ac & bc & c^2 \end{bmatrix}$.

5.2. Orthogonal Bases 178

5.2.1. $\mathbf{x}_R = \frac{15}{9} \begin{bmatrix} 2 \\ 1 \\ 2 \end{bmatrix} + \frac{-2}{2} \begin{bmatrix} 1 \\ 0 \\ -1 \end{bmatrix} = \frac{1}{3} \begin{bmatrix} 7 \\ 5 \\ 13 \end{bmatrix}$.

5.2.2. $A = \begin{bmatrix} u_1 & v_1 \\ u_2 & v_2 \\ u_3 & v_3 \end{bmatrix}$ and $A^T = \begin{bmatrix} u_1 & u_2 & u_3 \\ v_1 & v_2 & v_3 \end{bmatrix}$. Thus $P = AA^T = P(\boldsymbol{u}, \boldsymbol{v})$.

5.2.7. The columns of $A = \frac{1}{3} \begin{bmatrix} -1 & 2 & 2 \\ 2 & -1 & 2 \\ 2 & 2 & -1 \end{bmatrix}$.

CHAPTER 6 DETERMINANTS

6.1. Definition and Basic Properties 185

6.1.2. $\begin{vmatrix} 0 & 1 & 2 \\ 4 & 0 & 3 \\ 3 & 2 & 1 \end{vmatrix} = (-1) \begin{vmatrix} 3 & 2 & 1 \\ 4 & 0 & 3 \\ 0 & 1 & 2 \end{vmatrix} = (-1) \begin{vmatrix} 3 & 2 & 1 \\ 1 & -2 & 2 \\ 0 & 1 & 2 \end{vmatrix}$

$= (-1)^2 \begin{vmatrix} 1 & -2 & 2 \\ 3 & 2 & 1 \\ 0 & 1 & 2 \end{vmatrix} = (-1)^2 \begin{vmatrix} 1 & -2 & 2 \\ 0 & 8 & -5 \\ 0 & 1 & 2 \end{vmatrix} = (-1)^3 \begin{vmatrix} 1 & -2 & 2 \\ 0 & 1 & 2 \\ 0 & 8 & -5 \end{vmatrix}$

$= (-1)^3 \begin{vmatrix} 1 & -2 & 2 \\ 0 & 1 & 2 \\ 0 & 0 & -21 \end{vmatrix} = 21.$

6.1.10. A and B are similar if there exists an invertible matrix S such that $B = S^{-1}AS$. In that case,

$$\det(B) = \det(S^{-1}) \det(A) \det(S) = \frac{1}{\det(S)} \det(A) \det(S) = \det(A).$$

6.1.11. If AB is invertible, then $\det(AB) = \det(A) \det(B) \neq 0$ and so neither $\det(A)$ nor $\det(B)$ is zero, which implies that A and B are invertible.

Conversely, if A and B are invertible, then neither $\det(A)$ nor $\det(B)$ is zero, and so $\det(AB) = \det(A)\det(B) \neq 0$, which implies that AB is invertible.

6.2. Further Properties of Determinants 197

6.2.2. Expansion along the first row:

$$\begin{vmatrix} 0 & 1 & 2 \\ 4 & 0 & 3 \\ 3 & 2 & 1 \end{vmatrix} = 0\begin{vmatrix} 0 & 3 \\ 2 & 1 \end{vmatrix} - 1\begin{vmatrix} 4 & 3 \\ 3 & 1 \end{vmatrix} + 2\begin{vmatrix} 4 & 0 \\ 3 & 2 \end{vmatrix}$$

$$= 0(0-6) - 1(4-9) + 2(8-0) = 21.$$

6.2.7. $x_1 = \begin{vmatrix} 1 & 1 & 2 \\ 2 & 0 & 4 \\ 3 & 3 & 1 \end{vmatrix} \div \begin{vmatrix} 1 & 1 & 2 \\ 2 & 0 & 4 \\ 0 & 3 & 1 \end{vmatrix}$

$$= \frac{(0+12+12)-(0+12+2)}{(0+0+12)-(0+12+2)} = \frac{10}{-2} = -5,$$

$x_2 = 0,\ x_3 = 3.$

6.2.9. $A = \begin{bmatrix} 1 & 3 \\ 2 & 1 \end{bmatrix}$ implies $A_{11} = 1$, $A_{12} = -2$, $A_{21} = -3$, $A_{22} = 1$, and so $\operatorname{adj}(A) = \begin{bmatrix} 1 & -3 \\ -2 & 1 \end{bmatrix}$. Hence $A^{-1} = \frac{1}{|A|}\operatorname{adj}(A) = \frac{1}{-5}\begin{bmatrix} 1 & -3 \\ -2 & 1 \end{bmatrix}$.

6.2.12. If we apply Theorem 6.2.3 to A^{-1} in place of A, we get $(A^{-1})^{-1} = \frac{\operatorname{adj}(A^{-1})}{|A^{-1}|} = \frac{\operatorname{adj}(A^{-1})}{1/|A|}$, and so $\operatorname{adj}(A^{-1}) = \frac{(A^{-1})^{-1}}{|A|} = (|A|A^{-1})^{-1} = (\operatorname{adj}(A))^{-1}$.

6.2.15. If we expand the determinant along the first row, then we obtain a linear combination of the elements of that row, which can be rearranged into the given form because the coefficient of y is a nonzero determinant (see Exercise 6.1.14). If we substitute the coordinates of any of the given points for x, y, then two rows become equal, which makes the determinant vanish. Thus the given points lie on the parabola.

6.3. The Cross Product of Vectors in $\mathbb{R}^3$ 207

6.3.2. We may choose $\boldsymbol{n} = \overrightarrow{AB} \times \overrightarrow{AC} = \begin{vmatrix} \boldsymbol{i} & \boldsymbol{j} & \boldsymbol{k} \\ -1 & 0 & 1 \\ 2 & 1 & 0 \end{vmatrix} = -\boldsymbol{i} + 2\boldsymbol{j} - \boldsymbol{k}$ as the normal vector of the plane, and so the plane's equation can be written $\boldsymbol{n} \cdot (\boldsymbol{r} - \boldsymbol{a}) = 0$; that is, as $-1(x-1) + 2(y+1) - 1(z-2) = 0$.

CHAPTER 7 EIGENVALUES AND EIGENVECTORS

7.1. Basic Properties 217

7.1.4. $\lambda_1 = \lambda_2 = 0$, $\mathbf{s} = t(0, 1)^T$.

7.1.5. $\lambda_1 = 1$, $\lambda_2 = 2$, $\lambda_3 = 3$, $\mathbf{s}_1 = t(-1, 0, 1)^T$, $\mathbf{s}_2 = t(0, 1, 0)^T$, $\mathbf{s}_3 = t(1, 0, 1)^T$.

7.1.7. $\lambda_1 = 0$, $\lambda_2 = 1$, $\lambda_3 = 3$, $\mathbf{s}_1 = t(1, 1, 1)^T$, $\mathbf{s}_2 = t(1, 1, 0)^T$, $\mathbf{s}_3 = t(-1, 0, 2)^T$.

7.1.8. $\lambda_1 = \lambda_2 = 1$, $\lambda_3 = \lambda_4 = 2$, $\mathbf{s}_1 = s(1, 0, 0, 0)^T + t(0, 1, 0, 0)^T$, $\mathbf{s}_2 = s(0, 1, 1, 0)^T + t(1, 1, 0, 1)^T$.

7.1.10. If $B = A + cI$, then B has an eigenvalue $\lambda_B = \lambda_A + c$ for each eigenvalue of A and with the same eigenvectors.

7.2. Diagonalization of Matrices 227

7.2.5.
$$A^{100} = \begin{bmatrix} -1/\sqrt{2} & 0 & 1/\sqrt{2} \\ 0 & 1 & 0 \\ 1/\sqrt{2} & 0 & 1/\sqrt{2} \end{bmatrix} \begin{bmatrix} 1^{100} & 0 & 0 \\ 0 & 2^{100} & 0 \\ 0 & 0 & 3^{100} \end{bmatrix} \begin{bmatrix} -1/\sqrt{2} & 0 & 1/\sqrt{2} \\ 0 & 1 & 0 \\ 1/\sqrt{2} & 0 & 1/\sqrt{2} \end{bmatrix}$$
$$= \frac{1}{2}\begin{bmatrix} 3^{100}+1 & 0 & 3^{100}-1 \\ 0 & 2^{101} & 0 \\ 3^{100}-1 & 0 & 3^{100}+1 \end{bmatrix}.$$

7.2.6.
$$A^4 = \begin{bmatrix} 1 & 0 & 0 & 15 \\ 0 & 1 & 15 & 15 \\ 0 & 0 & 16 & 0 \\ 0 & 0 & 0 & 16 \end{bmatrix}.$$

7.2.7. If $B = SAS^{-1}$, then $|B - \lambda I| = |S^{-1}AS - \lambda I| = |S^{-1}AS - S^{-1}\lambda IS| = |S^{-1}(A - \lambda I)S| = |S^{-1}||(A - \lambda I)||S| = |A - \lambda I|$ because $|S^{-1}| = 1/|S|$.

7.3. Principal Axes 258

7.3.4. The matrix of this quadratic form is
$$A = \frac{1}{4}\begin{bmatrix} 3 & -1 & -1 \\ -1 & 3 & -1 \\ -1 & -1 & 3 \end{bmatrix}.$$

The eigenvalues are $\lambda_1 = \lambda_2 = 1$ and $\lambda_3 = 1/4$, and an orthogonal matrix of corresponding eigenvectors is
$$S = \frac{1}{\sqrt{6}}\begin{bmatrix} \sqrt{3} & 1 & \sqrt{2} \\ -\sqrt{3} & 1 & \sqrt{2} \\ 0 & -2 & \sqrt{2} \end{bmatrix}.$$

In the basis S the equation of the ellipsoid takes on the standard form
$$y_1^2 + y_2^2 + \frac{y_3^2}{4} = 1.$$

This is an ellipsoid of revolution, centered at the origin, with major axis of half-length 2 pointing in the $\mathbf{s}_3 = \frac{1}{\sqrt{3}}(1, 1, 1)^T$ direction, and having a circle of radius 1 as its cross-section that contains the minor axes.

INDEX